Textbooks in Telecommunication Engineering

Series Editor

Tarek S. El-Bawab, Ph.D.
Boston, MA, USA

Telecommunication and networks have evolved to embrace all aspects of our everyday life, including smart cities and infrastructures, healthcare, banking and businesses, manufacturing, space and aviation, meteorology and climate change, oceans and marine life, Internet of Things, defense, homeland security, education, research, social media, entertainment, and many others. Network applications and services continue to expand, virtually without limits. Therefore, specialized telecommunication and network engineering programs are recognized as a necessity to accelerate the pace of advancement in this field, and to prepare a new generation of engineers for imminent needs in our modern life. These programs need curricula, courses, labs, and textbooks of their own.

The IEEE Communications Society's Telecommunication Engineering Education (TEE) movement, led by Tarek S. El-Bawab- the editor of this Series, resulted in recognition of this field of engineering by the Accreditation Board for Engineering and Technology (ABET) on November 1, 2014. This Springer Series was launched to capitalizes on this milestone. The Series goal is to produce high-quality textbooks to fulfill the education needs of telecommunication and network engineering, and to support the development of specialized undergraduate and graduate curricula in this regard. The Series also supports research in this field and helps prepare its scholars for global challenges that lay ahead. The Series have published innovative textbooks in areas of network science and engineering where textbooks have been rare. It is producing high-quality volumes featuring innovative presentation media, interactive content, and online resources for students and professors.

Book proposals are solicited in all topics of telecommunication and network engineering including, but not limited to: network architecture and protocols; traffic engineering; network design, dimensioning, modeling, measurements, and analytics; network management and softwarization; cybersecurity; synchronization and control; applications of artificial intelligence in telecommunications and networks; applications of data science in telecommunications and networks; network availability, reliability, protection, recovery and restoration; wireless communication systems; cellular technologies and networks (through 5G, 6G, and beyond); satellite and space communications and networks; optical communications and networks; heterogeneous networks; broadband access and free-space optical communications; MSO/cable networks; storage networks; optical interconnects; and data centers; social networks; transmission media and systems; switching and routing (from legacy to today's paradigms); network applications and services; telecom economics and business; telecom regulation and policies; standards and standardization; and laboratories.

Proposals of interest shall be for textbooks that can be used to develop university courses, either in full or in part. They should include recent advances in the field while capturing whatever fundamentals that are necessary for students to understand the bases of the topic and appreciate its evolution trends. Books in this series will provide high-quality illustrations, examples, end-of-chapters' problems/exercises and case studies.

This series is indexed in Scopus.

For further information and to submit proposals, please contact the Series Editor, Dr. Tarek S. El-Bawab, telbawab@ieee.org; or Mary James, Executive Editor at Springer, mary.james@springer.com.

Jie Hu • Kun Yang

Integrated Data and Energy Transfer in Wireless Networks

Springer

Jie Hu
University of Electronic Science and Tec
Chengdu, China

Kun Yang
Wivenhoe Park
University of Essex
Colchester, UK

ISSN 2524-4345 ISSN 2524-4353 (electronic)
Textbooks in Telecommunication Engineering
ISBN 978-3-031-60513-0 ISBN 978-3-031-60514-7 (eBook)
https://doi.org/10.1007/978-3-031-60514-7

This Springer imprint is published by the registered company Springer Nature Switzerland AG
The registered company address is: Gewerbestrasse 11, 6330 Cham, Switzerland

Preface

In order to satisfy the power thirst of communication devices in the imminent sixth-generation (6G) and Internet of Things (IoT) era, wireless charging techniques have attracted much attention both from the academic and industrial communities. Although inductive coupling and magnetic resonance based charging techniques are indeed capable of supplying energy in a wireless manner, they tend to restrict the freedom of movement. By contrast, radio frequency (RF) signals are capable of supplying energy over distances, which are gradually inclining closer to our ultimate goal—charging anytime and anywhere. Furthermore, transmitters capable of emitting RF signals have been widely deployed in TV towers, cellular base stations, and WiFi access points. This communication infrastructure may indeed be employed also for wireless energy transfer (WET). Therefore, no extra investment in dedicated WET infrastructure is required. However, allowing RF signal-based wireless energy transfer (WET) may impair the wireless data transfer (WDT) operating in the same spectrum. Hence, it is crucial to coordinate and balance WET and WDT for integrated data and energy transfer (IDET). Therefore, in 6G mobile communication systems, low-power IoT devices can gain energy from network infrastructure, which creates energy self-sustainable 6G. This textbook aims for providing a landscape picture of IDET.

To this end, we firstly provide an overview of IDET in wireless networks in Chap. 1. We will look into the energy shortage of the electronic devices, compare the popular wireless charging techniques with one another, and highlight the RF signal-based WET and its distinctive features against the conventional wireless communication in the same spectral bands. Then we will describe the ubiquitous architecture of IDET-enabled wireless networks.

In Chap. 2, we focus on the information theoretical essence of both WET and WDT. We will commence the chapter with the basics of classic information theory, which will be then extended to include both WET and WDT and their interaction. Unary codes are then invoked to enable coding-level control of IDET with various modulation schemes.

In Chap. 3, we also introduce the fundamentals of classic modulation schemes. Then, we will describe how to design modulated waveform for IDET in a multi-user scenario. Moreover, multi-sinusoidal waveforms, orthogonal-frequency-division-multiplexing (OFDM) waveforms, and orthogonal-time-frequency-space (OTFS) waveforms are designed for both WET and WDT. Some basic skills about fractional optimization and other powerful optimization tools are also involved.

In Chap. 4, we move on to the radio-front of the IDET transceiver with multiple antennas. A general tutorial about all kinds of beamforming techniques is given at the beginning. Then a typical example of mmWave-based IDET transceiver design is provided, which includes a joint design of analogue beamforming and digital beamforming.

In Chap. 5, we generally outline the background of resource management in wireless networks, which is also along with an introduction of convex optimization. Then, we exemplify the resource management in an IDET-enabled wireless network, which jointly designs multi-dimensional resource allocation in full-duplex wireless powered communication networks.

In Chap. 6, we design a novel medium-access-control (MAC) layer protocol for scheduling multiple IoT devices in a distributed manner for both data transmission and energy replenishment. Therefore, energy-consuming central control is avoided. As a powerful mathematical tool, Markov chains are relied upon for modeling the behavior of IoT devices.

In Chap. 7, we design the deployment of hybrid base stations by considering mobilities of data and energy users. Therefore, when moving around in downtown, their requirement for both data transmissions and energy replenishing can be appropriately satisfied.

In Chap. 8, we prototype a mobile IDET station. It is capable of touring to IoT devices for transferring wireless energy to them. This mobile IDET station can accurately estimate the angle of IoT devices. Therefore, it may adjust its antenna to form an efficient energy beam for recharging the corresponding IoT devices. This chapter may help readers to build up their own IDET prototype for further research.

This textbook is for the undergraduate/postgraduate students to be familiar with this cutting-edge technique. It is also suitable for researchers from both the academia and industry so as to push the DEIN a step closer to the practical implementation.

We would like to thank Professor Tarek El-Bawab for his invaluable comments on the structure of the book and for his outstanding editorial organization of the book series.

Chengdu, China Jie Hu
Colchester, UK Kun Yang

Contents

Chapter 1
Introduction

1.1 Energy Dilemma for Electronic Devices

According to the prediction of the classic Moore's Law, the density of transistors in an integrated circuit doubles approximately every 2 years, which have been fuelling the spectacular proliferation of electronic devices since the 1960s. Furthermore, consumer electronic devices are becoming shirt-pocket-sized and mobile. These devices are normally powered by embedded batteries. However, as their functions become ever more sophisticated, their thirst for abundant energy is not matched by the slow progress of the batteries' capacity. The situation in the communication industry is even more daunting. Since the rollout of the fifth-generation (5G) cellular system and of the Internet of Things (IoT) is just around the corner, people's appetite for super-high data transmission rates, for high density of connectivity and for high mobilities, will indeed be satisfied to a large extent. A major portion of the future mobile data traffic will be constituted by novel types of services, including high-definition stereoscopic video streams, augmented/virtual reality, holographic tele-presence, cloud desktops, online games, etc. All these services require the user terminals to be implemented with high computing capabilities for real-time signal processing, which may quickly drain the embedded batteries. Furthermore, sensors will be deployed in every corner of the future smart cities [1]. These sensors monitor the environment and upload sensing results to central servers [2]. The lifespan of sensors and of sensing networks largely depends on the sensors' battery capacity. Regularly replacing the batteries may be an unrealistic or tedious task. Accordingly, new sources of energy have to be explored to prolong the depletion period of conventional batteries in order to relieve the energy concerns of various communication devices.

1.2 Near-Field Wireless Energy Transfer

Nowadays, resonant inductive coupling [3] and magnetic resonance coupling [4] have emerged for remotely charging electronic devices in the near-field. Resonant inductive coupling based wireless charging relies on the magnetic coupling that delivers electrical energy between two coils tuned to resonate at the same frequency. This technique has already been commercialized for some home electronic appliances [5], such as mobile phones, electric toothbrushes, smart watches, etc. However, the coupling coils only support near-field wireless energy transfer (WET) over a distance spanning from a few millimetres to a few centimetres [6], while achieving a WET efficiency as high as 56.7%, when operating at a frequency of 508 kHz [7]. Furthermore, resonant inductive coupling requires strict alignment of the coupling coils. Even a small misalignment may result in dramatic reduction of the WET efficiency [8]. As a result, during the charging process, the electronic appliances cannot be freely moved.

By contrast, magnetic resonance coupling [9] delivers electrical energy between two resonators by exploiting evanescent-wave coupling. This technique has already been adopted for charging the electric vehicles due to its high WET efficiency [10]. For example, magnetic resonance coupling is capable of achieving a WET efficiency of 90% over a distance of 0.75 m [11]. Both its WET efficiency and its charging distance are much higher than that of the resonant inductive coupling. However, magnetic resonance coupling still belongs to the category of near-field wireless charging, since its power transfer efficiency dramatically reduces to 30%, when the distance is increased to 2.25 m [11]. Nonetheless, magnetic resonance coupling does not require strict alignment between the rechargeable device and the energy source. Hence, during the charging process, the electronic appliances may be moved within the charging area [12]. Furthermore, a multiple-input multiple-output (MIMO)

J. Hu, K. Yang, *Integrated Data and Energy Transfer in Wireless Networks*, Textbooks in Telecommunication Engineering,
https://doi.org/10.1007/978-3-031-60514-7_1

system, which has already been widely adopted for improving the performance of the wireless communication, can also be introduced into the magnetic resonance coupling based WET system for the sake of further enhancing the WET efficiency [13, 14].

1.3 Energy Harvesting Aided Communications

Energy harvesting aided communications (EHAC) is a technology that uses environmental energy harvesters (such as solar cells, wind turbines, RF, etc.) to collect energy and provide power to devices for communication. EHAC technology can reduce the dependence of communication devices on the traditional power grid, achieving green and sustainable communication. At the same time, EHAC can also improve the coverage and reliability of communication devices, reduce the frequency of battery replacement, and extend the device's lifespan. EHAC technology has been widely used in low-power wireless communication fields such as the Internet of Things and sensor networks and is one of the future development trends of wireless communication [15].

There have been several outstanding surveys and tutorials in the related research topics, such as wireless charging and EHAC. Specifically, Niotaki et al. [16] have investigated solar/electromagnetic energy harvesting and wireless power transmission, covering the design of hybrid solar–electromagnetic harvesters, the circuits of direct current combiners, and the efficient solar-to-electromagnetic converters. Furthermore, novel circuit topologies were presented for the sake of minimizing the sensitivity of rectifier circuits. Lu et al. [17] have summarized the latest progress both in near-field inductive coupling and in magnetic resonance based wireless charging techniques, as well as in far-field RF-based WET techniques. Moreover, the technical standards of wireless charging and their recent advances in networked applications were also investigated. However, both [16] and [17] were mainly focused on circuit design for incorporating wireless charging into electronic devices, but they did not consider the integration of wireless charging and wireless communication. Apart from the existing survey and tutorials on WET, some further endeavors [18–20] have been invested in characterizing the landscape of EHAC. Specifically, Ulukus et al. [18] have summarized the state of the art in wireless networks relying on EHAC nodes, including their information theoretic performance limits, transmission scheduling policies, resource allocation, medium access control, and networking issues. Additionally, the energy cooperation between communication nodes and the energy harvesting and energy consumption models have also been reviewed. Another comprehensive survey [19] has been contributed by Ku et al., dealing with the challenges incurred by the intermittent random nature of the renewable energy arrival for improving EHAC systems. Furthermore, Lu et al. [20] have focused their attention on specific RF energy harvesting techniques and on their integration with wireless communication. Both the associated circuits, the receiver architecture and the resource allocation and protocol designs, have been reviewed in the context of RF energy harvesting aided wireless network design. However, energy harvesting is an operation adopted by the devices for acquiring additional energy from the environment; hence these devices can be regarded as passive energy receivers. As a result, scheduling and balancing energy and data flows of stochastic nature become the main challenge in EHAC. More specifically, RF energy harvesting aided wireless communication is mainly concerned with the operating strategy adopted by the receivers. For instance, the receivers may invoke efficient RF-DC converters for increasing the amount of energy harvested from RF signals. They may also decide how to split the received signals for simultaneously satisfying both the data and energy demands. However, RF energy harvesting aided wireless communication and networking ignores the evaluation and design of the entire RF-based WET chain, and it does not consider the joint design of careful integrated data and energy transmitters and receivers.

1.4 RF Signal Based Wireless Energy Transfer

In contrast to the abovementioned near-field WET techniques, the propagation of the RF signals is capable of supporting far-field wireless energy transfer [21]. The history of the RF signal based wireless energy transfer dates back to 1960, when the first long-distance WET system was established by William C. Brown [22, 23]. Brown jointly designed rectifiers and antennas for energy receivers, which is now widely known as rectennas. They are capable of efficiently converting the AC energy carried by the RF signals to direct current (DC) energy. His RF signal based WET system was validated by remotely powering a model helicopter from the ground in 1964 [22, 23]. In the 1970s and 1980s, intense efforts were invested into the research of RF signal based WET, which was largely motivated by the intention of developing a solar-powered satellite [24, 25]. In this system, a satellite may harvest energy from sunlight in outer space and beam the energy back to ground

Table 1.1 Main features of different WET techniques

Technique	Range	Direc.	Frequency	Antenna	Application
RF signals	Long	High	MHz–GHz	Parabolic dishes, rectennas, phased arrays	Solar-powered satellite, drone aircraft, IoT devices, portable devices RFID, smart cards, etc.
Magnetic resonant coupling	Middle	Low	kHz–GHz	Tuned wire coils, lumped element resonators	Portable devices, biomedical implants, electric vehicles, RFID, smartcard, etc.
Inductive coupling	Short	Low	Hz–MHz	Wire coils	Stovetops, industrial heaters and small electric appliances, such as electric toothbrush, razor, etc.

stations via the propagation of RF signals. Furthermore, the Jet Propulsion Laboratory of the National Aeronautics and Space Administration (NASA) led a project from 1969 to 1975, in which 30 kW of power was beamed over a distance of 1 mile at a 84% RF-DC efficiency [26].

There are three main technical challenges in the RF-based wireless energy transfer. Firstly, the long-distance propagation and adverse multipath fading may substantially attenuate the RF signals before they arrive at the receivers, which inevitably results in energy loss. Secondly, the energy carried by RF signals is of alternating current (AC) nature, which cannot be directly invoked for driving an electronic load. As a result, RF signals have to be converted into DC energy for any further use. However, some portion of energy is inevitably lost during this conversion process. For addressing the signal attenuation of wireless channels, the transmit beams have to be accurately aimed at the energy receivers [27], which requires the joint design of the transmit and receive antennas. For addressing the AC–DC conversion efficiency, the receive antennas have to be designed together with the rectifiers in order to achieve the impedance match for the sake of high-efficiency AC–DC conversion [28]. Last but not the least, the diffraction of the RF signals' waveform may expand the beam size. As a result, the receive antenna having limited size is not capable of capturing all the energy carried by the RF signals.

In general, the RF signal based WET has the following advantages over the near-field WET technique:

- Large coverage. Relying on the RF signals, energy can be transferred to receivers miles away.
- High flexibility. The angular selectivity transmit beam can be intelligently adjusted according to various WET requirements. For instance, a narrow beam can be invoked for realizing accurate and high-efficiency point-to-point WET, while a wide beam can be used for simultaneously charging multiple devices.
- More applications. RF signals can be leveraged for supplying a large amount of energy to energy-hungry appliances, such as solar-powered satellite systems. It can also be exploited for supplying energy to low-power devices, such as sensors and biomedical implants.
- Low investment. The transmitters of the RF signals have been deployed at every corner of the globe, such as radio broadcast stations, TV towers, cellular base stations, Wi-Fi access points, etc. The legacy of the communication infrastructure can all be exploited for radiating energy to electronic devices. Only limited extra investment is required for deploying energy transmitters in order to cover some blind spots.

The main features of different WET techniques are summarized in Table 1.1.

1.5 WET Versus WIT in the RF Spectral Band

Since RF signal based WET techniques require highly flexible beam directivity in order to satisfy diverse charging requests, the best spectral band for steering energy beams is in the range of 10 MHz to 100 GHz, which covers almost all the bands allocated for wireless communication services. For example, TV/radio broadcasting services operate in the band spanning from 40 to 220 MHz [29], the mobile cellular communication system operates in the spectral band spanning from 800 MHz to 3.7 GHz [30], while the Wi-Fi communication system operates in the spectral band spanning from 2.4 to 6 GHz [31]. Furthermore, as a key technique in the upcoming 5G era, millimetre wave (mmW) [32] may significantly increase the achievable throughput of the air interface, which operates in the spectral band ranging from 10 to 100 GHz.

Although they both operate in the same RF band, WET and WIT still have the following distinctive characteristics:

- They have different functional circuits. RF signals in the passband cannot be directly invoked for both the information decoding and the energy harvesting. For the information decoding, the RF signals in the passband have to be firstly converted to the baseband, since all the signal processing has to be accomplished in the baseband. By contrast, for the energy harvesting, the AC energy carried by the RF signals has to be converted to the DC energy first, since only the DC energy can be stored in batteries or drive electronic loads. Specifically, during the AC–DC conversion, the phase information carried by the RF signals is filtered.

- They require different absolute energy at receivers. The activation of the energy harvesting circuits requires a relatively high energy carried by the received RF signals, which is approximately on the order of -10 dBm. If the energy carried by the received RF signal does not achieve the required activation threshold, none of this energy can be harvested. By contrast, the successful information recovery relies on the energy ratio between the received RF signal and the noise plus interference, not on the absolute energy carried by the received RF signal. As a result, even a small amount of energy is capable of activating the information decoder, which is approximately on the order of -80 dBm.
- They have different coverage. The RF signals are attenuated by hostile wireless channels, such as the path loss, shadowing, and multipath fading. Since the energy harvesting requires a much higher absolute energy at the receivers than the information decoding, the range of WET is accordingly much shorter than that of WIT. Therefore, given the same set of transmitters and receivers, the resultant WET network has a different topology with the WIT network.
- They treat noise and interference differently. The interference and noise ubiquitously exist in any WIT system, which seriously impair the WIT performance. Mitigating the performance degradation induced by the interference and noise is a major challenge in the WIT system design. By contrast, WET systems may actually benefit from the interference and noise, since both of them are RF signals and they both carry useful energy. The interference and noise can be jointly harvested by the energy harvesting circuits, which may provide additional energy harvesting gains for the energy requesters.
- They have different definitions in energy efficiency. The energy efficiency of WET can be defined as the ratio of energy harvested by the receiver to the energy emitted by the transmitter, which can be formulated as

$$\eta_{WET} = \frac{1}{P_t} \cdot \rho \left(P_r + P_I + P_N\right) \quad \text{(Watt/Watt)}, \tag{1.1}$$

where ρ is the conversion rate from the received AC energy to the DC energy by considering a linear AC–DC converter. By contrast, in the community of green communications, the energy efficiency of WIT is defined as the ratio of spectral efficiency to energy consumption, which is evaluated in the unit of bps/Hz/Watt or bps/Hz/Joule. By exploiting the classic Shannon–Hartley theorem in an Additive White Gaussian Noise (AWGN) channel, the energy efficiency of WIT can be expressed as

$$\eta_{WIT} = \frac{1}{P_t} \cdot \log_2 \left(1 + \frac{P_r}{P_I + P_N}\right) \quad \text{(bps/Hz/Watt)}, \tag{1.2}$$

where P_t is the transmit power of the RF signal, P_r is the power received after the signal being attenuated by the hostile wireless channel, P_I is the aggregate interference power, and P_N is the noise power at the receiver.

In Fig. 1.1, we exemplify the energy efficiency of WET and that of WIT, which can be calculated by (1.2) and (1.1), respectively. Observe from Fig. 1.1a that in our setting, the energy efficiency of WET reduces from 1.1% but converges to 1%, which is due to the channel attenuation incurred by the path loss between the transmitter and the receiver pair. Observe from Fig. 1.1b that the energy efficiency of WIT gradually reduces from 35 bps/Hz/mW to 0 as the transmit power of the RF signal increases.

By contrast, WET and WIT operating in the same RF spectral band may compete for the precious resources in the air interface, and they may thus impair each other's performance to some extent. For example, WET requires that the RF signals carry a high power to the receivers for the efficient energy harvesting. However, the high-power RF signals of the WET system may impose excessive interference on the WIT receivers, which may thus significantly degrade the WIT performance attained. As a result, coordinating WET and WIT in the same RF band imposes critical challenges on the RF circuit design, on the transceiver design of the physical layer, on the resource scheduling/allocation schemes, and on the corresponding protocol design of the medium access control (MAC) layer. Furthermore, integrated data and energy transfer in the RF band also requires a joint networking concept for heterogeneous data and energy transceivers. All these challenging issues require novel Data and Energy Integrated Communication Networks (DEINs) [38].

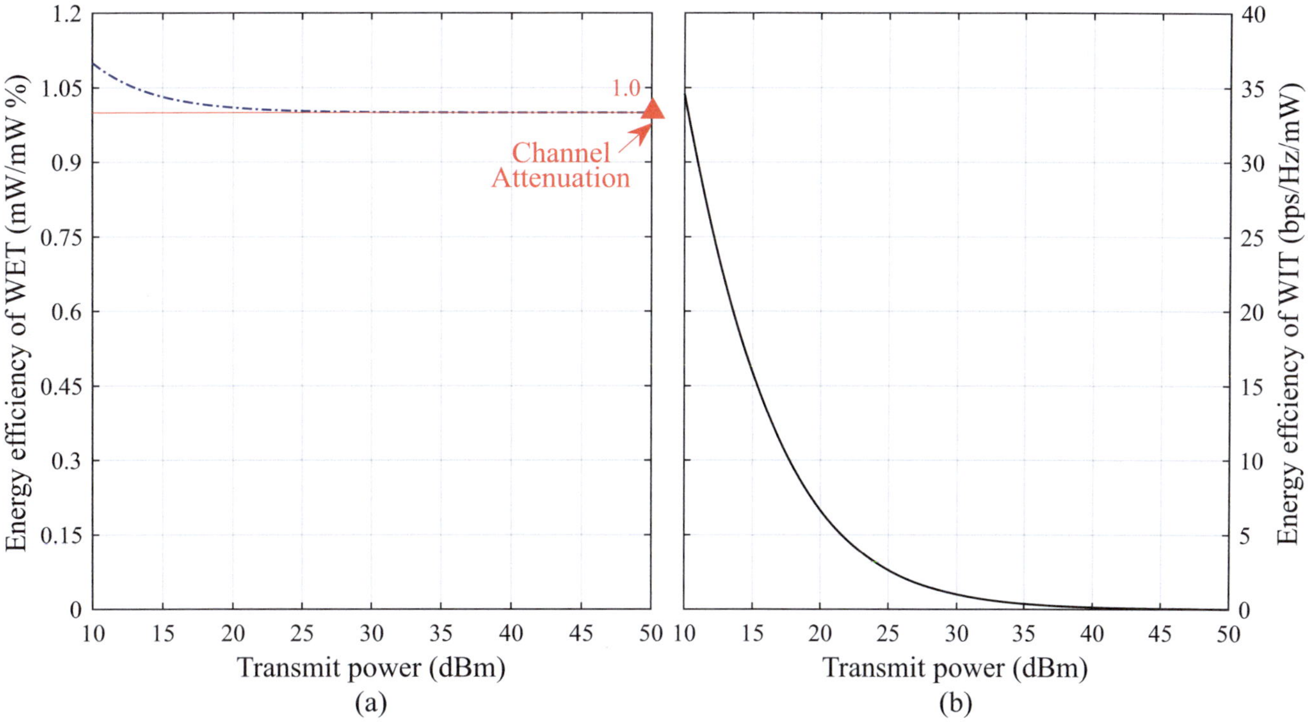

Fig. 1.1 Energy efficiency of WET (**a**) and that of WIT (**b**) against transmit power of RF signals. The noise power is $P_N = -94$ dBm, which is calculated by the power spectrum density of the thermal noise -174 dBm/Hz and 100 MHz of the RF signals' bandwidth. The aggregate interference power at the receiver is set to be $P_I = -20$ dBm, which appears in a heterogeneous cellular network with the highest probability [33]. The distance between a transmitter and a receiver pair is 10 m. The path loss is calculated by the model invoked in [34–37], where the path loss exponent is 2. No fading is assumed. The antenna gain in this example is set to be 40 dBi in order to counteract the path loss

1.6 Integrated Data and Energy Transfer

Integrated data and energy transfer (IDET), which integrates the traditionally separate two processes, i.e., wireless data transfer (WDT) and wireless energy transfer (WET), fulfilling co-transmission of data and energy. In particular, the energy transmission using radio frequency is for the purpose of energy harvesting (EH) rather than information decoding. One driving force of the advent of IDET is wireless big data, which comes from wireless sensors that produce a large number of small pieces of data. These sensors are typically powered by a battery that drains sooner of later and will have to be taken out and then replaced or recharged. EH has emerged as a technology to wirelessly charge batteries in a contactless way. WET and WDT operating in the same RF spectral band may impair each other's performance to some extent. For example, WET requires that the RF signals carry a high power to the receivers. However, the high-power RF signals of the WET system may impose excessive interference on the WDT receivers, which may substantially degrade the WDT performance attained. As a result, coordinating WET and WDT in the same RF band imposes critical challenges on the RF circuit design, on the IDET design of the physical layer, on the resource scheduling/allocation schemes, and on the corresponding protocol design of the medium access control (MAC) layer. It also requires a joint networking concept for heterogeneous data and energy transceivers. All these challenging issues require novel IDET technology [38].

Various state-of-the-art enabling techniques can be adopted from the toolbox of wireless communications and electronic engineering for efficiently coordinating WET and WDT in the same RF spectral bands, as illustrated by the outer shell of Fig. 1.2. The left side of the outer shell of Fig. 1.2 specifies techniques from the electronic engineering toolbox for enabling RF signal based WET of high efficiency. To elaborate:

- DC-RF Energy Converters are invoked by WET transmitters for converting the DC energy gleaned from the power grid or from the embedded batteries into RF signals. For instance, it is reported in [39] that a Doherty amplifier, which is realized by a pair of voltage mode power amplifiers and transformers, is capable of achieving as high as 45% of DC-RF conversion efficiency.
- The Airy Beam was conceived as early as 1979, which originated from a non-spreading Airy wave packet solution of the Schrodinger equation. The Airy beam was created and observed in both one- and two-dimensional configurations

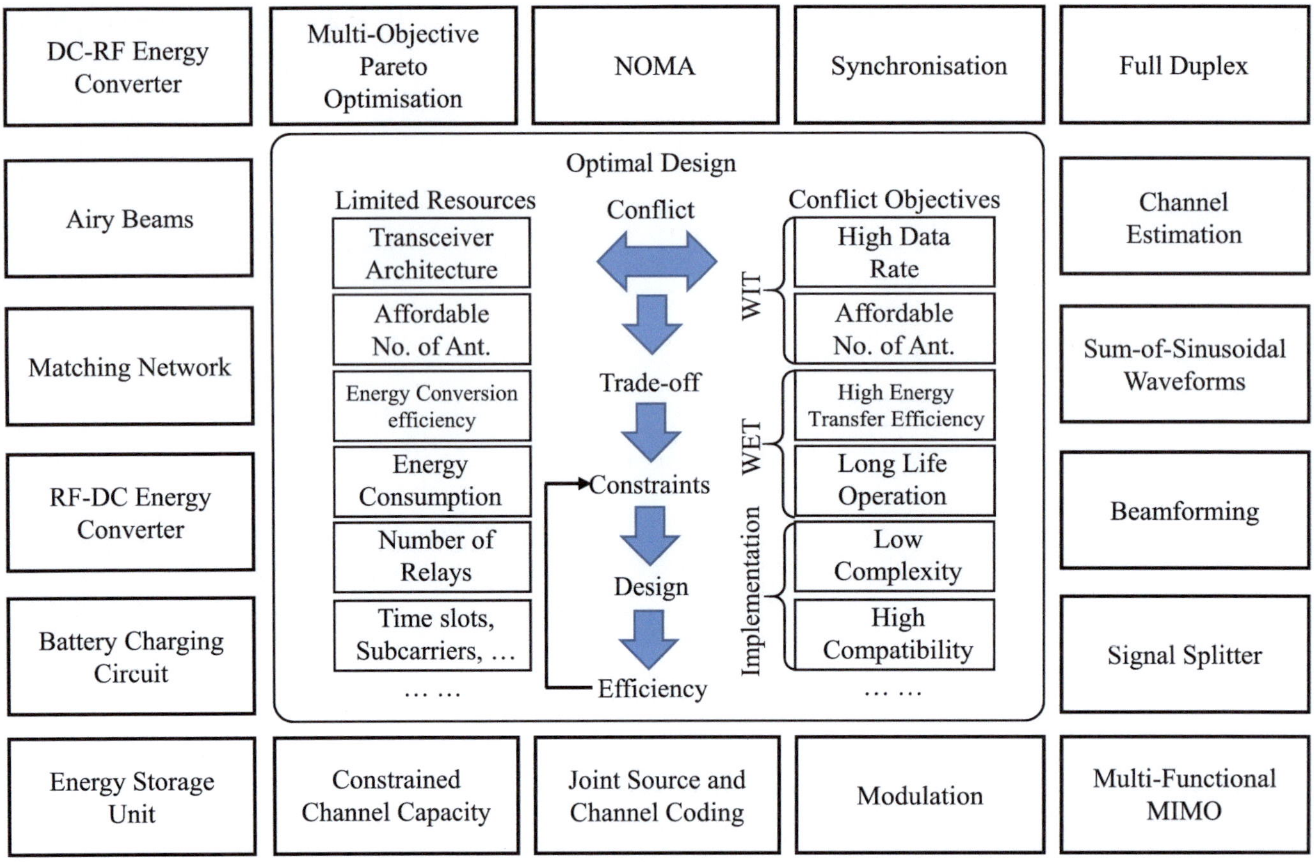

Fig. 1.2 Key techniques for coordinating WET and WIT in same RF spectral bands (outer shell) and optimal design principle of integrated WET and WIT (inner shell)

[40], and it has the following features: non-dispersing, self-accelerating, self-healing, and promptly auto-focusing, which constitutes as a potential solution of carrying energy to distant energy requesters. Some initial efforts [41] have been invested into generating Airy beams in the RF bands. Exploiting the compelling features of Airy beams for both the WDT and WET in the RF spectral bands calls for more research from both the electronic and communication communities.

- An accurate Impedance Matching Network is required between the receive antennas and the RF-DC converters in order to match the impedance of this pair of components [42]. Only when their impedance is accurately matched, we can suppress the energy leakage during the transfer process of RF signals from the receive antennas to the RF-DC converters.
- RF-DC Energy Converters play a salient role in energy reception. Due to its AC nature, energy carried by RF signals cannot be directly invoked for charging batteries or driving electronic loads. As a result, the RF energy has to be converted into DC energy for its further exploitation. The efficiency of RF-DC energy converters is naturally a nonlinear function of their input power. Suffice to say that a high RF-DC conversion efficiency can be achieved if the input power of the converter is high [43].
- Battery Charging Circuits are capable of adapting the resultant DC signals to the specific requirements of battery charging. Specifically, after the channel's attenuation and imperfect RF-DC conversion, only a low energy reaches the battery charging circuits. Efficiently exploiting this energy and amplifying the voltage for charging batteries is an essential task of the battery charging circuits. For instance, Hammed and Moez [44] designed a battery charging circuit capable of amplifying 170 mV of the input voltage to 2.8 V of output.
- Energy Storage Units are invoked for storing the DC energy converted from RF signals. Typically, lithium batteries are used in handheld electronic devices for storing and supplying energy, while super-capacitors are normally embedded into miniature IoT devices. A high storage density is required by energy storage units for storing energy extracted from RF signals.

The rest of the outer shell of Fig. 1.2 specifies the family of techniques used from wireless communication toolbox for enabling IDET within the same RF spectral bands:

- Constrained Channel Capacity is the terminology used for mutual information maximization under the constraint of particular output signals. This is in contrast to Shannon's classic channel capacity theory, which aims for maximizing the mutual information under the specific constraint of particular input signals. This information theoretic concept has been firstly proposed by Gastpar [45] for controlling the interference in the scenario of multiple transmitter and receiver pairs. A similar methodology can be adopted here for exploring the performance limits of data transmission under the constraint of having sufficient energy carried by the output signals, which can be harvested by the receiver.
- Joint Source and Channel Coding can be specifically designed for controlling integrated WET and WDT, as proposed by Tandon et al. [46]. Since the original messages are encoded into codewords carrying a different amount of energy, joint source and channel encoders are capable of beneficially controlling the energy transfer process. Therefore, given the characteristics of the information source, a carefully designed encoder is capable of responding to energy requests, while simultaneously satisfying the specific data transmission requirements.
- The Modulation Schemes have to be appropriately redesigned for supporting the IDET. In order to reduce hardware complexity of miniature devices, the information can also be modulated onto DC signals by pulse amplitude modulation, pulse width modulation, and pulse position modulation routinely adopted by optical communications [47]. As a result, the RF-DC energy converter becomes capable of replacing the function of the passband to baseband converter. However, the IDET is achieved at the cost of a data transmission rate reduction, since the RF-DC converters filter the frequency and phase of the RF signals; hence we have limited choices of dimensions to be modulated for data transmission.
- Multifunctional MIMOs are crucial for facilitating IDET as a benefit of their multiplexing, diversity beamforming gains [48]. Spatial multiplexing can be readily achieved by multifunctional MIMO systems in order to coordinate WET and WDT in different spatial domains, while the diversity gain attained by multiple antennas is capable of reducing the data transmission error probability, while simultaneously increasing the amount of energy transferred to the receivers. Furthermore, multifunctional MIMO systems are capable of forming narrow energy beams for supplying energy to a specific device. They are also capable of forming wide energy beams for supplying energy to multiple devices in a specific region.
- Signal Splitters are the most important modules in receiver architectures designed for IDET. After being received by the antennas, the RF signals are split into two portions by the signal splitters for dedicated energy and information reception. The time switcher [49] and power splitter [50] are capable of splitting the received RF signals in the time-domain and power-domain, respectively. The design of the signal splitter aims for avoiding energy leakage and for realizing adaptive adjustment.
- Beamforming is a key technique of compensating for the energy reduction during the RF signals' propagation, which is beneficial for both the dedicated WET and the IDET. There are two basic approaches for forming a directional beam, namely analog and digital beamforming. Analog beamforming is realized at the transmitter by invoking an analog phase shifter for each transmit antenna for the sake of appropriately phase-rotating the signal components impinging on all of the antennas. This allows us to form an angularly selective beam, hence compensating for the wireless channel's path-loss. However, the analog beamformer cannot adjust the amplitude weights of the signals transmitted by the antennas, which limits its capability of improving the performance of the dedicated WET and that of the IDET. By contrast, upon precoding the transmit symbols in the digital domain, the digital beamformer is capable of adjusting both the amplitude and the phase of the signals transmitted by the antennas. As a result, having an additional degree of freedom enables the digital beamforming to outperform its analog counterpart in terms of both the dedicated WET and the IDET. However, as the number of the antennas becomes high, it is impractical to invoke a dedicated RF chain for each antenna. As a result, hybrid beamforming [51] is required for constructing hardware-efficient IDET for future massive MIMO systems.
- Sum-of-Sinusoidal type Waveforms—such as OFDM waveforms—are capable of carrying a high amount of energy to the energy receivers [52]. Practical energy reception circuits can only be activated, if their instantaneous input power is higher than a threshold. Otherwise, none of the input power can be harvested. Therefore, the waveform of the received RF signals should have a high peak-to-average power ratio (PAPR) for dedicated WET [53], which may be readily achieved by accumulating a number of different frequency sinusoidal signals. However, accumulating a number of different frequency sinusoidal signals requires a considerable bandwidth. By contrast, a low PAPR is required for efficient WDT. These contradicting design principles can only be resolved by radical thinking in IDET.
- Channel Estimation has to be carried out for efficient exploitation and support of the antenna arrangement, of beamforming, and of waveform design. However, complex power-thirsty channel estimation algorithms are unsuitable for miniature IDET devices. Hence, the design of channel estimation should be energy-friendly, requiring both low complexity and low pilot overhead. For instance, a single-bit channel-quality feedback scheme has been proposed in [54] for relieving the burden of complex channel estimation at the receivers.
- Full-duplex communications allow a transmitter to simultaneously transmit and receive data in the same spectrum, which is potentially capable of doubling the bandwidth efficiency. The full duplex concept can also be exploited for supporting

integrated data and energy transfer. For instance, in [55], an IDET station was advocated, which is capable of transferring energy in the downlink and receiving data in the uplink. However, since WET normally requires a high transmit power, the self-interference imposed by the downlink WET on the uplink WDT has to be carefully dealt with. By contrast, the transmitter is also capable of transferring data in the downlink and receiving energy in the uplink. In this case, the self-interference can be recycled as an additional energy source for beneficially exploiting the energy emitted by the transmitter in [56].

- Synchronisation constitutes an essential part of all communications systems [57], and it typically entails establishing coarse initial synchronization of both the symbol timing and the carrier frequency of the transmitter and the receiver. Both the wireless channel and the oscillator's carrier phase offset impose undesirable phase rotation and hence a potential loss of synchronization between the transmitter and the receiver, which hence requires accurate tracking. Furthermore, the WET signals and WDT signals may be interleaved with each other in the time-domain. The receiver has to precisely control the timing of switching between the information decoding mode and the energy harvesting mode in order to recover the information requested from the WDT signals, while harvesting the requested amount of energy from the WET signals [58].
- Non-Orthogonal Multiple Access (NOMA) is a popular technique, which is capable of offering higher bandwidth efficiency than the orthogonal frequency-division multiple access (OFDMA) and the time-division multiple access (TDMA) techniques. Although the superimposed RF signals dedicated to different user equipments (UEs) may impose a high complexity on the receivers, constructively combining RF signals in a non-orthogonal manner is capable of beneficially concentrating the energy carried by the resultant RF signals. As a result, the efficiency of WET can be substantially improved. However, the accurate estimation of amplitude attenuation and phase shift of the channel is a prerequisite for constructively combining the RF signals at the transmitters.
- Multi-objective Optimization is also beneficial in terms of simultaneously satisfying the demands of both WET and WDT. IDET networks have to provide both data and energy services for the UEs. Multi-component objective functions have to be optimized in the design of all layers. However, finding the Pareto front of the multi-objective optimization requires substantial computing capabilities. The emerging quantum computing technique may efficiently solve multi-objective Pareto optimization problems. For instance, a powerful quantum computing aided algorithm has been invoked for simultaneously minimizing energy consumption and achieving load balancing in the routing design of ad hoc wireless networks [59].

Awareness of diverse types of contextual information [60], such as the specific activities of the devices, the geographic location information, the battery levels, the energy consumptions, and the channel coefficients, is essential for the efficient design of the IDET. The inner shell of Fig. 1.2 summarizes generic optimal design principles to be obeyed for coordinating WET and WDT in the same spectral band. Overall, in the design of IDET, the fundamental conflicts between the limited resources and unlimited quality of service (QoS) requirements have to be addressed. For instance, in an IDET system, the functionalities of the IDET transceivers are limited by their hardware. Only a limited number of antennas can be implemented, which results in limited diversity, multiplexing, and beamforming gains. Imperfect electronic modules of the transceivers suffer from considerable energy leakage and insufficient energy conversion. Only limited resources can be provided for IDET in the time-domain, frequency domain, power-domain, and spatial domain. However, a potentially unlimited QoS has to be satisfied in IDET networks. Specifically, WDT is always expected to achieve a high data rate and low transmission latency, while WET aims for achieving a high energy transfer efficiency for extending the battery recharge period of electronic devices. Additionally, the implementation of integrated WET and WDT requires a low complexity and a high compatibility. Furthermore, conflicts also exist between the different QoS requirements. For example, since both WDT and WET are supported by RF signals, they have to compete for limited resources for the sake of satisfying their QoS requirements. As a result, we have to strike a balance between the performance of WET and WDT. An integrated WET and WDT system of high performance also requires complex hardware implementation. We also have to carefully consider the trade-off between the achievable performance and the implementation complexity for arriving at an attractive IDET.

1.7 Ubiquitous Architecture of the DEIN

DEINs are naturally heterogeneous in terms of all their technical aspects. The ubiquitous architecture of the DEIN is exemplified in Fig. 1.3.

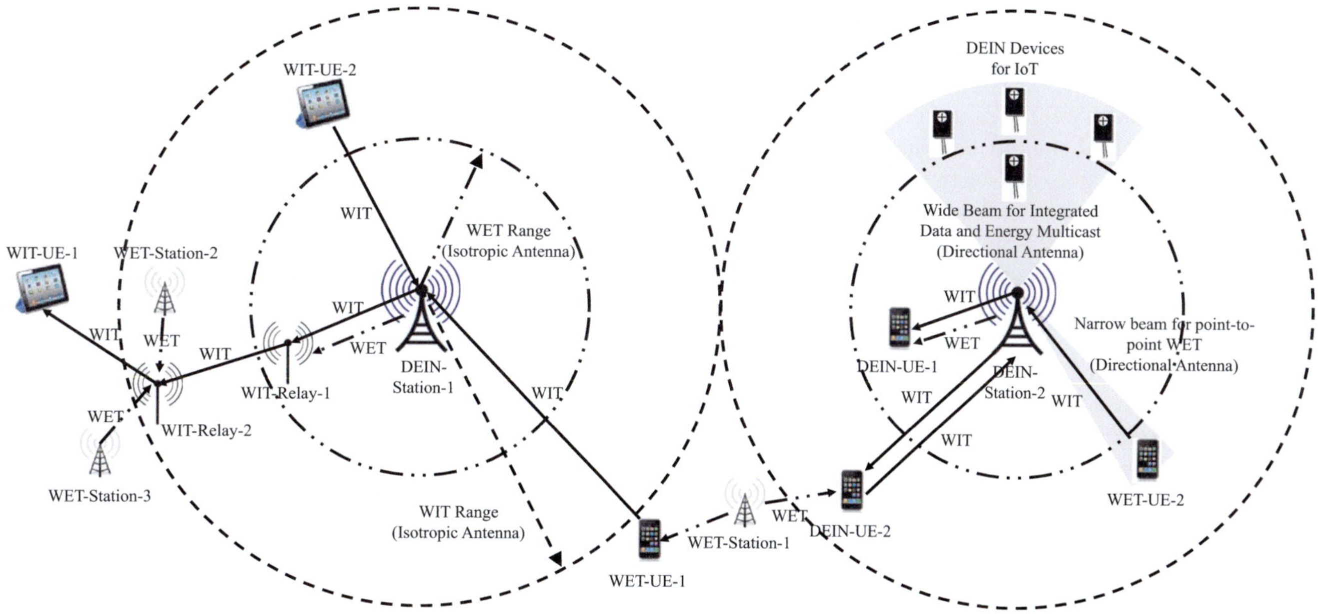

Fig. 1.3 Ubiquitous Architecture of Heterogeneous DEIN

1.7.1 Heterogeneous Infrastructure

First of all, there are various types of infrastructure elements in heterogeneous DEIN. As portrayed in Fig. 1.3, we have generally three basic types of infrastructure in DEINs, namely DEIN stations, WET stations, and WIT stations/relays. DEIN stations are capable of operating both as information transmitter and as energy transmitter for satisfying both of the user equipments' (UEs') data and energy requests. Thanks to their powerful functionalities, DEIN stations are also capable of realizing integrated data and energy transfer for the sake of increasing the spectrum efficiency of the congested RF band. Therefore, DEIN stations have to be connected to the core communication network, and they also have to be powered by stable energy sources, such as large solar energy harvesters and the power grid. As illustrated in Fig. 1.3, DEIN-Station-1 may satisfy the integrated data and energy requests from the IoT devices and those from DEIN-UE-1.

However, as we have discussed in Sect. 1.4, the reliable WET range is far shorter than the reliable WIT range, as exemplified in Fig. 1.3. As a result, some blind areas cannot be adequately covered by WET function of DEIN stations. Furthermore, some WET stations are also deployed in order to supply energy to the devices roaming in these blind areas. These WET stations are only connected to energy sources, but they do not have to be connected to the core communication network. As a result, they are dedicated to satisfying the UEs' charging request. For instance, as shown in Fig. 1.3, three WET stations are deployed in order to supply energy to the UEs beyond the WET range of the DEIN stations.

Apart from DEIN stations and WET stations, there are still many conventional communication stations in heterogeneous DEINs, namely the classic femto-cellular stations, pico-cellular stations, and macro-cellular stations. These communication stations have different levels of transmit power and coverage, which results in obvious heterogeneity in DEINs. Sometimes, low-cost relay stations are also deployed for forwarding the data packets to cell-edge UEs, as illustrated in Fig. 1.3. However, small cellular stations and relay stations are only capable of emitting RF signals at a limited power. They are not suitable for carrying out sophisticated WET tasks. Therefore, they are regarded as a dedicated communication infrastructure.

1.7.2 Heterogeneous User Equipment

Apart from the heterogeneous infrastructure, our DEINs have to accommodate both charging and communication requests from diverse types of UEs. We generally have three types of UEs in DEINs, namely the WIT UEs, the WET UEs, and the DEIN UEs, as exemplified in Fig. 1.3. WIT UEs only have downlink and uplink communication requests in DEINs. Since these UEs are always powered by stable energy sources, they do not request any wireless charging from the DEIN stations. Laptops and tablets are typical WIT UEs, which are either powered by high-capacity batteries or are connected to the power grid. For example, as illustrated in the right part of Fig. 1.3, WIT-UE-1 receives its requested data from DEIN-Station-2 with the aid of two WIT relay stations, while WIT-UE-2 may consume its own energy for powering its uplink information transmission.

By contrast, since WET UEs are not powered by stable energy sources, they have to request additional energy supply either from the DEIN stations or from the WET stations in order to support their basic functionalities, such as uplink information transmissions and energy-consuming computations. For instance, although WET-UE-1 is beyond the WIT range of DEIN-Station-1, it may still establish reliable uplink transmissions with DEIN-Station-2 by exploiting the additional energy received from WET-Station-1, as exemplified in Fig. 1.3. Similarly, the uplink transmission of WET-UE-2 toward DEIN-Station-2 is also powered by DEIN-Station-2 itself. Miniature-sized IoT devices are typical WET UEs, since their functionalities are limited by the amount of energy stored in their batteries.

Furthermore, some UEs simultaneously request data and energy transmissions, which are regarded as DEIN UEs. For instance, in the right cell of Fig. 1.3, DEIN-UE-1 simultaneously receives its requested data and energy from DEIN-Station-1, while DEIN-UE-2 also simultaneously requests both downlink data transmission and wireless charging. However, since DEIN-UE-2 is beyond the WET range of DEIN-Station-1, it can only receive the requested data from DEIN-Station-1, but it can receive energy from WET-Station-1. This energy may be exploited for supporting DEIN-UE-2's uplink data transmission to its associated DEIN-Station-1.

Sometimes, the functionalities of WIT relay stations are also limited by their energy supply, especially when the WIT relay stations rely on energy gleaned from batteries or harvested from renewable sources. As a result, they also need wireless charging from DEIN stations or WET stations for powering their data packet forwarding actions. As a result, WIT relay stations can also be regarded as special "DEIN UEs." As portrayed in the left cell of Fig. 1.3, both data and energy are simultaneously transferred from DEIN-Station-2 to WIT-Relay-1. The energy harvested by WIT-Relay-1 may be further exploited for forwarding the data packets to the next hop. Since WIT-Relay-2 is beyond the WET range of DEIN-Station-2, it has to request WET from the nearby WET-Station-2 and WET-Station-3. After receiving the data packets from WIT-Relay-1 and gleaning sufficient energy from the WET stations, the data packets are finally forwarded to their destination WIT-UE-2.

1.7.3 Heterogeneous Techniques for WIT and WET

Our DEIN architecture has to accommodate both the WET and WIT in the same RF spectral band. Although the WET and WIT both rely on the RF signal, they still have distinctive features, as summarized in Table 1.1. Therefore, in order to satisfy the UEs' information and energy requests, the coexistence of WET and WIT in the DEIN results in natural heterogeneity.

In order to guarantee the seamless WIT coverage, different techniques have to be invoked. As exemplified in Fig. 1.3, when the omnidirectional antennas are adopted, the boundary of a DEIN cell is determined by the WIT range of a DEIN station. As a result, the UEs residing within the WIT range of a DEIN station may receive their requested information via a single-hop cellular link. Furthermore, these UEs are also capable of uploading information to their associated DEIN stations. Observe from Fig. 1.3 that WIT-UE-1, DEIN-UE-1, and DEIN-UE2 all receive their requested information from the downlink WIT of their associated DEIN stations, while WET-UE-2 and DEIN-UE-2 both upload their information to their associated DEIN-Station-1. By contrast, in order to satisfy the information request of a UE sitting beyond the WIT range of a DEIN station, multiple relay stations have to be relied upon for forwarding the information from the DEIN station to the requester or in a reverse direction via the multi-hop transmissions, such as the downlink transmission from DEIN-Station-1 to WIT-UE-2 of Fig. 1.3. In addition, by exploiting the extra energy supplied by the WET stations, a UE beyond the WIT range of a DEIN station is also capable of uploading data to the DEIN station, such as the uplink transmission of WET-UE-2 in Fig. 1.3, which is powered by WET-Station-1.

If we further look into the wireless charging actions in DEINs, various WET techniques have to be invoked for satisfying diverse charging requirements. As illustrated in Fig. 1.3, a DEIN station's WET range is much shorter than its WIT range, when the omnidirectional antenna is adopted. The reason is that for the successful WET, the energy harvesting circuit of the receiver can only be activated by a high received energy. As a result, the WET is more sensitive to the wireless channel attenuation, which is dominated by the path loss. If a WET UE resides within the WET range of a DEIN station, it may successfully harvest energy from the RF signal emitted by this DEIN station. By contrast, when a WET UE is beyond the WET range, it has to request energy from its nearby WET station. If directional antennas are adopted by the DEIN stations, they may form a narrow energy beam for charging the WET UE beyond the WET range, which is characterized by the omnidirectional antennas. As exemplified in the right DEIN cell of Fig. 1.3, WET-UE-2 is still capable of receiving energy from the dedicated narrow energy beam formed by DEIN-Station-2.

Furthermore, IoT devices will be pervasively deployed in the near future. Our heterogeneous DEINs are also responsible for satisfying both of their communication and energy demands. IoT devices normally are clustered in a specific area in order to jointly carry out their tasks. As a result, for the sake of satisfying the charging requests of the multiple IoT devices, DEIN stations may form wide-angle energy beams for covering the cluster of requesters. This technique may be regarded as energy multicast. Moreover, this wide beam is also capable of transferring information and energy together to the multiple requesters.

Directional antennas may enable a DEIN station to focus its energy in the main-lobe, which substantially increases the long-range WET efficiency in the direction of the main-lobe. However, the resultant energy loss in the side-lobes may significantly reduce the WET efficiency in other directions.

References

1. Zanella, A., Bui, N., Castellani, A., Vangelista, L., & Zorzi, M. (2014). Internet of Things for smart cities. *IEEE Internet of Things Journal, 1*(1), 22–32. ISSN: 2327-4662.
2. Alrawais, A., Alhothaily, A., Hu, C., & Cheng, X. (2017). Fog computing for the Internet of Things: Security and privacy issues. *IEEE Internet Computing, 21*(2), 34–42. ISSN: 1089-7801.
3. Choi, B. H., Thai, V. X., Lee, E. S., Kim, J. H., & Rim, C. T. Dipole-coil-based wide-range inductive power transfer systems for wireless sensors. *IEEE Transactions on Industrial Electronics, 63*(5), 3158–3167 (2016). ISSN: 0278-0046.
4. Jiwariyavej, V., Imura, T., & Hori, Y. (2015). Coupling coefficients estimation of wireless power transfer system via magnetic resonance coupling using information from either side of the system. *IEEE Journal of Emerging and Selected Topics in Power Electronics, 3*(1), 191–200. ISSN: 2168-6777.
5. Riehl, P. S., Satyamoorthy, A., Akram, H., Yen, Y. C., Yang, J. C., Juan, B., Lee, C. M., Lin, F. C., Muratov, V., Plumb, W., & Tustin, P. F. (2015). Wireless power systems for mobile devices supporting inductive and resonant operating modes. *IEEE Transactions on Microwave Theory and Techniques, 63*(3), 780–790. ISSN: 0018-9480.
6. Pinuela, M., Yates, D. C., Lucyszyn, S., & Mitcheson, P. D. (2013). Maximizing DC-to-load efficiency for inductive power transfer. *IEEE Transactions on Power Electronics, 28*(5), 2437–2447. ISSN: 0885-8993.
7. Liu, H. (2011). Maximizing efficiency of wireless power transfer With resonant inductive coupling.
8. Zheng, C., Ma, H., Lai, J. S., & Zhang, L. (2015). Design considerations to reduce gap variation and misalignment effects for the inductive power transfer system. *IEEE Transactions on Power Electronics, 30*(11), 6108–6119. ISSN: 0885-8993.
9. Yan, Z., Li, Y., Zhang, C., & Yang, Q. (2014). Influence factors analysis and improvement method on efficiency of wireless power transfer via coupled magnetic resonance. *IEEE Transactions on Magnetics, 50*(4), 1–4. ISSN: 0018-9464.
10. Miller, J. M., Onar, O. C., & Chinthavali, M. (2015). Primary-side power flow control of wireless power transfer for electric vehicle Charging. *IEEE Journal of Emerging and Selected Topics in Power Electronics, 3*(1), 147–162. ISSN: 2168-6777.
11. Kurs, A., Karalis, A., Moffatt, R., Joannopoulos, J. D., Fisher, P., & Soljačić, M. (2007). Wireless power transfer via strongly coupled magnetic resonances. Science, 317(5834), 83–86. ISSN: 0036-8075. eprint: http://science.sciencemag.org/content/317/5834/83.full.pdf.
12. Hwang, H., Moon, J., Lee, B., Jeong, C. H., & Kim, S. W. (2014). An analysis of magnetic resonance coupling effects on wireless power transfer by coil inductance and placement. IEEE Transactions on Consumer Electronics, 60(2), 203–209. ISSN: 0098-3063.
13. Nguyen, M. Q., Plesa, D., Rao, S., & Chiao, J. C. (2014). A multi-input and multi-output wireless energy transfer system. In *2014 IEEE MTT-S International Microwave Symposium (IMS2014)* (pp. 1–3)
14. Nguyen, M. Q., Chou, Y., Plesa, D., Rao, S., & Chiao, J. C. (2015). Multiple-inputs and multiple-outputs wireless power combining and delivering systems. *IEEE Transactions on Power Electronics, 30*(11), 6254–6263. ISSN: 0885-8993.
15. Hu, J., Kun, Y., Guangjun, W., & Lajos, H. (2018). Integrated data and energy communication network: A comprehensive survey. *IEEE Communications Surveys Tutorials, 20*(4), 3169–3219.
16. Niotaki, K., Collado, A., Georgiadis, A., Kim, S., & Tentzeris, M. M. (2014). Solar/electromagnetic energy harvesting and wireless power transmission. *Proceedings of the IEEE, 102*(11), 1712–1722. ISSN: 0018-9219.
17. Lu, X., Wang, P., Niyato, D., Kim, D. I., & Han, Z. (2016). Wireless charging technologies: Fundamentals, standards, and network applications. *IEEE Communications Surveys Tutorials, 18*(2), 1413–1452. ISSN: 1553-877X.
18. Ulukus, S., Yener, A., Erkip, E., Simeone, O., Zorzi, M., Grover, P., & Huang, K. (2015). Energy harvesting wireless communications: A review of recent advances. *IEEE Journal on Selected Areas in Communications, 33*(3), 360–381. ISSN: 0733-8716.
19. Ku, M. L., Li, W., Chen, Y., & Liu, K. J. R. (2016). Advances in energy harvesting communications: Past, present, and future challenges. *IEEE Communications Surveys Tutorials, 18*(2), 1384–1412. ISSN: 1553-877X.
20. Lu, X., Wang, P., Niyato, D., Kim, D. I., & Han, Z. (2015). Wireless networks with RF energy harvesting: A contemporary survey. *IEEE Communications Surveys Tutorials, 17*(2), 757–789. ISSN: 1553-877X.
21. Kim, J. H., Yu, H. Y., & Cha, C. (2014). Efficiency enhancement using beam forming array antenna for microwave-based wireless energy transfer. In *2014 IEEE Wireless Power Transfer Conference* (pp. 288–291).
22. Brown, W. C., & George, R. H. (1964). Rectification of microwave power. *IEEE Spectrum, 1*(10), 92–97. ISSN: 0018-9235.
23. Brown, W. C. (1984). The history of power transmission by radio waves. *IEEE Transactions on Microwave Theory and Techniques, 32*(9), 1230–1242. ISSN: 0018-9480.
24. Brown, W. C. (1981). Status of the microwave power transmission components for the solar power satellite. *IEEE Transactions on Microwave Theory and Techniques, 29*(12), 1319–1327. ISSN: 0018-9480.

25. Brown, W. (1984). Recent advances in key microwave components that impact the design and deployment of the solar power satellite system. In *1984 Antennas and Propagation Society International Symposium* (Vol. 22, pp. 339–340).
26. Dickinson, R. M. (1976). Performance of a high-power, 2.388-GHz receiving array in wireless power transmission over 1.54 km. In *1976 IEEE-MTT-S International Microwave Symposium* (pp. 139–141).
27. Yedavalli, P. S., Riihonen, T., Wang, X., & Rabaey, J. M. (2017). Far-field RF wireless power transfer with blind adaptive beamforming for Internet of Things devices. *IEEE Access, 5*, 1743–1752. ISSN: 2169-3536.
28. Mitani, T., Kawashima, S., & Nishimura, T. (2017). Analysis of voltage doubler behavior of 2.45-GHz voltage doubler-type rectenna. *IEEE Transactions on Microwave Theory and Techniques, 65*(4), 1051–1057. ISSN: 0018-9480.
29. Moraitis, N., Vasileiou, P. N., Kakoyiannis, C. G., Marousis, A., Kanatas, A. G., & Constantinou, P. (2014). Radio planning of single-frequency networks for broadcasting digital TV in mixed-terrain regions. *IEEE Antennas and Propagation Magazine, 56*(6), 123–141. ISSN: 1045-9243.
30. Morelli, M., & Moretti, M. (2017). A maximum likelihood approach for SSS detection in LTE systems. *IEEE Transactions on Wireless Communications, 16*(4), 2423–2433. ISSN: 1536-1276.
31. Sanabria-Russo, L., Barcelo, J., Bellalta, B., & Gringoli, F. (2017). A high efficiency MAC protocol for WLANs: Providing fairness in dense scenarios. *IEEE/ACM Transactions on Networking, 25*(1), 492–505. ISSN: 1063-6692.
32. Ford, R., Zhang, M., Mezzavilla, M., Dutta, S., Rangan, S., & Zorzi, M. (2017). Achieving ultra-low latency in 5G millimeter wave cellular networks. *IEEE Communications Magazine, 55*(3), 196–203. ISSN: 0163-6804.
33. Zhang, T., An, L., Chen, Y., & Chai, K. K. (2014). Aggregate interference statistical modeling and user outage analysis of heterogeneous cellular networks. In *2014 IEEE International Conference on Communications (ICC)* (pp. 1260–1265).
34. Hu, J., Yang, L. L., & Hanzo, L. (2015). Distributed multistage cooperative-social-multicast-aided content dissemination in random mobile networks. *IEEE Transactions on Vehicular Technology, 64*(7), 3075–3089. ISSN: 0018-9545.
35. Hu, J., Yang, L. L., Poor, H. V., & Hanzo, L. (2015). Bridging the social and wireless networking divide: Information dissemination in integrated cellular and opportunistic networks. *IEEE Access, 3*, 1809–1848. ISSN: 2169-3536.
36. Hu, J., Yang, L. L., & Hanzo, L. (2016). Delay analysis of social group multicast-aided content dissemination in cellular system. *IEEE Transactions on Communications, 64*(4), 1660–1673. ISSN: 0090-6778.
37. Hu, J., Yang, L. L., & Hanzo, L. (2017). Energy-efficient cross-layer design of wireless mesh networks for content sharing in online social networks. *IEEE Transactions on Vehicular Technology, PP*(99), 1. ISSN: 0018-9545.
38. Yang, K., Yu, Q., Leng, S., Fan, B., & Wu, F. (2016). Data and energy integrated communication networks for wireless big data. *IEEE Access, 4*, 713–723. ISSN: 2169-3536.
39. Lee, H., Kwon, J., Lim, W., Lee, W., Kang, H., Hwang, K. C., Lee, K. Y., Park, C. S., & Yang, Y. (2017). Optimized current of the peaking amplifier for two-stage Doherty power amplifier. *IEEE Transactions on Microwave Theory and Techniques, 65*(1), 209–217. ISSN: 0018-9480.
40. Siviloglou, G. A., Broky, J., Dogariu, A., & Christodoulides, D. N. (2007). Observation of accelerating airy beams. *Physical Review Letters, 99*, 213901.
41. Chremmos, I. D., Fikioris, G., & Efremidis, N. K. (2013). Accelerating and abruptly-autofocusing beam waves in the fresnel zone of antenna arrays. *IEEE Transactions on Antennas and Propagation, 61*(10), 5048–5056. ISSN: 0018-926X.
42. Firrao, E. L., Annema, A. J., van Vliet, F. E., & Nauta, B. (2017). Hardware implementation overhead of switchable matching networks. *IEEE Transactions on Circuits and Systems I: Regular Papers, 64*(5), 1152–1163. ISSN: 1549-8328.
43. Charthad, J., Dolatsha, N., Rekhi, A., & Arbabian, A. (2016). System-level analysis of far-field radio frequency power delivery for mm-sized sensor nodes. *IEEE Transactions on Circuits and Systems I: Regular Papers, 63*(2), 300–311. ISSN: 1549-8328.
44. Hameed, Z., & Moez, K. (2014). Hybrid forward and backward threshold compensated RF DC power converter for RF energy harvesting. *IEEE Journal on Emerging and Selected Topics in Circuits and Systems, 4*(3), 335–343. ISSN: 2156-3357.
45. Gastpar, M. (2007). On capacity under receive and spatial spectrum-sharing constraints. *IEEE Transactions on Information Theory, 53*(2), 471–487. ISSN: 0018-9448.
46. Tandon, A., Motani, M., & Varshney, L. R. (2014). On code design for simultaneous energy and information transfer. In *2014 Information Theory and Applications Workshop (ITA)* (pp. 1–6). IEEE. ISBN: 978-1-4799-3589-5.
47. Wang, J. Y., Yang, Z., Wang, Y., & Chen, M. (2016). On the performance of spatial modulation-based optical wireless communications. *IEEE Photonics Technology Letters, 28*(19), 2094–2097. ISSN: 1041-1135.
48. Hanzo, L., El-Hajjar, M., & Alamri, O. (2011). Near-capacity wireless transceivers and cooperative communications in the MIMO era: Evolution of standards, waveform design, and future perspectives. *Proceedings of the IEEE, 99*(8), 1343–1385. ISSN: 0018-9219.
49. Dong, Y., Hossain, M. J., & Cheng, J. (2016). Joint power control and time switching for SWIPT systems with heterogeneous QoS requirements. *IEEE Communications Letters, 20*(2), 328–331. ISSN: 1089-7798.
50. Zhou, F., Li, Z., Cheng, J., Li, Q., & Si, J. (2017). Robust AN-aided beamforming and power splitting design for secure MISO cognitive radio with SWIPT. *IEEE Transactions on Wireless Communications, 16*(4), 2450–2464. ISSN: 1536-1276.
51. Sohrabi, F., & Yu, W. (2017). Hybrid analog and digital beamforming for mmWave OFDM large-scale antenna arrays. *IEEE Journal on Selected Areas in Communications, PP*(99), 1. ISSN: 0733-8716.
52. Clerckx, B. (2018). Wireless information and power transfer: Nonlinearity, waveform design, and rate-energy tradeoff. *IEEE Transactions on Signal Processing, 66*(4), 847–862.
53. Popovski, P., Fouladgar, A. M., & Simeone, O. (2013). Interactive joint transfer of energy and information. *IEEE Transactions on Communications, 61*(5), 2086–2097. ISSN: 0090-6778.
54. Xu, J., & Zhang, R. (2014). Energy beamforming with one-bit feedback. *IEEE Transactions on Signal Processing, 62*(20), 5370–5381. ISSN: 1053-587X.
55. Hu, J., Xue, Y., Yu, Q., & Yang, K. (2017). A joint time allocation and UE scheduling algorithm for full-duplex wireless powered communication networks. In *2017 IEEE 86th Vehicular Technology Conference (VTC-Fall)* (pp. 1–5)
56. Maso, M., Liu, C. F., Lee, C. H., Quek, T. Q. S., & Cardoso, L. S. (2015). Energy-recycling full-duplex radios for next-generation networks. *IEEE Journal on Selected Areas in Communications, 33*(12), 2948–2962. ISSN: 0733-8716.
57. Venmani, D. P., Le Moult, O., Deletre, F., Lagadec, Y., & Morlon, Y. (2016). On the role of network synchronization for future cellular networks: An operator's perspective. *IEEE Communications Magazine, 54*(9), 58–64.

58. Varshney, L. R. (2012). On energy/information cross-layer architectures. In *2012 IEEE International Symposium on Information Theory Proceedings* (pp. 1356–1360). IEEE. ISBN: 978-1-4673-2579-0.
59. Alanis, D., Hu, J., Botsinis, P., Babar, Z., Ng, S. X., & Hanzo, L. (2016). Quantum-assisted joint multi-objective routing and load balancing for socially-aware networks. *IEEE Access, 4*, 9993–10028. ISSN: 2169-3536.
60. Wu, J., Bisio, I., Gniady, C., Hossain, E., Valla, M., & Li, H. (2014). Context-aware networking and communications: Part 1 [guest editorial]. *IEEE Communications Magazine,52*(6), 14–15.

Chapter 2
Coding Controlled Integrated Data and Energy Transfer

2.1 Classic Information Theory

2.1.1 Entropy

First introduce the concept of entropy, which is a measure of the uncertainty of a random variable. Let X be a discrete random variable with a value space $\mathcal{X}$, and a probability density function (PDF) $p(x) = Pr(X = x), x \in \mathcal{X}$.

Definition 2.1 The entropy of a discrete random variable X is defined as:

$$H(X) = -\sum_{x \in \mathcal{X}} p(x) \log p(x), \tag{2.1}$$

where the base used for log is 2 and the units of entropy are bits. If not specifically stated, the log base is chosen to be 2 in this book. Note that entropy is actually a general function of the distribution of the random variable X and is not dependent on the actual value of X, but only on its probability distribution.

2.1.1.1 Joint Entropy and Conditional Entropy

The entropy of a single random variable was defined in Sect. 2.1.1. Now, the definition is extended to the case of two random variables.

Definition 2.2 For a pair of discrete random variables (X, Y) that obey a joint distribution of $p(x, y)$, the joint entropy $H(X, Y)$ is defined as:

$$H(X, Y) = -\sum_{x \in \mathcal{X}} \sum_{y \in \mathcal{Y}} p(x, y) \log p(x, y). \tag{2.2}$$

Definition 2.3 If $(X, Y) \sim p(x, y)$, the conditional entropy $H(X|Y)$ is defined as:

$$\begin{aligned} H(X|Y) &= \sum_{x \in \mathcal{X}} p(x) H(Y|X = x) \\ &= -\sum_{x \in \mathcal{X}} p(x) \sum_{y \in \mathcal{Y}} p(y|x) \log p(y|x) \\ &= -\sum_{x \subset \mathcal{X}} \sum_{y \subset \mathcal{Y}} p(x, y) \log p(y|x). \end{aligned} \tag{2.3}$$

J. Hu, K. Yang, *Integrated Data and Energy Transfer in Wireless Networks*, Textbooks in Telecommunication Engineering,
https://doi.org/10.1007/978-3-031-60514-7_2

This naturalness of the definition of joint and conditional entropy is captured by the fact that the entropy of a pair of random variables is equal to the entropy of one of the random variables plus the conditional entropy of the other random variable. The proof of this is given in the following theorem.

Theorem 2.1 (Chain Rule)

$$H(Y|X) = H(X) + H(Y|X) \tag{2.4}$$

Proof

$$\begin{aligned}
H(X, Y) &= -\sum_{x\in\mathcal{X}}\sum_{y\in\mathcal{Y}} p(x, y) \log p(x, y) \\
&= -\sum_{x\in\mathcal{X}}\sum_{y\in\mathcal{Y}} p(x, y) \log p(x)p(y|x) \\
&= -\sum_{x\in\mathcal{X}}\sum_{y\in\mathcal{Y}} p(x, y) \log p(x) - \sum_{x\in\mathcal{X}}\sum_{y\in\mathcal{Y}} p(x, y) \log p(y|x) \\
&= -\sum_{x\in\mathcal{X}} p(x) \log p(x) - \sum_{x\in\mathcal{X}}\sum_{y\in\mathcal{Y}} p(x, y) \log p(y|x) \\
&= H(X) + H(Y|X)
\end{aligned} \tag{2.5}$$

2.1.1.2 Relative Entropy and Mutual Information

Entropy is a measure of the uncertainty of a random variable; it is also a measure of the amount of information required to describe a random variable in an average sense. In this subsection two related concepts will be introduced: relative entropy and mutual information.

Relative entropy is a measure of the distance between two random distributions. Relative entropy $D(p||q)$ measures the nullity when the true distribution is p and the assumed distribution is q. For example, if the true distribution of a random variable is known to be p, a code of average description length $H(p)$ can be constructed. However, if a code for distribution q is used, then $H(p) + D(p||q)$ bits are required in an average sense to describe this random variable.

Definition 2.4 The relative entropy or Kullback–Leibler distance between two probability density functions as $p(x)$ and $q(x)$ is defined as:

$$\begin{aligned}
D(p||q) &= \sum_{x\in\mathcal{X}} p(x) \log \frac{p(x)}{q(x)} \\
&= E_p \log \frac{p(X)}{q(X)}.
\end{aligned} \tag{2.6}$$

Coming now to mutual information, it is a measure of the amount of information that a random variable contains about another random variable. The mutual information is also the amount of reduction in the uncertainty of the original random variable given the knowledge of the other random variable.

Definition 2.5 Consider two random variables X and Y with joint probability density functions $p(x, y)$, whose marginal probability density functions are $p(x)$ and $p(y)$, respectively. The mutual information $I(X; Y)$ is the relative entropy between the joint distribution $p(x, y)$ and the product distribution $p(x)p(y)$, i.e.:

$$\begin{aligned}
I(X; Y) &= \sum_{x\in\mathcal{X}}\sum_{y\in\mathcal{Y}} p(x, y) \log \frac{p(x, y)}{p(x)p(y)} \\
&= D(p(x, y)||p(x)p(y)) \\
&= E_{p(x,y)} \log \frac{p(X, Y)}{p(X)p(Y)}.
\end{aligned} \tag{2.7}$$

It is possible to extend this definition to the case of continuous random variables as well, using differential entropy to represent the entropy of a continuous random variable.

2.1.2 *Mutual Information*

The mutual information $I(X;Y)$ can be rewritten as:

$$\begin{aligned}
I(X;Y) &= \sum_{x,y} p(x,y)\log\frac{p(x,y)}{p(x)p(y)} \\
&= \sum_{x,y} p(x,y)\log\frac{p(x|y)}{p(x)} \\
&= -\sum_{x,y} p(x,y)\log p(x) + \sum_{x,y} p(x,y)\log p(x|y) \\
&= -\sum_{x} p(x)\log p(x) - (-\sum_{x,y} p(x,y)\log p(x|y)) \\
&= H(X) - H(X|Y).
\end{aligned} \tag{2.8}$$

The mutual information $I(X;Y)$ is therefore a reduction in the uncertainty of X given knowledge of Y. Symmetrically, one obtains:

$$I(X;Y) = H(Y) - H(Y|X). \tag{2.9}$$

Therefore, the amount of information that X contains Y is equivalent to the amount of information that Y contains X. From Theorem 2.1, we have:

$$I(X;Y) = H(X) + H(Y) - H(X,Y). \tag{2.10}$$

2.1.2.1 The Chain Rule of Entropy and Mutual Information

The entropy of a set of random variables is equal to the sum of the conditional entropies.

Theorem 2.2 (The Chain Rule of Entropy)

$$H(X_1, X_2 \ldots, X_n) = \sum_{i=1}^{n} H(X_i|X_{i-1}, \ldots, X_1) \tag{2.11}$$

Proof

$$\begin{aligned}
H(X_1, X_2 \ldots, X_n) &= -\sum_{x_1,x_2,\ldots,x_n} p(x_1,x_2,\ldots,x_n)\log p(x_1,x_2,\ldots,x_n) \\
&= -\sum_{x_1,x_2,\ldots,x_n} p(x_1,x_2,\ldots,x_n)\log\prod_{i=1}^{n} p(x_i|x_{i-1},\ldots,x_1) \\
&= -\sum_{x_1,x_2,\ldots,x_n}\sum_{i=1}^{n} p(x_1,x_2,\ldots,x_n)\log p(x_i|x_{i-1},\ldots,x_1) \\
&= -\sum_{i=1}^{n}\sum_{x_1,x_2,\ldots,x_i} p(x_1,x_2,\ldots,x_i)\log p(x_i|x_{i-1},\ldots,x_1) \\
&= \sum_{i=1}^{n} H(X_i|X_{i-1},\ldots,X_1)
\end{aligned} \tag{2.12}$$

The following defines conditional mutual information, which is the reduction in uncertainty about X due to knowledge of Y at a given Z.

Definition 2.6 The conditional mutual information of the random variables X and Y given the random variable Z is defined as:

$$\begin{aligned} I(X;Y|Z) &= H(X|Z) - H(X|Y,Z) \\ &= E_{p(x,y,z)} \log \frac{p(X,Y|Z)}{p(X|Z)p(Y|Z)}. \end{aligned} \tag{2.13}$$

Mutual information also satisfies the chain rule.

Theorem 2.3 (The Chain Rule of Mutual Information)

$$I(X_1, X_2, \ldots, X_N; Y) = \sum_{i=1}^{n} I(X_i; Y|X_{i-1}, X_{i-2}, \ldots, X_1) \tag{2.14}$$

Proof

$$\begin{aligned} I(X_1, X_2 \ldots, X_n; Y) &= H(X_1, X_2 \ldots, X_n) - H(X_1, X_2 \ldots, X_n|Y) \\ &= \sum_{i=1}^{n} H(X_i|X_{i-1} \ldots, X_1) - \sum_{i=1}^{n} H(X_i|X_{i-1} \ldots, X_1, Y) \\ &= \sum_{i=1}^{n} I(X_i; Y|X_1, X_2 \ldots, X_{i-1}). \end{aligned} \tag{2.15}$$

2.1.3 Lossless Source Coding

There are two important issues for the source: one is the calculation of the amount of information output from the source, and the other is the problem of how to represent the source output more efficiently. In order to improve communication efficiency, it is often necessary to process the source. One way of processing is source coding.

For any source, one wants to transmit all the information to the receiver without any reservations, i.e., to achieve lossless transmission. For this, it is first necessary to implement lossless source coding.

A discrete source outputs a message that is a random sequence of discrete symbols.

$$\begin{aligned} \mathbf{X} &= (X_1 X_2 \ldots X_l \ldots X_L) \\ X_l &\in a_1, a_2, \ldots, a_i, \ldots, a_n \end{aligned} \tag{2.16}$$

Source coding is the transformation of a random sequence of symbols from the source output into a code sequence.

$$\begin{aligned} \mathbf{Y} &= (Y_1 Y_2 \ldots Y_k \ldots Y_K) \\ Y_k &\in b_1, b_2, \ldots, b_j, \ldots, b_m \end{aligned} \tag{2.17}$$

To achieve distortion-free transmission, a one-to-one correspondence between source messages and code sequences is required. There are two methods of distortion-free discrete source coding, fixed-length and variable length. In the former case, the code word length K is fixed and the corresponding coding theorem is also known as the fixed-length coding theorem. In the latter case, K is variable and the corresponding coding theorem is also known as the variable-length coding theorem.

2.1.3.1 Fixed-Length Coding Theorem

A smooth memoryless sequence of symbols $X_1X_2 \ldots X_l \ldots X_L$ consisting of L symbols, each with entropy $H(X)$, can be encoded with fixed length by K symbols $Y_1Y_2 \ldots Y_k \ldots Y_K$, for any $\sigma > 0$, $\varepsilon > 0$, as long as

$$\frac{K}{L} \log m \geq H(X) + \varepsilon. \tag{2.18}$$

Then, when L is large enough, it must be possible to make the decoding error less than σ. Conversely, when:

$$\frac{K}{L} \log m \leq H(X) - 2\varepsilon. \tag{2.19}$$

At this point, the decoding error must be finite, and when L is large enough, the decoding must be wrong.

In the inequality, m represents the possible values of each symbol in the code sequence. If the m symbols are equally probable, the information content of a single symbol is $\log m$. For fixed-length coding, the length of each code word is K, so the code word is always m^K. If the source is smooth and memoryless, the average symbol entropy is $\log m^K$. Thus the source entropy $H(X)$ is a bound, and when the information rate of the encoder output exceeds this bound, it can be decoded without distortion-free decoding when the information rate at the output of the encoder exceeds this bound, but not otherwise.

2.1.3.2 Variable-Length Coding Theorem

If the symbol entropy of a discrete memoryless source is $H(X)$ and the source symbol is coded with m-element variable length, there must exist a distortion-free coding method whose average code word length $\hat{K}$ satisfies the following inequality:

$$1 + \frac{H(X)}{\log m} > \hat{K} \geq \frac{H(X)}{\log m}. \tag{2.20}$$

The average information rate satisfies the following inequality:

$$H(X) \leq R < H(X) + \varepsilon, \tag{2.21}$$

where ε represents any positive number.

2.1.4 Lossy Source Coding

The actual discrete source has a finite number of values and can be coded one-to-one to achieve lossless transmission of the source. Continuous sources, on the other hand, have an infinite number of values and cannot be coded one-to-one, so distortion is bound to occur. In practical communication systems, however, distortion-free transmission of information is not required, as long as the distortion is limited to a certain range, i.e., the fidelity criterion is satisfied, the quality of communication can be guaranteed.

Let the sequence of output random variables of a discrete smooth memoryless source be $\mathbf{X} = (X_1X_2 \ldots X_L)$. If the information rate distortion function of this source is $R(D)$, and a finite distortion function is selected. For any allowed average distortion $D \geq 0$, and any small ε, if the information rate satisfies:

$$R > R(D). \tag{2.22}$$

Provided that the source sequence length L is sufficiently long, there must exist a coding method C such that the average distortion after decoding satisfies:

$$\hat{D}(C) \leq D + \varepsilon \tag{2.23}$$

conversely, if $R < R(D)$. Then, regardless of the coding method used, there will necessarily be:

$$\hat{D}(C) > D. \tag{2.24}$$

It means that the average distortion of the decoding code must be greater than the allowed distortion. This is known as the distortion-limited source coding theorem, where the information rate distortion function is the minimum average amount of information necessary to find the reproduction of the source message, i.e., the minimum value of the average mutual information, under the condition that the fidelity criterion is satisfied, and its expression is:

$$R(D) = \min I(X; Y). \tag{2.25}$$

2.1.5 *Channel Coding*

Information transmitted over the channel is inevitably subject to random noise or human interference, and the purpose of channel coding is to reduce interference and improve the reliability of information transmission.

If there is a discrete memoryless smooth channel with a channel capacity of C and an input sequence of length L, provided that the transmitted information rate satisfies the following expression:

$$R < C. \tag{2.26}$$

It is always possible to find an encoding for which the decoding error probability $P_e < \varepsilon$ when L is sufficiently long, ε is any positive number greater than 0. Conversely, when $R > C$, P_e must be greater than 0 for any encoding.

The channel capacity is a critical value at which information can be transmitted almost distortion-free in a channel with interference, provided that the information transmission rate does not exceed this critical value, otherwise distortion will occur.

2.2 Information Theory for Integrated Data and Energy Transfer

The WET and the WIT entail several conflicting specifications, when they are coordinated in the same RF spectral band. As a result, theoretical investigations have to be carried out in order to reveal the underlying relationship between the WET and WIT in IDENs, which may provide researchers and engineers with further valuable insights on improving the system-level performance of IDENs. In this section, we will explore the information theoretical essence for integrated WET and WIT.

Note that the information theoretical exploration remain valid not only for the integrated WET and WIT operating in the RF spectral band, but also for all other integrated information and energy transfer scenarios, such as the power line communication [1] and the power over Ethernet technique [2]. Classic information theoretical channel capacity analysis has been dedicated to maximizing the mutual information under the constraint of specific input signals. By contrast, the pioneering work of Gastpar [3] has attempted to maximize the mutual information under the constraint of specific output signals, which aims for controlling the interference imposed on a communication pair by other peers. This piece of work may provide us with valuable hints for finding the performance limits of integrated information and energy transfer.

2.2.1 *Discrete-Input-Discrete-Output Memoryless Channel*

We first consider the classic Discrete-Input-Discrete-Output Memoryless Channel (DIDO-MC) of Fig. 2.1. If a DIDO-MC has a single input symbol x and a single output symbol y, the transition probability of this DIDO-MC channel may be expressed by the probability function of $p_{\mathcal{Y}|\mathcal{X}}(y|x)$. All the legitimate values of the input symbol x constitute the input codebook $\mathcal{X}$, while all the legitimate values of the output symbol y constitute the output codebook $\mathcal{Y}$. Furthermore, given a specific output symbol y, the energy carried by it can be represented by the non-negative function $g(y)$.

Let us now move on to the n-dimensional random input, which is expressed by the vector $\mathbf{X}^n = (X_1, X_2, \cdots, X_n)$. All the random symbols in the vector $\mathbf{X}^n$ are independent of one another. If a sample of the n dimensional random input is $\mathbf{x}^n = (x_1, x_2, \cdots, x_n)$, its corresponding occurrence probability can be expressed as $p_{\mathcal{X}^n}(\mathbf{x}^n) = \prod_{i=1}^{n} p_{\mathcal{X}}(x_i)$, where $\mathcal{X}^n$ represents the n-dimensional codebook containing all the possible values of the random input $\mathbf{X}^n$ and $p_{\mathcal{X}}(x_i)$ represents

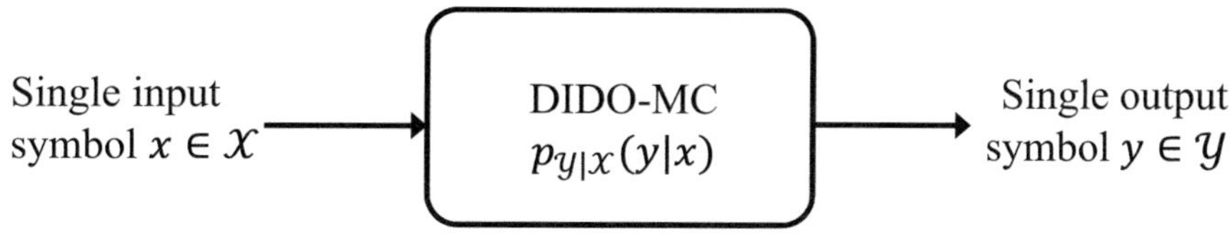

Fig. 2.1 Discrete-input-discrete-output memoryless channel

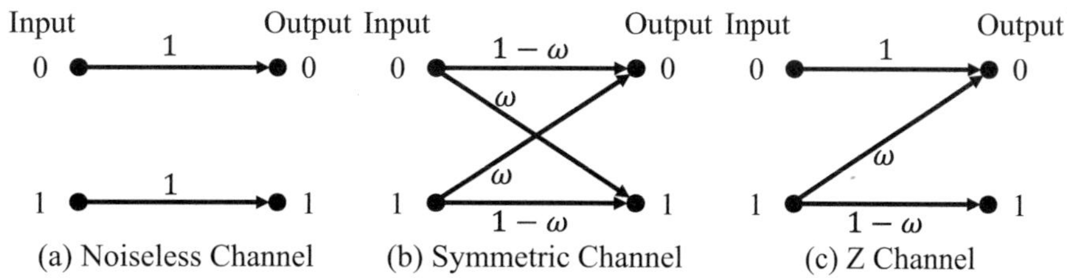

Fig. 2.2 Three typical binary channels

the probability of the symbol x_i being generated. When the input sample $\mathbf{x}^n$ is transmitted by the information source, the corresponding output at the information destination is denoted by the vector $\mathbf{y}^n = (y_1, y_2, \cdots, y_n)$. This sequence of symbols occurs with a probability of $p_{\mathcal{Y}^n}(\mathbf{y}^n) = \prod_{i=1}^n p_{\mathcal{Y}}(y_i)$, where $\mathcal{Y}^n$ represents the n-dimensional codebook containing all the possible values of the random input $\mathbf{Y}^n$ and $p_{\mathcal{Y}}(y_i) = \sum_{x_i \in \mathcal{X}} p_{\mathcal{X}}(x_i) p_{\mathcal{Y}|\mathcal{X}}(y_i|x_i)$ represents the probability of the symbol y_i being received by the information destination. The energy carried by the sequence $\mathbf{y}^n$ can be calculated by the non-negative function $g(\mathbf{y}^n)$. Assuming a random output sequence $\mathbf{Y}^n = (Y_1, Y_2, \cdots, Y_n)$, the average energy carried by this n-dimensional sequence can be formulated as $\mathbb{E}[g(\mathbf{Y}^n)] = \sum_{\mathbf{y}^n \in \mathcal{Y}^n} g(\mathbf{y}^n) \cdot p_{\mathcal{Y}^n}(\mathbf{y}^n)$. As a result, the trade-off between the information transfer and the energy transfer can be formulated as the following optimization problem:

$$\text{Objective: } \max_{p_{\mathcal{X}^n}(\mathbf{x}^n)} I(\mathbf{X}^n; \mathbf{Y}^n), \tag{2.27}$$

$$\text{Subject to: } \frac{1}{n} \cdot \mathbb{E}[g(\mathbf{Y}^n)] \geq \beta, \tag{2.27a}$$

where $I(\mathbf{X}^n; \mathbf{Y}^n)$ is the average mutual information between the n-dimensional input symbol sequence and its output counterpart. Given the transition probabilities $p_{\mathcal{Y}|\mathcal{X}}(\mathbf{Y}|\mathbf{X})$ of the DMC, the optimization problem (2.27) aims for finding the optimal n-dimensional information source, which is represented by the probabilities $p_{\mathcal{X}^n}(\mathbf{x}^n)$ of the symbol sequences being generated. The constraint (2.27a) suggests that the average energy carried by a single output symbol has to be higher than a threshold β in order to satisfy the energy transfer requirement.

Substituting the optimal $p_{\mathcal{X}^n}(\mathbf{x}^n)$ into the objective (2.27), we may derive the channel capacity $C_n(\beta)$, when the input is an n-dimensional symbol sequence. Note that $C_n(\beta)$ is a function of the energy constraint β. Furthermore, the normalized channel capacity subject to the energy constraint β can be further formulated as

$$C(\beta) = \sup_n \frac{1}{n} \cdot C_n(\beta), \quad \text{[bit/symbol]}, \tag{2.28}$$

which may be regarded as the rate-energy function. Observe from (2.28) that the rate-energy function is a natural extension of the classic channel capacity concept. This function only depends on a channel's statistical property and on the requirement of the energy transfer, but it does not rely on the information source.

The rate-energy functions of some simple binary channels will now be studied for illustrating the above information theoretical methodology. In binary channels, the energy carried by the output symbol can be defined as follows. If the output symbol is "0", its energy is $g(0) = 0$. If the output symbol is "1", its energy is $g(1) = 1$.

As illustrated in Fig. 2.2a, in a noiseless channel, the binary symbol "0" or "1" can be delivered from the information source to the destination without any distortion. By exploiting the maximum entropy theory of the information source, the closed-form rate-energy function of the noiseless channel can be expressed as [4]

$$C(\beta) = \begin{cases} \log 2, & 0 \le \beta \le \frac{1}{2}, \\ H_2(\beta), & \frac{1}{2} < \beta \le 1, \end{cases} \tag{2.29}$$

where $H_2(\beta) = -\beta \log \beta - (1 - \beta) \log(1 - \beta)$ is the binary entropy function with respect to β.

As portrayed in Fig. 2.2b, in a symmetric channel, the probabilities of the input symbols "0" and "1" being delivered to the destination without any distortion are both ω, while the probability of the input symbol 1 being converted to the output symbol 0 and the probability of the input symbol 0 being converted to the output symbol "1" are both $(1 - \omega)$. The closed-form rate-energy function of the symmetric channel can be formulated as [4]

$$C(\beta) = \begin{cases} \log 2 - H_2(\omega), & 0 \le \beta \le \frac{1}{2} \\ H_2(\beta) - H_2(\omega), & \frac{1}{2} < \beta \le 1 - \omega, \end{cases} \tag{2.30}$$

where $H_2(\beta)$ and $H_2(\omega)$ are binary entropy functions with respect to β and ω, respectively. According to classic information theory without any consideration of WET, the average mutual information can be maximized when the binary symbols generated by the binary information source have equal probabilities. In this case, the maximum energy delivered to the information destination is only $\frac{1}{2}$ unit in average. If the energy threshold at the destination is higher than $\frac{1}{2}$, the information source has to increase the probability of transmitting symbol "1". If the source only transmits symbol "1", the destination may receive the maximum energy $\beta_{\max} = (1 - \omega)$.

As shown in Fig. 2.2c, in a Z channel, the input symbol "0" can be delivered to the destination without any error probability. However, the input symbol "1" may be converted to the output symbol "0" with a probability of ω, while it remains the same at the output with a probability of $(1 - \omega)$. The closed-form rate-energy function of the Z channel can be formulated as [4]

$$C(\beta) = \begin{cases} \log \left(1 - \omega^{\frac{1}{1-\omega}} + \omega^{\frac{\omega}{1-\omega}}\right), & 0 \le \beta \le (1 - \omega)\pi^* \\ H_2(\beta) - \frac{\beta}{1-\omega} H_2(\omega), & (1 - \omega)\pi^* < \beta \le 1 - \omega, \end{cases} \tag{2.31}$$

where the variable π^* is given by [4]

$$\pi^* = \frac{\omega^{\frac{\omega}{1-\omega}}}{1 + (1 - \omega)\omega^{\frac{\omega}{1-\omega}}}. \tag{2.32}$$

We plot the rate-energy functions of these binary channels in Fig. 2.3. Observe from Fig. 2.3 that given the same energy threshold β, naturally the noiseless channel has the highest information rate, since the information symbols can be transmitted without any error probability over this channel. Furthermore, we also find that given the same energy threshold β, the Z channel may achieve a higher information rate than the symmetric channel, since the Z channel produces fewer errors than the symmetric channel.

2.2.2 *Continuous-Input-Continuous-Output Memoryless Channel*

A similar methodology can also be applied in a continuous-input-continuous-output memoryless channel (CICO-MC), as shown in Fig. 2.4. Both the input message and the output message are continuous random variables, defined as X and Y, respectively. The transition probability density function of a CICO-MC can be defined as $q_{Y|X}(y|x)$, when the information source randomly generates the continuous input message x and the corresponding continuous output message y is received by the information destination. All the legitimate values of the random input X constitute the space $\mathcal{X}$, while the space $\mathcal{Y}$ contains all the legitimate values of Y. The energy carried by a continuous output message y can be represented by the function $g(y)$. Given the transition probability density function $q_{Y|X}(y|x)$ of the CICO-MC, the average energy $\mathbb{E}[g(y)]$ is dependent on the cumulative distribution function $F_X(x)$ of the continuous random input X.

Therefore, the rate-energy function of the CICO-MC can be formulated by the following optimization problem:

$$\text{Objective: } \max_{F_X(x)} I(X; Y), \tag{2.33}$$

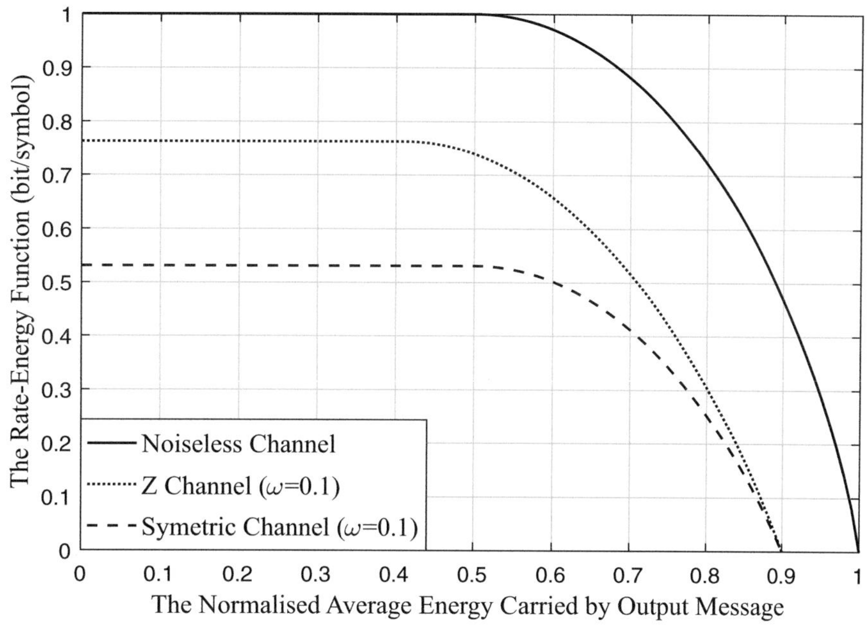

Fig. 2.3 Rate-energy functions for three binary channels

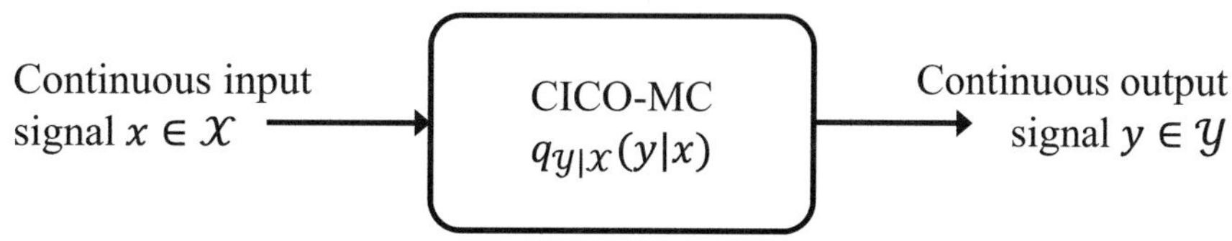

Fig. 2.4 Continuous-input-continuous-output memoryless channel

$$\text{Subject to: } \mathbb{E}[g(y)] \geq \beta, \tag{2.33a}$$

where $I(X; Y)$ is the average mutual information between the continuous random input X and the continuous output Y, whereas β is the minimum energy transfer requirement. The optimization problem (2.33) aims for finding a test information source X^*, whose cumulative distribution function (CDF) $F_{X^*}(x)$ is capable of maximizing the average mutual information $I(X; Y)$, while satisfying the minimum energy requirement β. Upon substituting the optimal CDF $F_{X^*}(x)$ into (2.33), we may derive the rate-energy function of the CICO-MC. Assuming that the input message falls into the region $[-\alpha, \alpha]$, the rate-energy function can be defined as $C(\alpha, \beta)$.

Let us now study the AWGN channel as an example of the CICO-MC. In the AWGN channel, the Gaussian distributed noise N has a mean of zero and a variance of σ_N^2. The output of the AWGN channel can be expressed as $Y = X + N$, whose average energy can be formulated as $\mathbb{E}[g(y)] = \mathbb{E}[Y^2] = \mu_x^2 + \sigma_x^2 + \sigma_N^2$, where μ_x is the mean of the random input X and σ_x^2 is its variance.

In the case of the AWGN channel, the optimization problem (2.33) can be expressed as

$$\text{Objective: } \max_{F_X(x)} H_c(Y) - H_c(N) = H_c(Y) - \frac{1}{2}\log 2\pi e \sigma_N^2 \tag{2.34}$$

$$\text{Subject to: } \mathbb{E}[g(y)] = \mu_x^2 + \sigma_x^2 + \sigma_N^2 \geq \beta. \tag{2.34a}$$

Naturally, we have to maximize the continuous entropy $H_c(Y)$ for the random output Y. Since the average power $P_Y = \mathbb{E}[g(y)]$ of the output message is constrained both by the average power of the input message as well as by the power of the noise, $H_c(Y)$ becomes maximum, when Y follows a Gaussian distribution having a zero mean and a variance of $\sigma_y^2 = P_Y$. Since the noise N is Gaussian distributed, the random input message X has to be a Gaussian distributed random variable having a zero mean as well. According to the constraint (2.34a), the variance of X should obey $\sigma_x^2 \geq (\beta - \sigma_N^2)$. Since the variance of the output message Y is $\sigma_y^2 = \sigma_x^2 + \sigma_N^2$, the rate-energy function of the AWGN channel can thus be expressed as

$$
\begin{aligned}
C_{AWGN}(\beta) &= H_{c,\max}(Y) - \frac{1}{2}\log 2\pi e\sigma_N^2 \\
&= \frac{1}{2}\log 2\pi e\sigma_y^2 - \frac{1}{2}\log 2\pi e\sigma_N^2 \\
&= \frac{1}{2}\log\frac{\sigma_x^2+\sigma_N^2}{\sigma_N^2} = \frac{1}{2}\log\frac{\beta}{\sigma_N^2}.
\end{aligned} \tag{2.35}
$$

2.2.3 *Joint Energy and Information Coding*

Channel coding, also known as the forward error correction coding, incorporates redundant bits into the original information stream at the transmitter side in order to correct the errors imposed by the hostile wireless channel. With the aid of the redundant bits, the receiver becomes capable of automatically correcting the erroneous bits, hence recovering the original information bits. As stated by information theory, channel coding allows us to communicate at a vanishingly low error probability at near-capacity rates. In the scenario of IDEN, every bit carries energy. As a result, redundant bits are not only capable of correcting errors, but they are also capable of delivering energy to the information destinations. Moreover, the information sources typically do not generate symbols with equal probabilities. For instance, according to the statistics, the letter "e" appears in modern English with a probability of 12.7%. By exploiting the specific probability of the information symbols, we are capable of designing more efficient joint source and channel coding schemes for achieving efficient information transmission as well as energy delivery.

Different symbols transmitted by the transmitter may carry different amount of energy, when the amplitude of the RF signal is exploited for conveying modulated information, such as amplitude shift keying (ASK) and quadrature amplitude modulation (QAM). Let us take the simplest on-off keying (OOK) modulation scheme as an example. In the OOK scheme, the information bit "1" is represented by the presence of the carrier waveform $A\sin(2\pi ft)$, where A is the amplitude and f is the frequency of the carrier waveform. As a result, when the bit "1" is transmitted, the amount of energy carried by its modulated waveform should be expressed as $g(1) = \int_0^T A^2\sin^2(2\pi ft)dt = \frac{A^2}{2} - \frac{A^2}{8\pi f}\sin(4\pi fT)$, where T is the symbol duration. If T is the period of the carrier waveform, we have $g(1) = \frac{A^2}{2}$. Furthermore, in the OOK modulation scheme, the information bit "0" is represented by the absence of the carrier waveform. Therefore, the amount of energy carried by the bit "0" is $g(0) = 0$. The OOK modulated binary symbols 0 and 1 are exemplified by Fig. 2.5. Similar analysis can be exploited for evaluating the energy carried by a specific symbol, when other ASK and QAM modulation schemes are invoked. The attenuation of the hostile channel may result in the distortion of information transmission and the uncertainty of energy transfer.

As a result, the arrival of the wireless energy at the battery of the receiver can be modeled by a stochastic process, rather than a deterministic one. Moreover, the depletion of energy from the battery may also be modeled by a stochastic process, which usually depends on the specific action of the IDEN terminal at the receiver side. If the IDEN receiver is a sensor, the energy in the battery may be dissipated by carrying out sensing tasks or by uploading sensing results. If the IDEN receiver is a mobile device, the energy in the battery may be dissipated also by supporting its computation and communication tasks. Therefore, the energy stored in the battery at the receiver side can be modeled by a dynamic queueing process, as shown in the stylised Fig. 2.5, whose statistical properties can be evaluated by mathematical tools, such as the classic Markov queueing theory [5], large deviation theory [6], etc. In order to support the routine operation of the IDEN terminal, on one hand, we have to set up a minimum energy threshold and ensure that the energy stored in the battery is always above this threshold. In a mathematical language, the terminal may tolerate that the underflow probability p_{uf} defining the likelihood of the energy

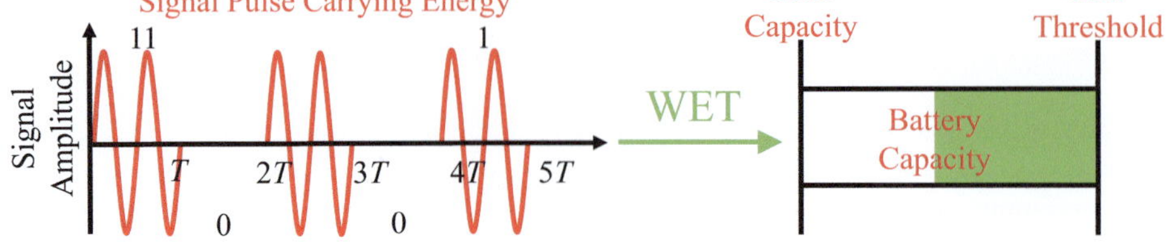

Fig. 2.5 On-off keying modulated RF signals for WET

stored in the battery being lower than the minimum threshold should be higher than a predefined value of P_{uf}. On the other hand, in order to avoid the energy waste, when the energy stored in the battery approaches its capacity, the energy transfer should be reduced in order to avoid the waste of energy and overcharging. In a mathematical term, the terminal may tolerate that the probability p_{of} of the energy supply overflowing the battery capacity should be lower than a predefined value of P_{of}.

Therefore, we have to design an appropriate coding scheme subject to these battery-status related WET requirements in order to maximize the information transmission rate. This optimization problem can be formulated as

$$\text{Objective:} \quad \max_{p(x), x \in \mathcal{X}} I(X; Y), \tag{2.36}$$

$$\text{Subject to: } p_{of} \leq P_{of}, \, p_{uf} \leq P_{uf}, \tag{2.36a}$$

where X and Y represent the random input message and output message, respectively, $p(x)$ is the probability of a specific codeword x in the codebook $\mathcal{X}$ being invoked, while $I(X; Y)$ is the average mutual information between the input message X and the output message Y.

There are some initial attempts in the literature on joint energy and information coding design. Babero et al. [7] studied the family of constrained codes for power transfer, modulation and error control in the context of the reader-to-tag channels in near-field passive radio frequency identification (RFID) systems. In their coding scheme, a block of information consists of the power content bits and the information content bits in order to achieve integrated data and energy transfer. However, the coding scheme of [7] was designed for a channel between a pair of inductive coupling coils, which is incapable of addressing the practical issues of the RF signal based IDEN. In [8], the constrained run-length limited codes were invoked for IDEN by replacing conventional unconstrained capacity achieving codes. According to the simulation results of Fouladgar et al. [8], the codeword structure enabling the integrated data and energy transfer can be adjusted more readily to the receiver's energy consumption pattern than the family of classic unstructured codes. Tandon et al. [9] proposed constant sub-block composition codes for the real-time simultaneous information and energy transfer. In their scheme, all sub-blocks have the same fixed composition of the energy bits and information bits, which is carefully designed for striking a trade-off between the information rate and the amount of energy transferred. Moreover, in [10], the same group of authors provided a necessary and sufficient condition for the sub-block length of the constant sub-block composition code in order to avoid any WET outage. Then they proposed a generalized sub-block energy-constrained codes by ensuring that sufficient energy is transferred within every sub-block duration [11]. Specifically, they presented a tight lower bound for the constant sub-block composition code sequences on the average energy per symbol within a sliding time window.

2.3 Unary Coding for Integrated Data and Energy Transfer

2.3.1 Unary Coding Design for SWIPT with OOK

In the era of Internet of Things (IoT), massively deployed low-power devices may account for a major portion of wireless connectivity [12] for data collecting, data processing, data downloading, and data uploading. However, frequent operations may quickly drain embedded batteries having limited capacity. Moreover, a lot of IoT devices are deployed in hardly reachable places. Replacing their batteries may substantially increase maintenance expenses. RF signal based WPT is capable of delivering controllable, reliable and flexible energy to remote low-power IoT devices [8]. However, WIT has already resided in RF bands. Introducing WPT in the same band may severely degrade WIT performance, since they may compete for precious radio resources to fulfil their distinctive targets [13]. Therefore, carefully coordinating both WIT and WPT in RF bands yields an emerging technique of SWIPT [14].

A great deal of efforts have been invested in SWIPT spanning from signal and transceiver designs to medium-access-control strategies as well as resource allocation and networking schemes [14–23]. For example, Clerckx et al. [15] designed the optimal multi-sinusoidal signal for dedicated WPT, while Zeng et al. [16] focused on the design of multi-sinusoidal SWIPT signal. Zhou et al. [14] made the very first contribution to receiver architecture of SWIPT, which adopted either a power splitter, a time switcher or a current splitter for simultaneous reception of information and energy. Moreover, Zhang et al. [17] optimized transmit beamformer for a MIMO aided SWIPT transmitter. Furthermore, Lv et al. [18] proposed an optimal time-domain resource allocation scheme for a time-division-multiple-access (TDMA) aided multi-user SWIPT system, which maximized sum-throughput and fair-throughput of users' uplink transmissions, respectively. Based on [18], Yang et al. implemented multiple antennas at a SWIPT transmitter, which results in a joint resource allocation scheme [19] in the spatial-, time- and power-domain. Moreover, Gautam et al. [20] studied joint relay selection and subcarrier allocation in an OFDMA aided multi-user SWIPT system. In order to support the "ad hoc" access of batteryless devices, Zhao et al. proposed several enhanced carrier-sensing-multiple-access and collision avoidance (CSMA/CA) protocols and analyzed their attainable performance in [21] and [22]. Moreover, Fouladgar et al. [24] investigated the coordination of distributed wireless nodes for achieving the required quality of service (QoS) of SWIPT. Furthermore, Zhao et al. [23] studied an optimal deployment scheme of SWIPT transmitters by considering social characteristics of mobile SWIPT users. The impact of discrete messages on the SWIPT was discussed in [25], which firstly provided a design guide of the SWIPT coding and modulation.

Surprisingly, the research on the information theoretical essence of SWIPT has been comparatively slower, since Varshney [4] first studied the trade-off between maximum mutual information and energy harvesting requirement in different channels. This seminal work demonstrated that SWIPT performance can be controlled by adjusting the codeword structure, which provided the theoretical fundamental for coding controlled SWIPT. Furthermore, Varshney [26] also proposed a cross-layer architecture for jointly optimizing both the energy and information delivery from the information theoretical perspective, where a powerline communication system was exemplified as a typical integrated information and energy transfer system. Popovski et al. [27] derived both the inner and outer bounds of information transmission rate in a two-way noiseless channel, where the energy contained in the received signal was reused for powering the successive communications. Then, Yin et al. [28] studied a variable-length coding design for SWIPT in a binary noiseless channel. Moreover, Tandon et al. [11] divided a codeword into several sub-blocks for satisfying the receiver's real-time energy demands. The optimal structure of codewords was found for maximizing the mutual information in a binary symmetric channel (BSC). However, both of [4] and [11] only considered memoryless information source in their analysis but ignored the correlation among the binary signs in the output sequence of a practical encoder. Therefore, these works optimized transmit probabilities of binary signs in order to achieve the maximum SWIPT performance without considering a practical coding scheme.

Conventional source encoders generate equi-probable binary signs for maximizing codewords' capability of information transmission. However, codewords having equi-probable binary signs are only capable of carrying a fixed amount of energy. Therefore, conventional source encoders cannot satisfy diverse energy requests of receivers. *Constrained code* [7] has a certain degree of freedom to adjust codeword structure for satisfying specific WPT requirements without sacrificing WIT efficiency.

As a typical constrained code, a run-length-limited (RLL) encoder was designed by Fouladgar et al. [8] for minimizing battery overflow/underflow probability at the receiver subject to a constraint on the achievable mutual information. Moreover, Tandon et al. [29] found the capacity lower bound of BSC and Z channel, when a (d, ∞) RLL code was adopted. However, they only used the minimum run-length d to ensure the minimum energy delivery requirement, which largely overlooks devices' long-term energy requirements.

Furthermore, unary code has been widely used for source coding [30–32] in wireless communication, due to its low encoding and decoding complexity. Specifically, unary code was used by Babar et al. [30] for joint throughput and dimming control in visible light communication system. However, the SWIPT performance of unary code in the RF band is not well understood from the information theoretical perspective.

Against this background, our novel contributions are summarized as below:

- We propose a *practical* unary coding aided SWIPT transceiver and analyzed its performance from the information theoretical perspective. The correlation among the information signs generated by the unary encoder has been taken into account in the performance analysis.
- By considering infinite battery capacity, the codeword distribution is optimized for maximizing the average energy harvested from a single sign received by an energy receiver (ER), while satisfying the mutual information requirement of an information receiver (IR).
- By also considering finite battery capacity, the optimal codeword distribution is found for minimizing the battery overflow/underflow probability of an ER, while satisfying the mutual information requirement of an IR.
- The optimal coding design is obtained by exploiting a low-complexity Genetic Algorithm (GA), which achieves almost the same performance of the exhaustive search (ES).

All the notations are summarized in Table 2.1.

2.3.1.1 System Model

As portrayed in Fig. 2.6, a SWIPT transmitter is constituted by the following functional modules:

Table 2.1 Notation list

Notation	Definition
$\mathbb{X}/\widehat{\mathbb{X}}$	Original/coded information source
$\mathbb{Y}/\widehat{\mathbb{Y}}$	Original/coded information destination
$\mathcal{X}_m/\widehat{\mathcal{X}}_k$	m-th message/k-th codeword
$p(\mathcal{X}_m)/p(\widehat{\mathcal{X}}_k)$	Probability of m-th message/k-th codeword
$\mathcal{X}/\widehat{\mathcal{X}}$	Message/code book
$\mathbf{P}(\mathbb{X})/\mathbf{P}(\widehat{\mathbb{X}})$	Message/Codeword Distribution
ω_I/ω_E	Crossover prob. of WIT/WPT channel
K	Level of Unary code
$\mathbf{S}_N$	State sequence of Markov chain
$\mathbf{X}_N/\mathbf{Y}_N$	Channel input/output binary sign sequence
q_n	Transition prob. from state n to $(n+1)$
$\mathbf{Q}$	Matrix of state transition probabilities
π_n	Steady-state probability of state n
$\widehat{\mathbb{E}}$	Energy destination
$\mathcal{E}(\widehat{\mathbb{E}})$	Energy harvested per received binary sign
$I_{\text{inf}}(\widehat{\mathbb{X}};\widehat{\mathbb{Y}})$	Inf of mutual information between $\widehat{\mathbb{X}}$ and $\widehat{\mathbb{Y}}$
$B(t)$	Energy level in battery at the t-th instant
$\lambda(t)/\mu(t)$	Energy arrival/departure at the t-th instant
$\xi_{B,B+b}$	Trans. prob. from energy level B to $B+b$
$\widehat{\pi}_B$	Steady-state probability of energy level B
$B_{\max}/B_{\min}$	Upper/lower bound of battery energy level
p_{of}/p_{uf}	Battery overflow/underflow probability

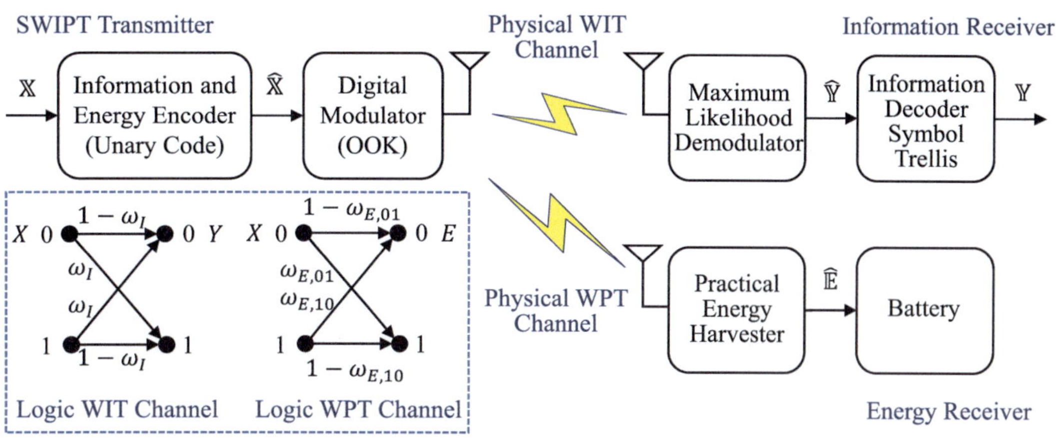

Fig. 2.6 Practical transceiver for coding controlled SWIPT

- *Information and Energy Encoder* adopts the K-level unary coding for mapping M messages $\mathcal{X} = \{\mathcal{X}_1, \cdots, \mathcal{X}_M\}$ randomly generated by the *information source* $\mathbb{X}$ onto K codewords $\{\widehat{\mathcal{X}}_1, \cdots, \widehat{\mathcal{X}}_K\}$, where the k-th codeword $\widehat{\mathcal{X}}_k$ has $(k-1)$ signs "1" followed by a sign "0", as portrayed in Table 2.2. The resultant *coded information source* is denoted as $\widehat{\mathbb{X}}$. Given a specific message distribution $\mathbf{P}(\mathbb{X}) = [p(\mathcal{X}_1), \cdots, p(\mathcal{X}_M)]$ of $\mathbb{X}$, we may obtain an arbitrary codeword distribution $\mathbf{P}(\widehat{\mathbb{X}}) = \left[p(\widehat{\mathcal{X}}_1), \cdots, p(\widehat{\mathcal{X}}_K)\right]$ by exploiting the equation $\mathbf{P}(\widehat{\mathbb{X}}) = \mathbf{P}(\mathbb{X}) \times \mathbf{P}(\widehat{\mathbb{X}}|\mathbb{X})$, where the matrix $\mathbf{P}(\widehat{\mathbb{X}}|\mathbb{X}) = \left[p(\widehat{\mathcal{X}}_k|\mathcal{X}_m)\right]_{M\times K}$ is carefully chosen. The entry $p(\widehat{\mathcal{X}}_k|\mathcal{X}_m)$ is the probability of mapping the message $\mathcal{X}_m$ onto the codeword $\widehat{\mathcal{X}}_k$.[1]

[1] By carefully adjusting the entries in the encoding matrix $\mathbf{P}(\widehat{\mathbb{X}}|\mathbb{X})$, we may achieve either a lossless or a lossy source coding scheme.

Table 2.2 K-level unary code

Messages	Message Prob.	Codewords	Codeword Prob.
$\mathcal{X}_1$	$p(\mathcal{X}_1)$	$\widehat{\mathcal{X}}_1 = 0$	$p(\widehat{\mathcal{X}}_1)$
$\mathcal{X}_2$	$p(\mathcal{X}_2)$	$\widehat{\mathcal{X}}_2 = 10$	$p(\widehat{\mathcal{X}}_2)$
$\mathcal{X}_3$	$p(\mathcal{X}_3)$	$\widehat{\mathcal{X}}_3 = 110$	$p(\widehat{\mathcal{X}}_3)$
$\vdots$	$\vdots$	$\vdots$	$\vdots$
$\mathcal{X}_M$	$p(\mathcal{X}_M)$	$\widehat{\mathcal{X}}_K = \underbrace{11\cdots 1}_{(K-1)\text{ sign}}\ 0$	$p(\widehat{\mathcal{X}}_K)$

- *Digital Modulator* modulates the binary codewords onto the analogue RF signals. Amplitude based modulator allocates different power to the codewords. In order to highlight the impact of the codeword structure on the SWIPT performance, we adopt an OOK based digital modulator. Binary sign "1" is represented by the presence of the RF signal $\sqrt{P_t}\sin(2\pi f_c t)$, where P_t and f_c denote the transmit power and the carrier frequency, respectively. By contrast, sign "0" is represented by the absence of the RF signal.

Discussion Traditional information encoder aims for maximizing the mutual information, which ignores the QoS of WPT performance. By contrast, information and energy encoder achieves a balance between the WIT and WPT performance. The resultant optimal codeword distribution may sacrifice some WIT performance in order to satisfy the QoS of WPT.

An information receiver (IR) consists of the following modules, as illustrated in Fig. 2.6:

- *Digital Demodulator* demodulates the information carried by the received RF signal by adopting the maximum likelihood (ML) principle. The resultant binary sign sequences constitute the coded information destination $\widehat{\mathbb{Y}}$.
- *Information Decoder* recovers the original message from the received codeword by exploiting the symbol-level trellis [30–32], which yields the *information destination* $\mathbb{Y}$.

Due to the broadcast nature of wireless channels, the RF signal transmitted by the SWIPT transmitter can also be received by an energy receiver (ER) for energy harvesting, which consists of the following modules, as illustrated in Fig. 2.6:

- *Energy Harvester* converts the RF signal received to the direct current (DC). A single energy unit can be harvested, when the input RF power is higher than P_{th}. Binary notation $E = 1$ represents that the ER harvests a single energy unit, while $E = 0$ represents no energy harvested, which constitutes an energy destination $\widehat{\mathbb{E}}$.
- *Battery* is charged by the DC. We consider both infinite- and finite-capacity based batteries in the optimal coding design.

Discussion The IR and ER are either co-located or separated for simultaneous information reception and energy harvesting.

WIT Channel Modeling: A logic WIT channel connecting the SWIPT transmitter and the IR consists of an OOK based modulator, a physical wireless channel and an ML based demodulator. The RF signal received by the IR is either $r_I = A_I + z_I$, if the binary sign $X = 1$ is input to the channel, or $r_I = z_I$, if the channel input is $X = 0$. Specifically, $A_I = \sqrt{P_t/\Omega(d_I)}$ is the amplitude of the modulated RF signal received by the IR, where $\Omega(d_I)$ is the path-loss since the IR is d_I m away from the SWIPT transmitter, while $z_I \sim \mathcal{N}(0, \sigma_I^2)$ is the random ambient RF interference at the IR.

By adopting the ML based demodulator, the crossover probability from the channel input $X = 0$ to the output $Y = 1$ and that from $X = 1$ to $Y = 0$ are both equal to $\omega_I = Q\left(\frac{A_I}{2\sigma_I}\right)$, where $Q(\cdot)$ is the tail distribution function of a standard Gaussian distribution. Therefore, the logic WIT channel can be modeled as a binary symmetric channel (BSC), as portrayed in Fig. 2.6.

WPT Channel Modeling: A logic WPT channel connecting the SWIPT transmitter and the ER consists of an OOK based modulator, a physical wireless channel and a practical energy harvester. The RF signal received by the ER is either $r_E = A_E + z_E$, if the binary sign $X = 1$ is input to the channel, or $r_E = z_E$, if the channel input is $X = 0$. Specifically, $A_E = \sqrt{P_t/\Omega(d_E)}$ is the amplitude of the modulated RF signal received by the IR, where $\Omega(d_E)$ is the path-loss since the ER is d_E m away from the SWIPT transmitter, while $z_E \sim \mathcal{N}(0, \sigma_E^2)$ is the random RF interference at the ER.

By considering a practical energy harvester, the crossover probability $\omega_{E,01}$ from the channel input $X = 0$ to the channel output $E = 1$ and that $\omega_{E,10}$ from $X = 1$ to $E = 0$ are expressed as

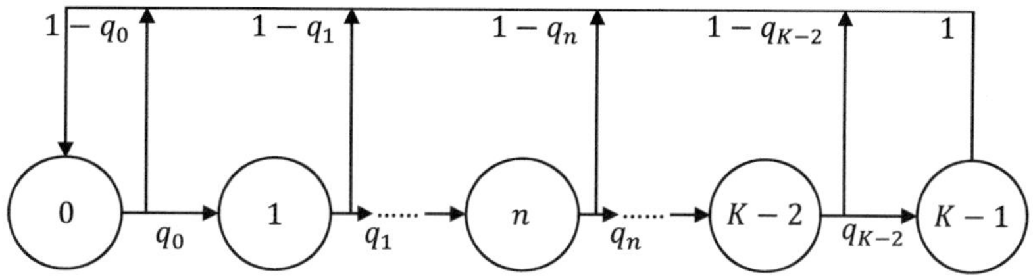

Fig. 2.7 Markov modeling of the unary coded information source $\widehat{\mathbb{X}}$

$$
\begin{cases}
\omega_{E,01} = 2Q\left(\frac{\sqrt{P_{th}}}{\sigma_E}\right), \\
\omega_{E,10} = 1 - Q\left(\frac{\sqrt{P_{th}}+A_E}{\sigma_E}\right) - Q\left(\frac{\sqrt{P_{th}}-A_E}{\sigma_E}\right),
\end{cases}
\tag{2.37}
$$

respectively. Therefore, the logic WPT channel can be modeled as a binary asymmetric channel (BAC), as portrayed in Fig. 2.6.

Discussion: Different from physical SWIPT channels adopted in the physical layer design [14–23], we conceive logic WIT and WPT channels by jointly considering the impact of modulation on WIT and the impact of energy harvester's sensitivity on WPT.

2.3.1.2 WIT Performance Analysis

The K-level unary encoded information source $\widehat{\mathbb{X}}$ sequentially sends a range of codewords chosen from the codebook $\widehat{\mathcal{X}}$ for simultaneously conveying information to the IR and transferring energy to the ER, which generates an N-sign sequence $\mathbf{X}_N = \{X_1, X_2, \cdots, X_N\}$. The correlation among the binary signs is modeled by a finite-state Markov chain, as portrayed in Fig. 2.7.

State n of this Markov chain represents that the unary coded information source $\widehat{\mathbb{X}}$ has already output a run of sign "1" having a length of m at the current instant, since $\widehat{\mathbb{X}}$ is in the process of sending an arbitrary codeword chosen from $\{\widehat{\mathcal{X}}_{n+1}, \widehat{\mathcal{X}}_{n+2}, \cdots, \widehat{\mathcal{X}}_K\}$. By considering the codeword structure of Table 2.2, if $\widehat{\mathbb{X}}$ is in the process of sending codewords $\{\widehat{\mathcal{X}}_{n+2}, \cdots .\widehat{\mathcal{X}}_K\}$, the next output sign is "1". The run-length of sign "1" is then updated to $(n+1)$, which results in the state transition from n to $(n+1)$. The corresponding state transition probability is expressed as

$$
q_n = \frac{p(\widehat{\mathcal{X}}_{n+2}) + \cdots + p(\widehat{\mathcal{X}}_K)}{p(\widehat{\mathcal{X}}_{n+1}) + p(\widehat{\mathcal{X}}_{n+2}) + \cdots + p(\widehat{\mathcal{X}}_K)},
\tag{2.38}
$$

for $0 \leq n \leq K-2$. The probability of the Markov chain transiting from state n back to state 0 is thus $(1 - q_n)$. Specifically, when the Markov chain stays at state $(K-1)$, it transits back to state 0 at the next instant for certain, since $(K-1)$ is the maximum allowable run-length of sign "1" for a K-level unary code, as shown in Table 2.2.

The matrix $\mathbf{Q}$ containing all the state transition probabilities is expressed as:

$$
\mathbf{Q} = \begin{bmatrix}
1-q_0 & q_0\ 0 & \cdots 0\ 0 \\
1-q_1 & 0\ \ q_1 & \cdots 0\ 0 \\
\vdots & \vdots\ \ \vdots & \ddots \vdots\ \vdots \\
1-q_{K-2} & 0\ \ 0 & \cdots 0\ q_{K-2} \\
1 & 0\ \ 0 & \cdots 0\ 0
\end{bmatrix}.
\tag{2.39}
$$

The stationary distribution $\boldsymbol{\pi} = \{\pi_0, \pi_1, \cdots, \pi_{K-1}\}$ of the Markov chain can be obtained by solving the following linear equations:

$$
\begin{cases}
\boldsymbol{\pi} \times \mathbf{Q} = \boldsymbol{\pi} \\
\boldsymbol{\pi} \times \mathbf{I}_{K\times 1} = 1
\end{cases},
\tag{2.40}
$$

where $\mathbf{I}_{K\times 1}$ is a $K \times 1$ column vector having all its entries equal to one.

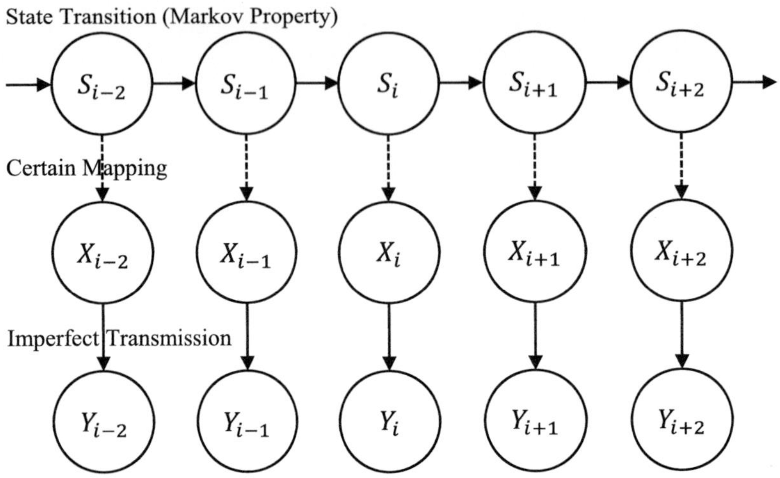

Fig. 2.8 Correlation among the state sequence and the sign sequences of the channel input/output

A state sequence $\mathbf{S}_N = \{S_1, \cdots, S_N\}$ can be mapped onto a sign sequence $\mathbf{X}_N = \{X_1, \cdots, X_N\}$, as portrayed in Fig. 2.8. For an example of a 4-level unary code, a 6-states sequence $\mathbf{S}_6 = \{1, 2, 0, 1, 2, 3, 0\}$ corresponds to a sign sequence $\mathbf{X}_6 = \{1, 1, 0, 1, 1, 1, 0\}$.

When the N-sign sequence $\mathbf{X}_N$ is input to the channel, the corresponding output sign sequence is $\mathbf{Y}_N = \{Y_1, Y_2, \cdots, Y_N\}$ after the imperfect transmission, as exemplified in Fig. 2.8. The mutual information between the coded source $\widehat{\mathbb{X}}$ and the coded destination $\widehat{\mathbb{Y}}$ can be formulated as

$$I(\widehat{\mathbb{X}}; \widehat{\mathbb{Y}}) = H(\widehat{\mathbb{X}}) - H(\widehat{\mathbb{X}}|\widehat{\mathbb{Y}}). \tag{2.41}$$

The entropy $H(\widehat{\mathbb{X}})$ of the coded source can be expressed as

$$\begin{aligned} H(\widehat{\mathbb{X}}) &= \lim_{N\to\infty} \frac{H(\mathbf{X}_N)}{N} \overset{(a)}{=} \lim_{N\to\infty} \frac{H(\mathbf{S}_N)}{N} \\ &\overset{(b)}{=} \lim_{N\to\infty} H(S_N|S_1, S_2, \cdots, S_{N-1}) \\ &\overset{(c)}{=} \lim_{N\to\infty} H(S_N|S_{N-1}) \overset{(d)}{=} H(S_2|S_1). \end{aligned} \tag{2.42}$$

In Eq. (2.42), the equality (a) is derived due to the certain mapping between the sign sequence $\mathbf{X}_N$ and the state sequence $\mathbf{S}_N$, as exemplified in Fig. 2.8. The equality (b) is obtained by exploiting *Theorem 4.2.1* of [33]. The equality (c) is derived due to the Markov property of $\widehat{\mathbb{X}}$, while the derivation of the equality (d) is based on the stationary property of $\widehat{\mathbb{X}}$.

In order to obtain the conditional entropy $H(\widehat{\mathbb{X}}|\widehat{\mathbb{Y}})$, the following pair of Lemmas are introduced for investigating the correlation between $\mathbf{S}_N$ and $\mathbf{Y}_N$:

Lemma 2.1 *Given state S_i at the i-th instant, state S_{i+1} is independent of the channel output signs $\{Y_1, \cdots, Y_{i-1}, Y_i\}$ at the previous instants.*

Proof Please refer to 2.3.1.6.

Lemma 2.2 *State S_i at the i-th instant is correlated to the channel output signs $\{Y_{i+1}, Y_{i+2}, \ldots\}$ at the subsequent instants. If the state transition probabilities satisfy $\{q_n = q|n = 0, 1, \cdots, K-2\}$, where $q \in (0, 1)$ is a constant, S_i is independent of $\{Y_{i+1}, Y_{i+2}, \cdots\}$.*

Proof Please refer to 2.3.1.7.

Therefore, the conditional entropy $H(\widehat{\mathbb{X}}|\widehat{\mathbb{Y}})$ is calculated as

$$H(\widehat{\mathbb{X}}|\widehat{\mathbb{Y}}) = \lim_{N\to\infty} \frac{H(\mathbf{X}_N|\mathbf{Y}_N)}{N} = \lim_{N\to\infty} \frac{H(\mathbf{S}_N|\mathbf{Y}_N)}{N}$$

$$\overset{(e)}{=} \lim_{N\to\infty} \frac{1}{N}\Big[H(S_1|Y_1,\cdots,Y_N) + \sum_{i=2}^{N} H(S_i|S_{i-1},\cdots,S_1,Y_1,\cdots,Y_N)\Big]$$

$$\overset{(f)}{=} \lim_{N\to\infty} \frac{1}{N}\Big[H(S_1|Y_1,\cdots,Y_N) + \sum_{i=2}^{N} H(S_i|S_{i-1},Y_i,\cdots,Y_N)\Big]$$

$$\overset{(g)}{\leq} \lim_{N\to\infty} \frac{H(S_1|Y_1,\cdots,Y_N) + (N-1)H(S_2|S_1,Y_2)}{N}$$

$$= H(S_2|S_1,Y_2). \tag{2.43}$$

By exploiting the *chain rule for entropy* (*Theorem 2.5.1* in [33]), we obtain the equality (e) of Eq. (2.43). The equality (f) is derived by exploiting i) the Markov property and ii) *Lemma 2.1*. According to *Lemma 2.2*, State S_i is dependent of the channel output signs $\{Y_i, Y_{i+1}, \cdots, Y_N\}$, which results in the inequality (g).

By substituting Eqs. (2.42) and (2.43) into (2.41), the mutual information is finally derived as

$$I(\widehat{\mathbb{X}};\widehat{\mathbb{Y}}) \overset{(h)}{\geq} H(S_2|S_1) - H(S_2|S_1,Y_2)$$

$$= I(S_2;Y_2|S_1) = H(Y_2|S_1) - H(Y_2|S_2,S_1)$$

$$= -\sum_{s_1}\sum_{y_2} p(y_2,s_1)\log_2 p(y_2|s_1) + \sum_{s_1}\sum_{s_2}\sum_{y_2} p(y_2,s_2,s_1)\log_2 p(y_2|s_2s_1)$$

$$\overset{(i)}{=} \sum_{n=0}^{K-1} \pi_n \{H[q_n\omega_I + (1-q_n)(1-\omega_I)] - H(\omega_I)\}$$

$$\triangleq I_{\inf}(\widehat{\mathbb{X}};\widehat{\mathbb{Y}}), \tag{2.44}$$

where $H(x) = -x\log_2(x) - (1-x)\log_2(1-x)$ for $0 < x < 1$. The inequality (h) of Eq. (2.44) is obtained by substituting the upper bound $H(S_2|S_1,Y_2)$ of the conditional entropy $H(\widehat{\mathbb{X}}|\widehat{\mathbb{Y}})$. We further derive the equality (i) by exploiting the stationary distribution $\boldsymbol{\pi}$ of the Markov chain and the corresponding matrix $\mathbf{Q}$ of the transition probabilities. The lower bound of $I(\widehat{\mathbb{X}};\widehat{\mathbb{Y}})$ derived in Eq. (2.44) is attainable, according to *Lemma 2.2*. Therefore, it is the infimum of $I(\widehat{\mathbb{X}};\widehat{\mathbb{Y}})$ denoted as $I_{\inf}(\widehat{\mathbb{X}};\widehat{\mathbb{Y}})$, which is a function of the codeword distribution $\mathbf{P}(\widehat{\mathbb{X}})$, since entries in both $\mathbf{Q}$ and $\boldsymbol{\pi}$ are functions of $\mathbf{P}(\widehat{\mathbb{X}})$, according to Eqs. (2.38) and (2.40).

2.3.1.3 WPT Performance Analysis

When the battery has infinite capacity, we focus on the average energy harvested from a single sign received by the ER.

When the Markov chain of Fig. 2.7 stays at state n, the K-level unary coded source $\widehat{\mathbb{X}}$ may output a sign "1" with a probability of q_n, while it may output a sign "0" with a probability of $(1-q_n)$. Given the stationary distribution $\boldsymbol{\pi}$ derived by solving Eq. (2.40), the stationary probability of $\widehat{\mathbb{X}}$ outputting a sign "1" is expressed as $p_1 = \sum_{n=0}^{K-1}\pi_n q_n$. Similarly, the stationary probability of $\widehat{\mathbb{X}}$ outputting a sign "0" is expressed as $p_0 = \sum_{n=0}^{K-1}\pi_n(1-q_n)$. Therefore, after the imperfect transmission in the WPT channel of Fig. 2.6, the average energy harvested from a single sign received by the ER is formulated as

$$\mathcal{E}(\widehat{\mathbb{E}}) = p_1(1-\omega_{E,10}) + p_0\omega_{E,01}$$

$$= \sum_{n=0}^{K-1} \pi_n[q_n(1-\omega_{E,10}) + (1-q_n)\omega_{E,01}], \tag{2.45}$$

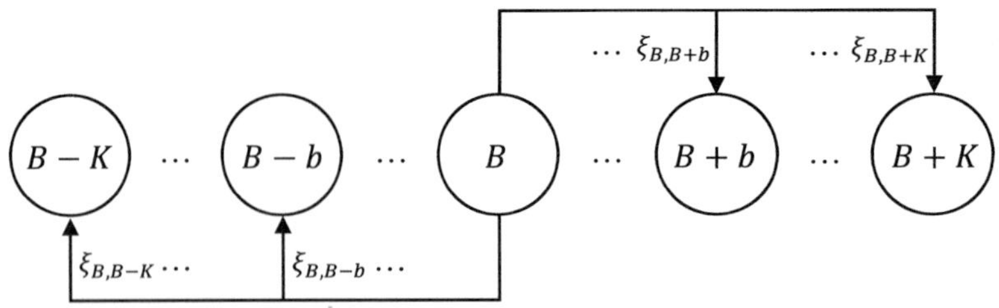

Fig. 2.9 Markov modeling of the battery queuing process $B(t)$

which is a function of the codeword distribution $\mathbf{P}(\widehat{\mathbb{X}})$, since entries in both $\mathbf{Q}$ and $\boldsymbol{\pi}$ are all functions of $\mathbf{P}(\widehat{\mathbb{X}})$, according to Eqs. (2.38) and (2.40).

When the battery has a finite capacity $B_{\max}$ and a warning threshold $B_{\min}$, we study both the battery overflow probability p_{of} and the battery underflow probability p_{uf} of the ER.

The instantaneous energy level can be modeled by the following queuing process:

$$B(t+1) = \begin{cases} \max\left\{B_{\min}, B(t)+\lambda(t)-\mu(t)\right\}, \\ \min\left\{B(t)+\lambda(t)-\mu(t), B_{\max}\right\}, \end{cases} \tag{2.46}$$

where $B(t)$ is the energy level of the battery at the t-th instant, $\lambda(t)$ is the amount of energy arriving at the battery of the ER at the t-th instant and $\mu(t)$ is the amount of energy departing from the battery at the t-th instant. The energy level $B(t+1)$ at the next instant is lower-bounded by $B_{\min}$ and upper-bounded by $B_{\max}$. At a single instant, an arbitrary unary codeword is sent by the SWIPT transmitter.

Specifically, the energy arrival $\lambda(t)$ is a stochastic process, which has the following randomness:

- A codeword $\widehat{\mathcal{X}_k}$ is randomly sent by the SWIPT transmitter with a probability of $p(\widehat{\mathcal{X}_k})$, which includes $(k-1)$ signs "1" carrying energy.
- The binary signs in $\widehat{\mathcal{X}_k}$ may be randomly flipped by the logic WPT channel of Fig. 2.6.

We assume that the probability of the ER consuming a single energy unit in a sign duration is ν. Therefore, if a codeword $\widehat{\mathcal{X}_k}$ is sent by the SWIPT transmitter at the t-th instant, the amount of energy $\mu(t)$ consumed by the ER obeys a Binomial distribution having parameters of (k, ν).

The queuing process $B(t)$ can be further modeled by another discrete Markov chain, as illustrated in Fig. 2.9. The states in this Markov chain span from $B_{\min}$ to $B_{\max}$, which represent all possible energy levels of the battery. We now focus on deriving state transition probabilities. The probability ξ_b of the energy level increment $b = \lambda(t) - \mu(t)$ is expressed as

$$\begin{aligned} \xi_b &= \sum_{\widehat{\mathcal{X}_k}} \sum_{\lambda} p(\mu(t) = \lambda(t) - b, \lambda(t) = \lambda, \widehat{\mathcal{X}_k}) \\ &= \sum_{k=1}^{K} p(\widehat{\mathcal{X}_k}) \sum_{\lambda} p(\mu(t) = \lambda - b | \lambda(t) = \lambda, \widehat{\mathcal{X}_k}) p(\lambda(t) = \lambda | \widehat{\mathcal{X}_k}). \end{aligned} \tag{2.47}$$

Given the unary codeword $\widehat{\mathcal{X}_k}$ sent at the t-th instant, the conditional probability of $\lambda(t) = \lambda$ is

$$\begin{aligned} p\left(\lambda(t) = \lambda | \widehat{\mathcal{X}_k}\right) = &\binom{k}{\lambda} \frac{(1-\omega_{E,10})^{\lambda-1} \omega_{E,10}^{k-1-\lambda}}{k} \\ &\cdot \left[(k-\lambda)(1-\omega_{E,10})(1-\omega_{E,01}) + \lambda \omega_{E,10} \omega_{E,01}\right], \end{aligned} \tag{2.48}$$

for $0 \le \lambda \le k$, and $p\left(\lambda(t) = \lambda | \widehat{\mathcal{X}_k}\right) = 0$, otherwise. Given the unary codeword $\widehat{\mathcal{X}_k}$ and $\lambda(t) = \lambda$, the conditional probability $p\left(\mu(t) = \lambda - b | \widehat{\mathcal{X}_k}, \lambda(t) = \lambda\right)$ is calculated as

$$p(\mu(t) = \lambda - b | \widehat{\mathcal{X}_k}, \lambda(t) = \lambda)$$

$$= \begin{cases} \binom{k}{\lambda - b} \nu^{\lambda-b}(1-\nu)^{k-\lambda+b}, & 0 \le \lambda - b \le k, \\ 0, & \text{otherwise.} \end{cases} \tag{2.49}$$

By substituting both Eqs. (2.48) and (2.49) into (2.47), we may obtain ξ_b as

$$\xi_b = \sum_{k=1}^{K} p(\widehat{\mathcal{X}}_k)\Phi(b,k), \text{ for } -K \le b \le K, \tag{2.50}$$

and $\xi_b = 0$, otherwise, where we have $\Phi(b,k)$ expressed in Eq. (2.51).

$$\Phi(b,k) = \begin{cases} \sum_{\lambda=b}^{k} \binom{k}{\lambda - b}\binom{k}{\lambda} \nu^{\lambda-b}(1-\nu)^{k-\lambda+b} \frac{(1-\omega_{E,10})^{\lambda-1}\omega_{E,10}^{k-1-\lambda}}{k} & \\ \times \left[(k-\lambda)(1-\omega_{E,10})(1-\omega_{E,01}) + \lambda\omega_{E,10}\omega_{E,01}\right], & 0 \le b \le k, \\ \sum_{\lambda=0}^{k+b} \binom{k}{\lambda - b}\binom{k}{\lambda} \nu^{\lambda-b}(1-\nu)^{k-\lambda+b} \frac{(1-\omega_{E,10})^{\lambda-1}\omega_{E,10}^{k-1-\lambda}}{k} & \\ \times \left[(k-\lambda)(1-\omega_{E,10})(1-\omega_{E,01}) + \lambda\omega_{E,10}\omega_{E,01}\right], & -k \le b \le 0, \\ 0, & \text{otherwise.} \end{cases} \tag{2.51}$$

Without loss of generality, we assume that the battery has an energy level of $B(t) = B$ at the t-th instant. Therefore, at the $(t+1)$-th instant, the energy level of the battery could be $B(t+1) = B + b$, which also represents the state transition from B to $(B+b)$ in the Markov chain of Fig. 2.9. The corresponding state transition probability $\xi_{B,B+b}$ can be then expressed as

$$\xi_{B,B+b} = \begin{cases} \xi_b, & \max(B_{\min} - B + 1, -K) \le b \le \min(B_{\max} - B - 1, K) \\ & \text{and } B_{\min} + 1 \le B \le B_{\max} - 1, \\ 0, & K < b \le B_{\max} - B \text{ and } B_{\min} \le B < B_{\max} - K, \\ 0, & B_{\min} - B \le b < -K \text{ and } B_{\min} + K < B \le B_{\max}. \end{cases} \tag{2.52}$$

If the current energy level B satisfies $B_{\max} - K \le B \le B_{\max}$ and the energy level increment b satisfies $B_{\max} - B \le b \le K$, the energy level of the battery at the end of this instant is constrained to $B_{\max}$, according to Eq. (2.46). Therefore, the state transition probability $\xi_{B,B_{\max}}$ is expressed as

$$\xi_{B,B_{\max}} = \sum_{b=B_{\max}-B}^{K} \xi_b, \text{ for } B_{\max} - K \le B \le B_{\max}. \tag{2.53}$$

Similarly, if the current energy level satisfies $B_{\min} \le B \le B_{\min} + K$ and the energy level increment b satisfies $-K \le b \le B_{\min} - B$, the energy level of the battery at the end of this instant is constrained to $B_{\min}$, according to Eq. (2.46). Therefore, the state transition probability $\xi_{B,B_{\min}}$ is expressed as

$$\xi_{B,B_{\min}} = \sum_{b=-K}^{B_{\min}-B} \xi_b, \text{ for } B_{\min} \le B \le B_{\min} + K. \tag{2.54}$$

With the aid of Eqs. (2.52)–(2.54), the state transition probability matrix $\Xi = [\xi_{i,j}]$ is found. By following the similar method of Eq. (2.40), the stationary distribution $\widehat{\pi} = \{\widehat{\pi}_{B_{\min}}, \cdots, \widehat{\pi}_B, \cdots, \widehat{\pi}_{B_{\max}}\}$ is obtained for all the states in the Markov chain of Fig. 2.9. Finally, the battery overflow probability p_{of} and the underflow probability p_{uf} are then formulated as

$$\begin{cases} p_{of} = \sum_{B=B_{\max}-K+1}^{B_{\max}} \widehat{\pi}_B \sum_{b=B_{\max}-B+1}^{K} \xi_b, \\ p_{uf} = \sum_{B=B_{\min}}^{B_{\min}+K-1} \widehat{\pi}_B \sum_{b=-K}^{B_{\min}-B-1} \xi_b. \end{cases} \tag{2.55}$$

Algorithm 2.1 ES algorithm for solving (P2.3.1)

Require: The crossover probability ω of the BSC; The probability increment ε for the ES; The minimum WIT requirement R_{th};
Ensure: The optimal codeword distribution $\mathbf{P}_1^*$; The maximum average energy carried per received sign $\mathcal{E}_{\max}(\widehat{\mathbb{E}})$;

Initialise a K-dimensional probability space for the exhaustive search: $\mathcal{P} \leftarrow \{m\varepsilon | m = 0, 1, \cdots, \lfloor 1/\varepsilon \rfloor\}^K$;
Initialise $E_{\max} \leftarrow 0$;
while $\mathcal{P}$ is not null **do**
  Let $\mathbf{P}_1 \leftarrow$ an arbitrary $K \times 1$ vector from $\mathcal{P}$;
  Calculate $I_{\inf}(\widehat{\mathbb{X}}; \widehat{\mathbb{Y}})$ by substituting $\mathbf{P}_1$ into (2.44);
  if $I_{\inf}(\widehat{\mathbb{X}}; \widehat{\mathbb{Y}}) \geq R_{th}$ **then**
    Calculate $\mathcal{E}(\widehat{\mathbb{E}})$ by substituting $\mathbf{P}_1$ into (2.45);
    if $\mathcal{E}(\widehat{\mathbb{E}}) > E_{\max}$ **then**
      Let $E_{\max} \leftarrow \mathcal{E}(\widehat{\mathbb{E}})$ and $\mathbf{P}_1^* = \mathbf{P}_1$;
    end if
  end if
  Let $\mathcal{P} \leftarrow \mathcal{P} - \mathbf{P}_1$;
end while
return $\mathcal{E}_{\max}(\widehat{\mathbb{E}}) \leftarrow E_{\max}$ and $\mathbf{P}_1^*$.

According to Eq. (2.50), both p_{of} and p_{uf} are the functions of the codeword distribution $\mathbf{P}(\widehat{\mathbb{X}})$.

2.3.1.4 Optimal Coding Design

We formulate the first optimal coding design by considering an infinite battery capacity at the ER, when the K-level unary code is adopted at the SWIPT transmitter:

$$\text{(P2.3.1): } \max_{\mathbf{P}(\widehat{\mathbb{X}})} \mathcal{E}(\widehat{\mathbb{E}}), \tag{2.56}$$

$$\text{s. t.: } I_{\inf}(\widehat{\mathbb{X}}; \widehat{\mathbb{Y}}) \geq R_{th}. \tag{2.56a}$$

The objective of (P2.3.1) is to maximize the average energy $\mathcal{E}(\widehat{\mathbb{E}})$ of Eq. (2.45) harvested from a single received sign by finding the optimal codeword distribution $\mathbf{P}_1^*(\widehat{\mathbb{X}})$, while satisfying the constraint (2.56a) that the mutual information's infinimum $I_{\inf}(\widehat{\mathbb{X}}; \widehat{\mathbb{Y}})$ of Eq. (2.44) should be higher than a rate threshold R_{th}. Since we have $R = I(\widehat{\mathbb{X}}; \widehat{\mathbb{Y}}) \geq I_{\inf}(\widehat{\mathbb{X}}; \widehat{\mathbb{Y}})$, the actual information transmission rate R is also higher than R_{th}. We use $\mathbf{P}_1^*$ to represent $\mathbf{P}_1^*(\widehat{\mathbb{X}})$ in Sect. V-B.

We then formulate the second optimal coding design by considering a finite battery capacity at the ER:

$$\text{(P2.3.2): } \min_{\mathbf{P}(\widehat{\mathbb{X}})} p_{of} (\text{or } p_{uf}), \tag{2.57}$$

$$\text{s. t.: } I_{\inf}(\widehat{\mathbb{X}}; \widehat{\mathbb{Y}}) \geq R_{th}. \tag{2.57a}$$

In contrast to (P2.3.1), we aim for minimizing the battery overflow/underflow probability in (P2.3.2) by finding the optimal codeword distribution $\mathbf{P}_2^*(\widehat{\mathbb{X}})$, while satisfying the minimum rate requirement R_{th}. We use $\mathbf{P}_2^*$ to represent $\mathbf{P}_2^*(\widehat{\mathbb{X}})$ in Sect. V-C.

The calculations of $I_{\inf}(\widehat{\mathbb{X}}; \widehat{\mathbb{Y}})$ and $\mathcal{E}(\widehat{\mathbb{E}})$ as well as p_{of} and p_{uf} all require the stationary distribution $\boldsymbol{\pi}$, which may only be obtained by numerically solving the system of linear equations (2.40), given a specific state transition probability matrix $\mathbf{Q}$. Therefore, the optimization problems (P2.3.1) and (P2.3.2) cannot be solved in polynomial time, when we have a high-level unary encoder.

We first provide an exhaustive search (ES) based algorithm for solving (P2.3.1), whose pseudocode is provided in Algorithm 2.1. Its complexity $\mathcal{O}(\lfloor 1/\varepsilon \rfloor^K)$ depends on the searching increment ε and the level K. If we increase ε for obtaining a more accurate result and K for implementing high-level unary codes, the complexity increases dramatically. Algorithm 2.1 can also be slightly adjusted for solving (P2.3.2), which is omitted here for the page limitation.

In order to reduce the computing complexity of the coding design but maintain its optimality, we then propose a GA aided coding design for solving the optimization problem (P2.3.2). By exploiting both crossover and mutation treatments on the "parents," GA aided coding design is capable of escaping from the local optimum, which is detailed in Algorithm 2.2. The main steps of Algorithm 2.2 can be summarized as below:

Algorithm 2.2 GA aided optimal coding design for solving the optimisation problem (P2.3.2)

Require: The crossover probability ω of the BSC; The minimum WIT requirement R_{th}; The maximum number of generations G; The population of a single generation N;

Ensure: The minimum overflow probabilities $p_{of,\min}$; The optimal codeword distribution $\mathbf{P}_2^*$;

1: Randomly initialise a generation $\mathcal{P} \leftarrow \{\mathbf{P}_{(j)} | j = 1, \cdots, N\}$, where $\mathbf{P}_{(j)}$ is a legitimate codeword distribution;
2: Initialise a generation index $g \leftarrow 1$ and the minimum overflow probability $p_{of,\min} \leftarrow 1$;
3: **while** $g \leq G$ **do**
4: Initialise the mutual information of the g-th generation $\mathcal{I} \leftarrow \{I_{\text{inf},(j)} | j = 1, \cdots, N\}$, where $I_{\text{inf},(j)}$ is calculated by substituting $\mathbf{P}_{(j)}$ into Eq. (2.44);
5: Initialise the overflow probabilities of the g-th generation $\mathbf{p}_{of} \leftarrow \{p_{of,(j)} \leftarrow 1 | j = 1, \cdots, N\}$;
6: Initialise the survival probabilities of the g-th generation $\mathbf{p}_{live} \leftarrow \{p_{live,(j)} \leftarrow 0 | j = 1, \cdots, N\}$;
7: Initialise the selected parental individuals $\mathcal{P}_{sel} \leftarrow \{\mathbf{P}_{sel,(j)} \leftarrow \mathbf{P}_{(j)} | j = 1, \cdots, N\}$;
8: **for** $\forall j = 1, \cdots, N$ **do**
9: **if** $I_{\text{inf},(j)} \geq R_{th}$ **then**
10: Update $p_{of,(j)}$ by substituting $\mathbf{P}_{(j)}$ into Eq. (2.55);
11: **else**
12: Update $p_{of,(j)} \leftarrow 1$;
13: **end if**
14: Update $p_{live,(j)} \leftarrow \dfrac{(1 - p_{of,(j)})}{\sum_j (1 - p_{of,(j)})}$;
15: **end for**
16: **if** $\min \mathbf{p}_{of} < p_{of,\min}$ **then**
17: Update $p_{of,\min} \leftarrow \min \mathbf{p}_{of}$ and $\mathbf{P}_2^* \leftarrow \arg\min_{\mathbf{P}_j} \mathbf{p}_{of}$;
18: **end if**
19: **for** $\forall j = 1, \cdots, N$ **do**
20: Generate ζ uniformly distributed in [0, 1];
21: Initialise a temporary index $n \leftarrow 1$;
22: **while** $\zeta - p_{\text{live},(n)} \geq 0$ **do**
23: Update $\zeta \leftarrow \zeta - p_{live,(n)}$ and $n \leftarrow n + 1$;
24: **end while**
25: Update $\mathbf{P}_{sel,(j)} \leftarrow \mathbf{P}_{(n)}$;
26: **end for**
27: Obtain a new generation $\mathcal{P}'$ by inputting $\mathcal{P}_{sel}$ into Algorithm 2.3 and update $\mathcal{P} \leftarrow \mathcal{P}'$;
28: **end while**
29: **return** $p_{of,\min}$ and $\mathbf{P}_2^*$.

- *Step 1*: Randomly generate N legitimate codeword distribution as a generation $\mathcal{P}$, as shown in Line 1 of Algorithm 2.2.
- *Step 2*: Calculate the corresponding mutual information and initialize overflow probabilities and the survival probabilities for all the codeword distributions in $\mathcal{P}$, as detailed in Lines 4-15 of Algorithm 2.2, while update the minimum overflow probability $p_{of,\min}$ expressed in the objective (2.57) of (P2.3.2), as shown in Lines 16-18 of Algorithm 2.2. If the mutual information of an individual codeword distribution violates the minimum rate requirement R_{th} of the (P2.3.2)'s constraint (2.57a), its corresponding overflow probabilities are set to 1. According to Line 14, this individual has a zero survival probability.
- *Step 3*: Randomly select individuals from $\mathcal{P}$ according to their survival probabilities $\mathbf{p}_{live}$ in order to generate the parental individuals $\mathcal{P}_{sel}$ for the sake of giving birth to the next generation, as detailed in Lines 19-26 of Algorithm 2.2. The individuals having lower overflow probabilities can be selected with higher probabilities, which is in line with the objective (2.57) of (P2.3.2).
- *Step 4*: Obtain a new generation $\mathcal{P}'$ by carrying out crossover and mutation on the parental individuals $\mathcal{P}_{sel}$ in Algorithm 2.3 and repeat from Step 2, until we obtain the minimum overflow probability $p_{of,\min}$ and the optimal codeword distribution $\mathbf{P}_2^*$ for (P2.3.2).

Furthermore, Algorithm 2.3 provides the details of crossover and mutation operations on the parental individuals. All the N child individuals are born by the crossover of randomly chosen parents from $\mathcal{P}_{sel}$, as shown in Lines 2-4 of Algorithm 2.3. Every child individual has a probability of ε to mutate, as illustrated in Lines 5-13 of Algorithm 2.3. Specifically, Lines 8-11 ensure that the mutated individual is a legitimate codeword distribution.

The complexity of the proposed GA aided optimal coding design is $\mathcal{O}(G^2)$, where G is the number of generations in Algorithm 2.2. The complexity is irrelevant to the level K of the unary code. Therefore, when a high-level unary code is adopted for the SWIPT, the GA aided optimal coding design is capable of substantially reducing the computational complexity. Furthermore, by substituting p_{of} in Algorithm 2.2 by p_{uf}, we can readily obtain the optimal codeword

Algorithm 2.3 Crossover and mutation algorithm

Require: A selected parental individuals $\mathcal{P}_{sel}$; A mutation probability ε.
Ensure: A new generation $\mathcal{P}'$;
Initialise a new generation $\mathcal{P}' \leftarrow \{\mathbf{P}'_{(j)} \leftarrow \mathbf{P}_{sel,(j)} | j = 1, \cdots, N\}$;
for $\forall j = 1, \cdots, N$ **do**
  Randomly choose a pair of individuals $\mathbf{P}_{sel,(m)}$ and $\mathbf{P}_{sel,(l)}$ from $\mathcal{P}_{sel}$;
  Update $\mathbf{P}'_{(j)} \leftarrow \frac{\mathbf{P}_{sel,(m)} + \mathbf{P}_{sel,(l)}}{2}$;
  Generate ζ uniformly distributed in [0, 1];
  if $\zeta \leq \varepsilon$ **then**
    Update $\mathbf{P}'_{(j)} \leftarrow \mathbf{P}'_{(j)} + \boldsymbol{\gamma}$, where $\boldsymbol{\gamma} \sim \mathcal{N}(0, \mathbf{I}) \in \mathcal{R}^{1\times K}$;
    if $\min \mathbf{P}'_{(j)} < 0$ **then**
      Update $\mathbf{P}'_{(j)} \leftarrow \mathbf{P}'_{(j)} - \min \mathbf{P}'_{(j)}$;
    end if
    Normalise $\mathbf{P}'_{(j)} \leftarrow \mathbf{P}'_{(j)} / \sum \mathbf{P}'_{(j)}$;
  end if
end for
return $\mathcal{P}'$.

distribution for minimizing the battery underflow probability. Moreover, Algorithm 2.2 can also be slightly adjusted for solving (P2.3.1), which is omitted here for the page limitation.

2.3.1.5 Numerical Results

In our simulation, we consider *co-located* IR and ER as a SWIPT receiver, whose distance from the SWIPT transmitter is $d_E = d_I = d = 9.4$ m. The path-loss from the SWIPT transmitter to the receiver is modeled as $\Omega(d) = G_t G_r \left(\frac{4\pi d f_c}{c}\right)^2$, where the antenna gains are set to $G_t = G_r = 1$, the carrier frequency is $f_c = 800$ MHz and c is the speed of the light. The transmit power of the SWIPT transmitter is set to $P_t = 20$ dBm, while the sensitivity of the energy harvester is $P_{th} = -30$ dBm. Generally, the power of the ambient RF interference on both IR and ER are $\sigma_I^2 = \sigma_E^2 = \sigma^2$ varying from -40 dBm to -25 dBm, since IR and ER are *co-located*.

First of all, we demonstrate the convergence of the GA aided coding design and compare its performance to that of the ES aided counterpart in Fig. 2.10. All the performance of Fig. 2.10 is attained by averaging over 50 random results. Observe from Fig. 2.10a and b that the GA aided coding design converges to the optimality in terms of both $\mathcal{E}_{\max}(\mathbb{E})$ and $p_{\min,uf}$, as the generation number increases. Furthermore, as shown in Fig. 2.10a, after 10 generations, the GA aided coding design with medium population $N = 100$ achieves the WPT performance of $\mathcal{E}_{\max}(\widehat{\mathbb{E}}) = 0.3461$ energy unit/sign, only 3.1% lower than that of the ES aided counterpart. Moreover, as shown in Fig. 2.10b, the minimum underflow probability $p_{\min,uf} = 0.0526$ is attained by the GA with $N = 100$ after 10 generations, which is 2.6% higher than that of the ES aided counterpart. Since the low-complexity GA aided coding design achieves almost the same performance with the ES, the rest of the numerical results are obtained by invoking GA.

We then investigate the SWIPT performance by considering an infinite battery capacity in Fig. 2.11, where the WPT performance is characterized by the average energy units harvested from a single received sign. Observe from both Fig. 2.11a and b that the maximum WPT performance $\mathcal{E}_{\max}(\widehat{\mathbb{E}})$ reduces, as we increase the WIT requirement R_{th}. Note that R_{th} cannot exceed the channel capacity due to the limited WIT capability of a specific codeword distribution. For instance, as illustrated in Fig. 2.11a, the maximum WIT capability of 2-level unary code is 0.47 bit/sign. When we have $R_{th} = 0.47$ bit/sign, the WPT performance reaches its lowest, only 0.111 energy unit/sign. We may further observe from Fig. 2.11a that the 8-level unary code has supreme SWIPT performance. Moreover, observe from Fig. 2.11b that the ambient RF interference power is helpful for WPT but it is harmful for WIT. For instance, when we have $\sigma^2 = -25$ dBm, the highest WPT performance is achieved, but the WIT performance of the 4-level unary code is very limited, which is lower than 0.033 bit/sign. Furthermore, when R_{th} is lower than 0.15 bit/sign, the WPT performance associated with $\sigma^2 = -35$ dBm is higher than that with $\sigma^2 = -40$ dBm. By contrast, if the R_{th} is between 0.15 and 0.64 bit/sign, the channel having $\sigma^2 = -40$ dBm is capable of transferring more energy to the SWIPT receiver.

We then consider finite battery capacity having $B_{\max} = 2$ energy units and $B_{\min} = 0$ energy units, while the ER consumes a single energy unity with a probability of $\nu = 0.5$ within a sign duration. Observe from Fig. 2.12a and b that both the battery underflow and overflow probabilities increases, as the WIT requirement becomes stringent. Since the unary codeword distribution has to be adjusted in order to satisfy the harsh WIT requirement, the battery underflow and overflow probabilities

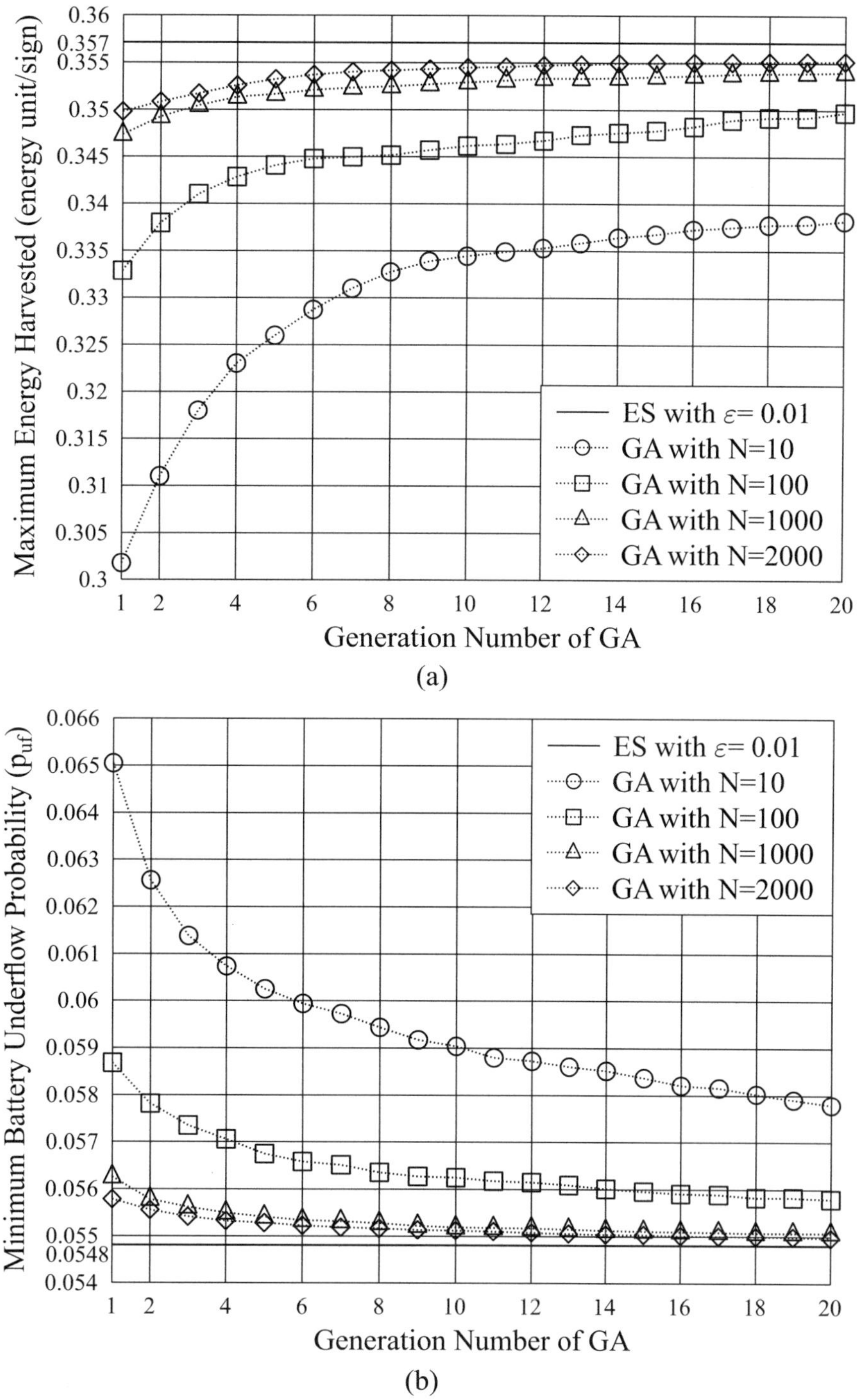

Fig. 2.10 Convergence of the GA aided coding design with different population $N = \{10, 100, 1000, 2000\}$, where the 4-level unary code is adopted, the ambient RF interference power is $\sigma^2 = -40$ dBm and the minimum WIT requirement is $R_{th} = 0.2$ bit/sign: **(a)** The maximum WPT performance $\mathcal{E}_{\max}(\widehat{\mathbb{E}})$ with infinite battery capacity and **(b)** the minimum underflow probability $p_{\min,uf}$ with finite battery capacity

are inevitably sacrificed during the coding design. Furthermore, as illustrated in Fig. 2.12a, a higher level of unary code has higher freedom for the sake of satisfying harsher WIT requirements and reaching better WPT performance. Moreover, observe from Fig. 2.12b that when the ambient RF interference power σ^2 becomes higher, the WIT performance becomes worse, since it impairs the information demodulation of the IR, while the WPT performance becomes better, since ambient RF interference provides more energy to the ER.

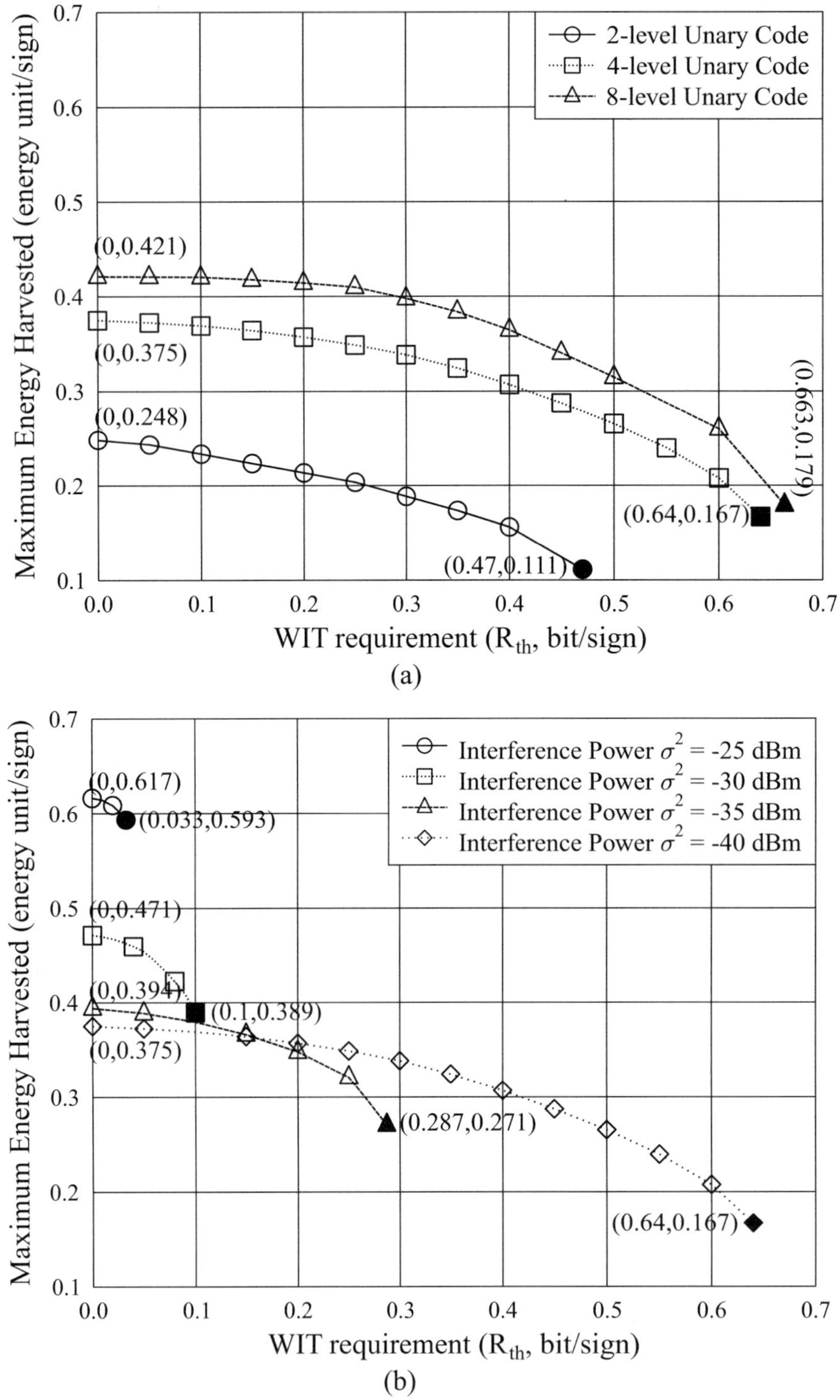

Fig. 2.11 The SWIPT performance with an infinite battery capacity: (**a**) (2,4,8)-level unary codes with the ambient RF interference power $\sigma^2 = -40$ dBm; (**b**) 4-level unary code with the ambient RF interference power $\sigma^2 = \{-25, -30, -35, -40\}$ dBm. The filled markers represent the conventional WIT coding schemes, which aims for maximizing the mutual information of Eq. (2.44) by ignoring the QoS requirement of WPT

2.3.1.6 Proof of Lemma 2.1

The Markov chain of Fig. 2.7 transits from the state S_i to S_{i+1} due to the coding principle of the unary encoder. By contrast, the transition from the state S_i to the sign Y_i is induced by the imperfect information transmission in the WIT channel of

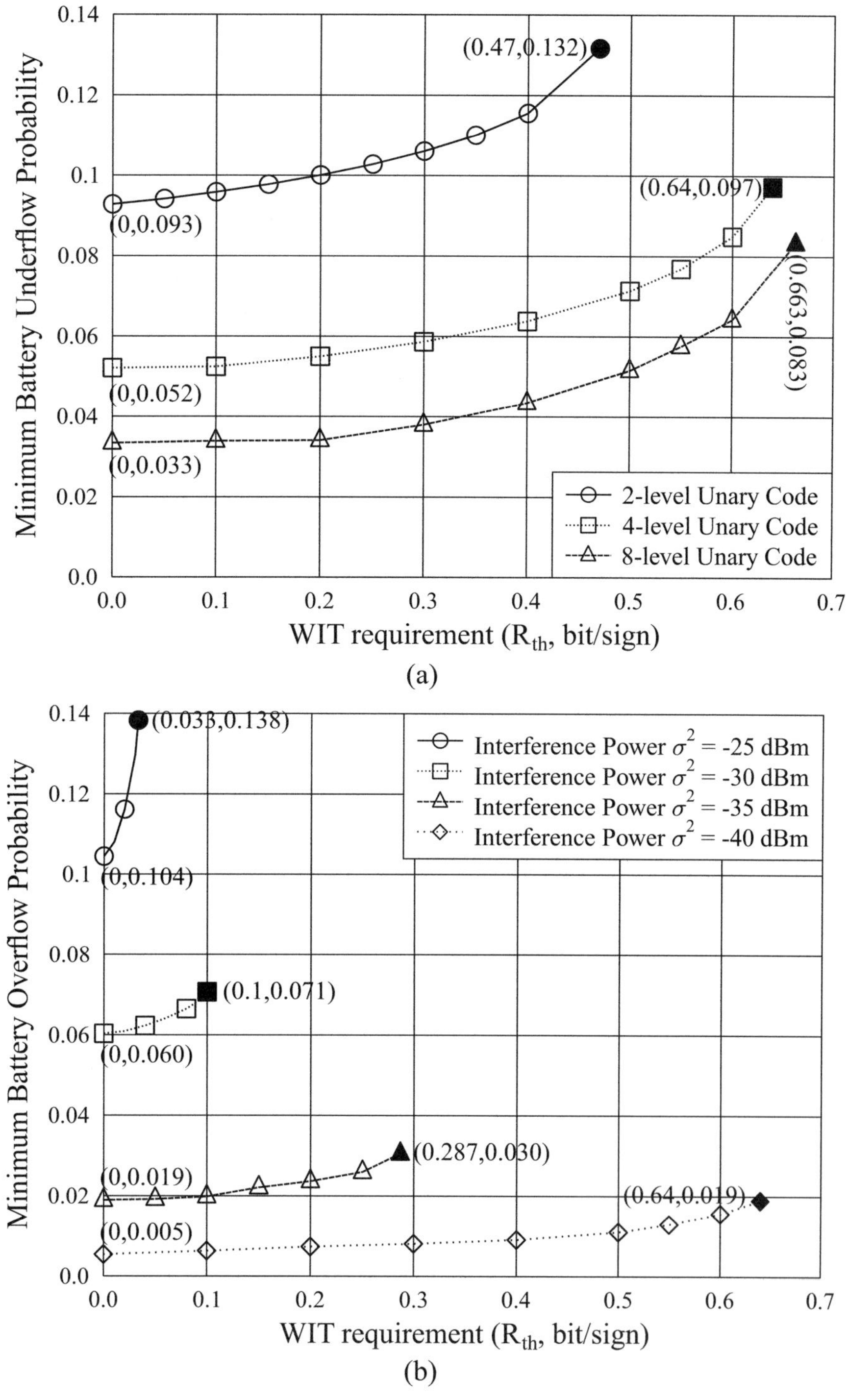

Fig. 2.12 The SWIPT performance with a finite battery capacity: **(a)** (2,4,8)-level unary codes with the ambient RF interference power $\sigma^2 = -40$ dBm; **(b)** 4-level unary code with the ambient RF interference power $\sigma^2 = \{-25, -30, -35, -40\}$ dBm. The filled markers represent the conventional WIT coding schemes, which aims for maximizing the mutual information of Eq. (2.44) by ignoring the QoS requirement of WPT

Fig. 2.6. Obviously, the transition process from the state S_i to the state S_{i+1} is physically uncorrelated to the transition process from the state S_i to Y_i, which can be expressed as the following equality:[2]

[2] The capital letter S_i represents a random variable and s_i represents a possible value of this random variable.

$$p(y_i, s_{i+1}|s_i) = p(y_i|s_i) \cdot p(s_{i+1}|s_i). \tag{2.58}$$

Moreover, the conditional probability $p(y_i, s_{i+1}|s_i)$ can be further expressed as

$$p(y_i, s_{i+1}|s_i) = p(y_i|s_{i+1}, s_i) \cdot p(s_{i+1}|s_i) = p(y_i|s_i) \cdot p(s_{i+1}|s_i), \tag{2.59}$$

where the second equality is derived by substituting Eq. (2.58) into Eq. (2.59). Therefore we have $p(y_i|s_{i+1}, s_i) = p(y_i|s_i)$. By multiplying $p(s_i)$ on both sides of this equation and summing them over all possible values of s_i, we may derive the following equalities[3]

$$\begin{aligned}
&\sum_{s_i} p(y_i|s_{i+1}, s_i) \cdot p(s_i) = \sum_{s_i} p(y_i|s_i) \cdot p(s_i) \\
&\Rightarrow \sum_{s_i} p(y_i, s_i|s_{i+1}) = \sum_{s_i} p(y_i, s_i) \Rightarrow p(y_i|s_{i+1}) = p(y_i) \\
&\Rightarrow p(y_i|s_{i+1}) \cdot p(s_{i+1}) = p(y_i) \cdot p(s_{i+1}) \\
&\Rightarrow p(y_i, s_{i+1}) = p(y_i) \cdot p(s_{i+1}),
\end{aligned} \tag{2.60}$$

which demonstrates that the random event $Y_i = y_i$ is independent of the random event $S_{i+1} = s_{i+1}$.

We may further prove that the channel output signs $\{Y_1, \cdots, Y_{i-1}\}$ are all uncorrelated to the state S_{i+1}. Firstly, $\{Y_1, \cdots, Y_{i-1}\}$ only depends on the corresponding states $\{S_1, \cdots, S_{i-1}\}$ at the same instants. Secondly, when S_i is given, the state S_{i+1} only depends on S_i but it is independent of all the previous states due to the Markov property. Therefore, *Lemma 2.1* is proved.

2.3.1.7 Proof of Lemma 2.2

The basic principle of proving this lemma is to find the inequalities of $\{p(S_i = s_i|Y_{i+n} = y_{i+n}) \neq p(S_i = s_i)|n = 1, 2, \cdots, \}$, where $s_i \in \{0, 1, \cdots, 2^k - 1\}$ and $y_{i+n} \in \{0, 1\}$.

(1) We first prove that the random event $S_i = s_i$ depends on the random event $Y_{i+1} = y_{i+1}$.

The joint probability of these two random events is $p(s_i, y_{i+1}) = p(s_i) \cdot p(y_{i+1}|s_i)$. Furthermore, the probability of the random event $Y_{i+1} = y_{i+1}$ occurring is $p(y_{i+1}) = \sum_{s_i} p(s_i) \cdot p(y_{i+1}|s_i)$. Therefore, the conditional probability $p(s_i|y_{i+1})$ can be obtained as

$$p(s_i|y_{i+1}) = \frac{p(s_i, y_{i+1})}{p(y_{i+1})} = \frac{p(s_i) \cdot p(y_{i+1}|s_i)}{\sum_{s_i} p(s_i) \cdot p(y_{i+1}|s_i)}. \tag{2.61}$$

When $y_{i+1} = 1$, the conditional probability $p(y_{i+1} = 1|s_i)$ of the WIT channel is expressed as

$$p(y_{i+1} = 1|s_i) = q_{s_i} \cdot (1 - \omega_I) + (1 - q_{s_i}) \cdot \omega_I, \tag{2.62}$$

where q_{s_i} is the transition probability from the state s_i to $(s_i + 1)$, which is also the probability of the unary encoder inputting a sign "1" to the WIT channel. This sign "1" then has a probability of $(1 - \omega_I)$ to be unchanged by the WIT channel. Moreover, the probability of a sign "0" being input to the WIT channel is $(1 - q_{s_i})$, while this sign is changed to sign "1" with a probability of ω_I. Substituting Eq. (2.62) into (2.61), we have

$$p(s_i|y_{i+1} = 1) = \frac{p(s_i) \cdot (q_{s_i} - 2q_{s_i}\omega_I + \omega_I)}{\sum_{s_i} p(s_i) \cdot (q_{s_i} - 2q_{s_i}\omega_I + \omega_I)}. \tag{2.63}$$

Observe from Eq. (2.63) that $\{q_{s_i} = q|s_i = 0, 1, \cdots, K - 1\}$ is a feasible solution to let $p(s_i|y_{i+1} = 1) = p(s_i)$, where $q \in [0, 1]$ is an arbitrary constant. Otherwise, we have $p(s_i|y_{i+1} = 1) \neq p(s_i)$.

When $y_{i+1} = 0$, the conditional probability $p(y_{i=1} = 0|s_i)$ of the WIT channel is expressed as

[3] $p(s_i)$ is the probability of the Markov state machine staying at the state s_i. When the Markov chain becomes stationary, we have $p(s_i) = \pi_{s_i}$.

$$p(y_{i=1}=0|s_i)=q_{s_i}\omega_I+(1-q_{s_i})(1-\omega_I). \tag{2.64}$$

This is because the unary encoder inputs a sign "1" to the WIT channel, when the Markov chain of Fig. 2.7 transits from s_i to s_{i+1} with a probability of q_{s_i}. This sign "1" has a probability of ω_I to be changed to "0" by the WIT channel. Moreover, the probability of a sign "0" being input is $(1-q_{s_i})$, while this sign is unchanged by the WIT channel with a probability of $(1-\omega_I)$. Substituting Eq. (2.64) into (2.61), we have

$$p(s_i|y_{i+1}=0)=\frac{p(s_i)\cdot[q_{s_i}\omega_I+(1-q_{s_i})(1-\omega_I)]}{\sum_{s_i}p(s_i)\cdot[q_{s_i}\omega_I+(1-q_{s_i})(1-\omega_I)]}. \tag{2.65}$$

Observe from both Eqs. (2.63) and (2.65) that $\{q_{s_i}=q|s_i=0,1,\cdots,K-1\}$ is a feasible solution to simultaneously let $p(s_i|y_{i+1}=1)=p(s_i)$ and $p(s_i|y_{i+1}=0)=p(s_i)$, which indicates that the random event $S_i=s_i$ is independent of $Y_{i+1}=y_{i+1}$. Otherwise, $S_i=s_i$ and $Y_{i+1}=y_{i+1}$ are correlated with each other, since we have $p(s_i|y_{i+1})\neq p(s_i)$.

(2) We then prove that the random event $S_i=s_i$ depends on the random event $Y_{i+2}=y_{i+2}$:

The joint probability of these two random events is $p(s_i,y_{i+2})=\sum_{s_{i+1}}\sum_{s_{i+2}}p(s_i,s_{i+1},s_{i+2})\cdot p(y_{i+2}|s_i,s_{i+1},s_{i+2})$. Furthermore, the probability of the random event $Y_{i+2}=y_{i+2}$ occurring is $p(y_{i+2})=\sum_{s_i}p(s_i,y_{i+2})$. Therefore, the conditional probability $p(s_i|y_{i+2})$ can be obtained as

$$p(s_i|y_{i+2})=\frac{p(s_i,y_{i+2})}{p(y_{i+2})}=\frac{p(s_i,y_{i+2})}{\sum_{s_i}p(s_i,y_{i+2})}. \tag{2.66}$$

When the output sign of the WIT channel at the $(i+2)$-th instant is $y_{i+2}=1$, the joint probability $p(s_i,y_{i+2}=1)$ can be formulated as

$$\begin{aligned}p(s_i,y_{i+2}=1)&=p(s_i,s_i+1,s_i+2)p(y_{i+2}=1|s_i,s_i+1,s_i+2)\\&+p(s_i,s_i+1,0)p(y_{i+2}=1|s_i,s_i+1,0)\\&+p(s_i,0,1)p(y_{i+2}=1|s_i,0,1)\\&+p(s_i,0,0)p(y_{i+2}=1|s_i,0,0)\\&=p(s_i)\left\{(1-\omega_I)[q_{s_i}q_{s_i+1}+(1-q_{s_i})q_0]\right.\\&\left.+\omega_I[q_{s_i}(1-q_{s_i+1})+(1-q_{s_i})(1-q_0)]\right\}.\end{aligned} \tag{2.67}$$

This is because when $s_{i+2}\neq 0$, the unary encoder inputs a sign "1" to the WIT channel, which is unchanged with a probability of $(1-\omega_I)$. When $s_{i+2}=0$, a sign "0" is input to the WIT channel, which is changed with a probability of ω_I. Substituting Eq. (2.67) into (2.66), we have

$$p(s_i|y_{i+2}=1)=\frac{p(s_i,y_{i+2}=1)}{\sum_{s_i}p(s_i,y_{i+2}=1)}. \tag{2.68}$$

Observe from Eqs. (2.67) and (2.68) that $\{q_{s_i}=q|s_i=0,1,\cdots,K-1\}$ is able to let $p(s_i|y_{i+2}=1)=p(s_i)$, where $q\in[0,1]$ is an arbitrary constant. Otherwise, we have $p(s_i|y_{i+2}=1)\neq p(s_i)$.

When the output sign of the WIT channel at the $(i+2)$-th instant is $y_{i+2}=0$, the joint probability $p(s_i,y_{i+2}=0)$ can be formulated as

$$\begin{aligned}p(s_i,y_{i+2}=0)&=p(s_i,s_i+1,s_i+2)p(y_{i+2}=0|s_i,s_i+1,s_i+2)\\&+p(s_i,s_i+1,0)p(y_{i+2}=0|s_i,s_i+1,0)\\&+p(s_i,0,1)p(y_{i+2}=0|s_i,0,1)\\&+p(s_i,0,0)p(y_{i+2}=0|s_i,0,0)\\&=p(s_i)[q_{s_i}q_{s_i+1}\omega_I+q_{s_i}(1-q_{s_i+1})(1-\omega_I)\\&+(1-q_{s_i})q_0\omega_I+(1-q_{s_i})(1-q_0)(1-\omega_I)].\end{aligned} \tag{2.69}$$

This is because when $s_{i+2} \neq 0$, the unary encoder inputs a sign "1" to the WIT channel, which is changed to "0" with a probability of ω_I. When $s_{i+2} = 0$, a sign "0" is input to the WIT channel, which is unchanged with a probability of ω_I. Substituting Eq. (2.69) into (2.66), we have

$$p(s_i|y_{i+2}=0) = \frac{p(s_i)[1-(1-\omega_I)(q_{s_i}q_{s_i+1}+q_0-q_{s_i}q_0)]}{\sum_{s_i} p(s_i)[1-(1-\omega_I)(q_{s_i}q_{s_i+1}+q_0-q_{s_i}q_0)]}. \tag{2.70}$$

Observe from Eqs. (2.67) and (2.68) that $\{q_{s_i} = q|s_i = 0, 1, \cdots, K-1\}$ is also a feasible solution to let $p(s_i|y_{i+2} = 0) = p(s_i)$, where $q \in [0, 1]$ is an arbitrary constant. Otherwise, we have $p(s_i|y_{i+2} = 0) \neq p(s_i)$. Therefore, the random event $S_i = s_i$ and $Y_{i+2} = y_{i+2}$ are independent, only if we have $\{q_{s_i} = q|s_i = 0, 1, \cdots, K\}$. Otherwise, they are correlated with each other.

Relying on the similar methodology, we may also prove that $S_i = s_i$ is correlated with $Y_{i+n} = y_{i+n}$ for $\forall n \geq 1$. Furthermore, $\{q_{s_i} = q|s_i = 0, 1, \cdots, K-1\}$ is a feasible solution to let $S_i = s_i$ simultaneously independent of $\{Y_{i+1} = y_{i+1}, Y_{i+2} = y_{i+2}, \cdots\}$. *Lemma 2.2* is proved.

2.3.2 Unary Coding Design for SWIPT with Practical M-QAM

The current analysis of WIT performance are based on the classic Shannon-Hartley channel capacity [34–36], which assumes Gaussian distributed transmit signals. This is far from practice, since modulated signals are practically transmitted with finite alphabet. The overestimated WIT performance may occupy more resources, which hence reduces attainable WPT performance in practice. Recently, very few works discussed the modulation and coding design for SWIPT. In [37], Zhao et al. studied a joint interleaving and modulation design in a NOMA-SWIPT system, where the energy interleaver and the constellation rotator were optimally designed for ensuring constructive combination of signals targeting different users. Chaehun et al. [38] proposed a multi-tone amplitude modulation scheme and the receiver architecture for SWIPT, where the energy harvested amount and the symbol error rate (SER) at the receiver were both improved. Other existing works about modulation design were studied in amplitude [39], frequency [40] or the tone index of multisine waveforms [41].

Coding design for information and energy transfer has also been studied in recent years. For instance, the integrated information and energy transfer performance of the run-length-limited (RLL) code [8] and sub-block-constrained code [11] were analyzed and optimized, respectively. In [8], Fouladgar et al. investigated an RLL code based SWIPT system. The battery overflow and underflow probabilities of the receiver was maximized. In [11], Tandon et al. adopted a sub-block energy-constrained code (SECC) to ensure that sufficient energy is transferred within every sub-block duration. A constant sub-block-composition codes, which is a subclass of SECC, was also studied to maximize the rate of information transfer while meeting the energy requirement. Moreover, in [42], Hu et al. studied an unary coding controlled SWIPT system, where the input distribution of unary codewords was optimized, in order to maximize WPT performance by considering infinite and finite battery capacity, while satisfying the WIT requirements.

However, conventional source and channel encoder tends to output equally distributed binary sings "1" and "0". In order to control the SWIPT performance in the coding level, an additional SWIPT encoder can be implemented at the transmitter for dynamically adjusting the distribution of the output binary signs and for attaining arbitrary WPT performance. Moreover, the impact of modulation [37–41] and coding [8, 11, 42] on SWIPT performance were separately studied. In fact, the binary codewords, which are generated by a SWIPT encoder, cannot solely determine the SWIPT performance. The binary signs have to be mapped onto modulated symbols carrying different amount of energy, which results in various WPT performance. Different modulators also have various bit-error-rate (BER) performance, which results in distinctive WIT performance. For instance, in [42], the unary coding controlled SWIPT is only feasible for on-off-keying (OOK) based modulator. In this chapter, we aim to investigate the SWIPT coding design by jointly considering the impact of practical modulation schemes.

Therefore, we aim for optimizing the codeword distribution of a unary coding aided SWIPT encoder for maximizing the SWIPT performance, when the binary codewords are modulated by a practical M-QAM based modulator. Our novel contributions are then summarized as follows:

- We proposed a novel transceiver architecture including an additional SWIPT encoder. The unary coding controlled SWIPT system was originally studied by jointly considering the impact of practical M-QAM on both the WIT and WPT performance.
- In a SWIPT system, discrete Markov chains are exploited for characterizing the correlation among the binary signs generated by the SWIPT encoder and for modeling the transmit pattern of the modulated symbols.

Table 2.3 The description of the notations

Notations	The meaning of notations
$\mathbb{X}$	Information source
$\overline{\mathbb{X}}$	Coded information source
$\mathbb{Y}$	Information destination
$\overline{\mathbb{Y}}$	Coded information destination
$\mathbf{X}_L$	Binary sequence generated by $\overline{\mathbb{X}}$
χ_j	j-th message of $\mathbb{X}$
$\overline{\chi}_k$	k-th codeword of $\overline{\mathbb{X}}$
Pr_x	Codeword distribution
Pr_c	State transition probabilities of coded Information source
Pr_s	State transition probabilities of modulators
s_m	m-th modulated symbol in M-QAM
$\boldsymbol{\zeta}_m$	Binary sequence corresponding to s_m
$\zeta_{m,n}$	n-th binary sign in $\boldsymbol{\zeta}_m$
$\lambda_m^{(b)}$	Backward run-length of "1" of s_m
$\lambda_m^{(f)}$	Forward run-length of "1" of s_m
ξ	Minimum Euclidean distance in a modulation constellation
$\kappa_{t,m}$	Transmit power of s_m
$\kappa_{r,m}$	Receive power of s_m
P_t	Average transmit power
P_r	Average receive power
ϵ_s	Symbol error rate
ϵ_b	Bit error rate

- Both mutual information between the encoded information source and the destination as well as the energy harvested by the SWIPT receiver are analyzed in semi-closed-form by considering a practical MISO channel with Rician fading.
- The codeword distribution of the SWIPT encoder is obtained by exploiting the genetic algorithm, which maximizes the average energy harvested by the SWIPT receiver, while guaranteeing a specific mutual information requirement.
- Our theoretical analysis and sub-optimal design are validated by numerical results, when various unary coding and modulation schemes are adopted. The advantage of the unary coding based SWIPT encoder and the trade-off between the WIT and WPT performance are also demonstrated.

Some important mathematical notations are summarized in Table 2.3.

Notations $\mathbb{C}$ denotes the set of complex numbers. The scalar variable is denoted in the form of x, while the vector and the matrix are denoted in the form of $\mathbf{x}$ and $\mathbf{X}$, respectively. $\mathbf{X}^*$ (or $\mathbf{x}^*$) denotes the conjugate transpose of the matrix $\mathbf{X}$ (or the vector $\mathbf{x}$). Moreover, $|x|$ represents the absolute value of a complex variable x, while $||\mathbf{x}||$ denotes the 2-norm of the vector $\mathbf{x}$.

2.3.2.1 System Model and transceiver Architecture

A SWIPT transceiver is illustrated in Fig. 2.13, which includes a SWIPT transmitter, a wireless channel and a SWIPT receiver. The SWIPT transmitter should be equipped with these functional modules, which are described as follows:

- *Source and channel encoder.* The source and channel encoder encodes the original messages into binary codewords $\mathbb{X}$, in order to improve the efficiency and the reliability of the information transmission. Conventional source coding techniques, such as Huffman code and L-Z code, as well as channel coding techniques, such as LDPC code and polar code, can be adopted by the source and channel encoder.
- *SWIPT encoder.* The output $\mathbb{X}$ of the source and channel encoder is then encoded into another binary codewords by a SWIPT encoder, which constitutes a binary coded information $\overline{\mathbb{X}}$. The unary code is then adopted by the SWIPT encoder.

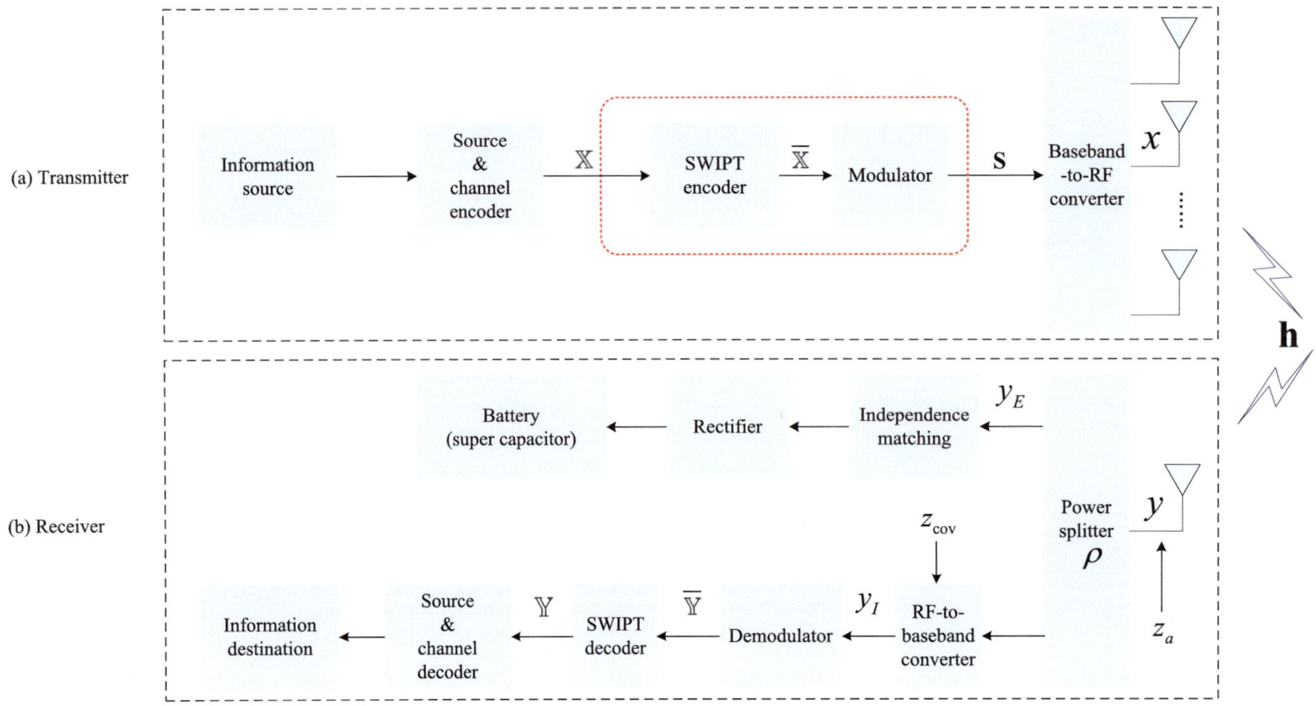

Fig. 2.13 Transceiver architecture

- *Modulator.* A digital modulator maps the binary sequences onto a range of modulated symbols **S**. Traditional M-QAM modulator is conceived at the SWIPT encoder.[4]
- *Baseband-to-RF converter.* The baseband-to-RF converter is responsible for converting the baseband signals to the RF band. An appropriate beamformer is adopted by the multi-antennas transmitter, in order to maximize the received power at the receiver.

After passing the baseband-to-RF converter, the wireless signals are transmitted by N_t antennas. A Rician block fading based multiple-input-single-output (MISO) channel is assumed between the SWIPT transmitter and receiver. The channel coefficient is expressed as

$$\mathbf{h} = \Omega \mathbf{g}, \tag{2.71}$$

where Ω is the large scale fading caused by path-loss, and $\mathbf{g} = [g_1, \cdots, g_{N_t}] \in \mathbb{C}^{1 \times N_t}$ represents the multipath fading of the MISO channel.

The SWIPT receiver has the following main functional modules:

- *Power splitter.* The power splitter enables the SWIPT receiver to operate energy harvesting and information decoding simultaneously by splitting the received RF signals into two portions according to a power splitting ratio ρ.
- *WIT modules.* One portion of the received RF signal is converted to the baseband signal y_I by an RF-to-baseband converter. The signal y_I is then demodulated by a digital demodulator into binary sequences, which constitutes a coded information destination $\overline{\mathbb{Y}}$. The resultant binary sequences are then decoded into $\mathbb{Y}$ by an SWIPT decoder, after which the original information is recovered by the source and channel decoder. Iterative decoding [43] can also be adopted at the receiver in order to improve the decoding efficiency and accuracy.[5]

[4] Note that any modulator is applicable in the following analysis. Without loss of generality, we select M-QAM, since it is widely adopted in modern digital communication systems.

[5] Note that the design of decoding algorithms is out of the scope in this chapter, since we mainly focus on the optimal coding design for improving the WIT and WPT performance. Any efficient decoding algorithms, including the hard and soft decoding approaches, can be adopted at the receiver.

- *WPT modules*. The other portion of the received signals flow into an energy harvesting circuit, including an independence matching and a rectifier. The RF signal is then transformed to direct current (DC). The DC energy is finally stored in a battery or a super-capacitor.

The signal received by the single antenna of the SWIPT receiver is expressed as

$$y = \mathbf{hw}x + z_a, \tag{2.72}$$

where x is the transmit RF signal corresponds to the baseband modulated symbol s having a maximum transmit power $P_{\max}$, $\mathbf{w}$ is the transmitter beamforming vector, and z_a represents the complex additive-white-Gaussian-noise (AWGN) imposed on the receive antenna, which has a zero mean and a variance of $\sigma_a/2$ per dimension. The maximum ratio based beamformer [44] has been proved to be optimal in a $N_t \times 1$ MISO system, where we have $\mathbf{w} = \mathbf{h}^*/||\mathbf{h}||$. A portion of $\sqrt{\rho}$ of the received RF signal is exploited for energy harvesting, which is expressed as

$$y_E = \sqrt{\rho}(\Omega||\mathbf{g}||x + z_a). \tag{2.73}$$

Moreover, the other portion $\sqrt{1-\rho}$ is relied upon for information demodulation and decoding, which is expressed as

$$y_I = \sqrt{1-\rho}(\Omega||\mathbf{g}||x + z_a) + z_{\text{cov}}, \tag{2.74}$$

where z_{cov} represents the AWGN imposed on the RF-to-baseband module, which has a zero mean and a variance of $\sigma_{\text{cov}}/2$ per dimension.

2.3.2.2 Markov Chains for Coding and Modulation

In this section, we calculate the probabilities of the run-length of the unary code as well as those of the transmit symbols by exploiting the coding level and the symbol-level Markov chain analysis. The joint impact of the codeword distribution and practical modulation is originally considered compared with other existing works.

Assuming that the source and channel encoder $\mathbb{X}$ randomly outputs a message chosen from a message book $\{\chi_1, \cdots, \chi_J\}$, a message is then encoded by one of the unary codewords chosen from a codebook $\{\overline{\chi}_1, \cdots, \overline{\chi}_K\}$. The k-th unary codeword $\overline{\chi}_k$ has $k-1$ binary signs "1" followed by a binary sign "0". We denote the message distribution of $\mathbb{X}$ as $\{p(\chi_1), \cdots, p(\chi_J)\}$, while the codeword distribution of the SWIPT encoder is denoted as $\mathbf{Pr\_x} = \{p(\overline{\chi}_1), \cdots, p(\overline{\chi}_K)\}$. We have $p(\overline{\chi}_k) = \sum_{j=1}^{J} \Pr(\overline{\chi}_k|\chi_j)p(\chi_j)$, where $\Pr(\overline{\chi}_k|\chi_j)$ represents the probability that the message χ_j is mapped onto the codeword $\overline{\chi}_k$ by the SWIPT encoder.

The resultant coded information source $\overline{\mathbb{X}}$ may generate a binary sequence $\mathbf{X}_L = \{X_1, \cdots, X_L\}$ having a length of L. We define a range of states $\mathbf{C}_L = \{C_1, \cdots, C_L\}$ for the coded information source $\overline{\mathbb{X}}$, where C_l represents the run-length of "1" after the transmission of the l-th binary sign in $\mathbf{X}_L$. Since the run-length of "1" has a maximum of $(K-1)$ when the K-level unary code is conceived, we have $C_l \in \{0, \cdots, k, \cdots, K-1\}$ for $\forall l = 1, \cdots, L$. Generally, there is always a definite mapping relationship between $\mathbf{X}_L$ and the corresponding state sequence $\mathbf{C}_L$. For example, if we have $L = 8$ and $\mathbf{X}_8 = \{0, 1, 1, 0, 1, 1, 1, 1\}$, the corresponding state sequence should be $\mathbf{C}_8 = \{0, 1, 2, 0, 1, 2, 3, 4\}$. The transitions of the states $C_l = k$ $(k = 0, \cdots, K-1)$ follow the Markov property, as portrayed in Fig. 2.14.

Theorem 2.4 *The state transition probabilities $Pr\_c(k'|k) = Pr\{C_{l+1} = k'|C_l = k\}$ of the coded information source $\overline{\mathbb{X}}$ are formulated as:*

$$Pr\_c(k'|k) = \begin{cases} p_k, & k \le K-2, k' = k+1, \\ 1-p_k, & k \le K-2, k' = 0, \\ 1, & k = K-1, k' = 0, \\ 0, & \textit{otherwise}, \end{cases} \tag{2.75}$$

where

$$p_k = \frac{p(\overline{\chi}_{k+2}) + \cdots + p(\overline{\chi}_K)}{p(\overline{\chi}_{k+1}) + p(\overline{\chi}_{k+2}) + \cdots + p(\overline{\chi}_K)}, k = 0, \cdots, K-2. \tag{2.76}$$

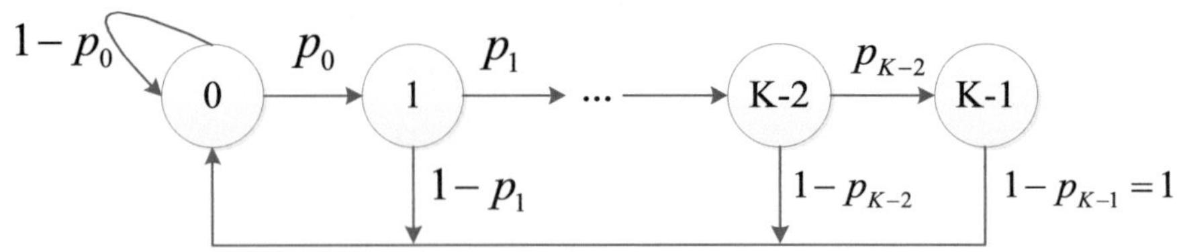

Fig. 2.14 Markov chain for a K-level unary coded information source $\overline{\mathbb{X}}$

Table 2.4 Mapping relationship between binary sequences and symbols and their characteristics for 16-QAM

m	ζ_m	$\lambda_m^{(b)}$ (←)	$\lambda_m^{(f)}$ (→)	m	ζ_m	$\lambda_m^{(b)}$ (←)	$\lambda_m^{(f)}$ (→)
1	0000	0	0	9	1000	0	1
2	0001	1	0	10	1001	1	1
3	0010	0	0	11	1010	0	1
4	0011	2	0	12	1011	2	1
5	0100	0	0	13	1100	0	2
6	0101	1	0	14	1101	1	2
7	0110	0	0	15	1110	0	3
8	0111	3	0	16	1111	4,5,···	4

Proof Please refer to Sect. 2.3.2.6 for detailed proof.

Let us denote the state transition probabilities as a $K \times K$ matrix **Pr_c**, whose element in the k-th row and in the k'-th column represents the probability of Pr_c$(k'-1|k-1)$. Therefore, stationary distribution $\boldsymbol{\pi}_c$ is then derived by solving the following equations

$$\begin{cases} \boldsymbol{\pi}_c = \boldsymbol{\pi}_c \cdot \mathbf{Pr\_c}, \\ \sum_{k=1}^{K} \pi_{c,k} = 1, \end{cases} \tag{2.77}$$

where $\pi_{c,k}$ is the k-th element in $\boldsymbol{\pi}_c$ representing the stationary probability of $C_l = k-1$.

By conceiving M-QAM as the digital modulator at the transmitter, the binary sequences generated by the SWIPT encoder are modulated as different symbols by obeying the Gray mapping rule. Given the m-th symbol s_m $(m = 1, \cdots, M)$ in the M-QAM constellation, its corresponding binary sequence is denoted as $\boldsymbol{\zeta}_m = \{\zeta_{m,1}, \cdots, \zeta_{m,n}, \cdots, \zeta_{m,N}\}$ $(n = 1, \cdots, N)$, where $\boldsymbol{\zeta}_m$ is the binary formation of the decimal number $(m-1)$ and $N = \log_2(M)$.

Given a specific symbol s_m and its corresponding binary sequence $\boldsymbol{\zeta}_m$, we define $\lambda_m^{(b)}$ as the number of successive binary sign "1" counting from $\zeta_{m,N}$ in a backward direction. For $\forall m = 1, \cdots, M-1$, we have $\lambda_m^{(b)} < N$, since there is at least a single sign "0" in the sequence $\boldsymbol{\zeta}_m$. For the symbol s_M, we have $\lambda_M^{(b)} \in \{N, N+1, \cdots, K-1\}$ if $K-1 \geq N$, which depends on the binary sequences of previously transmitted symbols. Furthermore, we define $\lambda_m^{(f)}$ as the number of successive binary sign "1" counting from $\zeta_{m,1}$ to $\zeta_{m,N}$ in a forward direction. The mapping relationship of the symbols in 16-QAM and their corresponding binary sequences $\boldsymbol{\zeta}_m$, $\lambda_m^{(b)}$, $\lambda_m^{(f)}$ are exemplified in Table 2.4.

Two cases of (a) $K = 4$, $N = 4$ and (b) $K = 6$, $N = 4$ are exemplified in Fig. 2.15 in order to help readers understand all these concepts. For instance, in Fig. 2.15a, the binary sequence of the modulated symbol s_{12} is $\boldsymbol{\zeta}_{12} = 1011$. Its backward run of "1" of its sub-sequence $\{\zeta_{12,3}, \zeta_{12,4}\} = 11$ has a length of $\lambda_{12}^{(b)}$, while the forward run of "1" of the sub-sequence $\{\zeta_{12,1}\} = 1$ has a length of $\lambda_{12}^{(f)}$. Observe from Fig. 2.15a that the maximum run-length of "1" is $K-1 = 3 < N = 4$, which indicates that the sequence $\boldsymbol{\zeta}_{16}$ and the corresponding symbol s_{16} are never output. Transmissions of a pair of consecutive modulated symbols also follow the Markov property, due to the correlation among binary signs output by the coded information source.

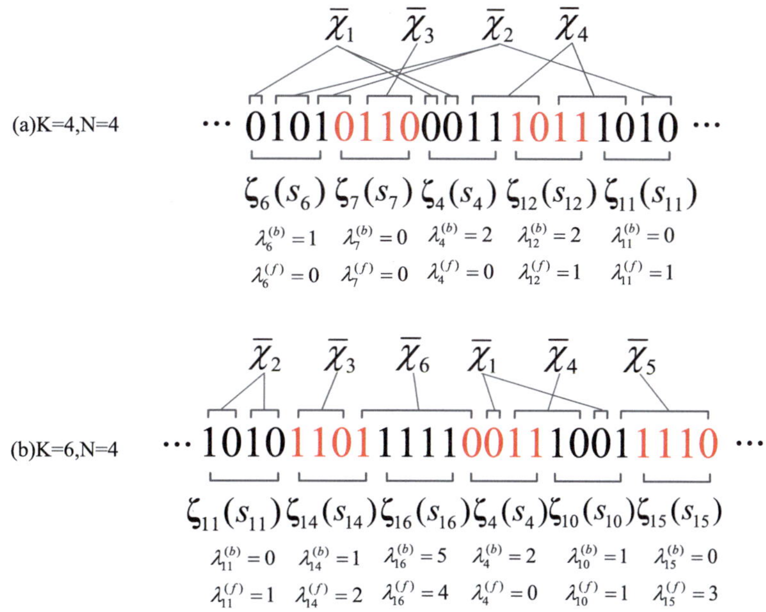

Fig. 2.15 Example of transmit binary sequences with two cases of $K = 4$, $N = 4$ and $K = 6$, $N = 4$

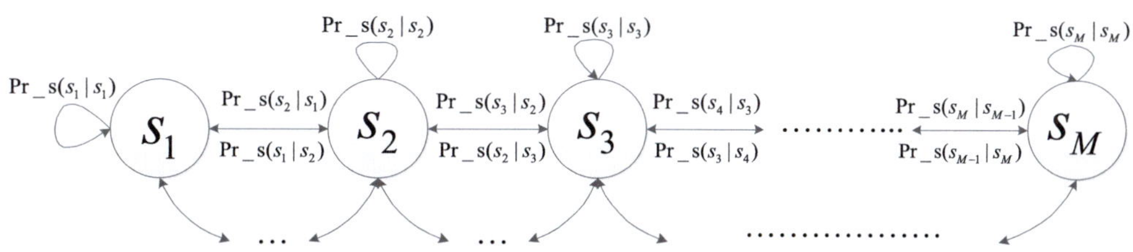

Fig. 2.16 Markov chain for the case of $K - 1 < N$

By contrast, in the case of Fig. 2.15b, we have $K - 1 = 5 > N = 4$. Therefore, the binary sequence $\boldsymbol{\zeta}_{16}$ can be output by the coded information source $\overline{\mathbb{X}}$. Observe from Fig. 2.15b that the backward run-length $\lambda_{16}^{(b)}$ of "1" of the binary sign $\boldsymbol{\zeta}_{16}$ also depends on the previous sequence $\boldsymbol{\zeta}_{14}$ in this example and its corresponding value is $\lambda_{16}^{(b)} = 5$. In this case, the probability that $\boldsymbol{\zeta}_4$ is output after $\boldsymbol{\zeta}_{16}$ also depends on $\boldsymbol{\zeta}_{14}$. As a result, when we have $K - 1 \geq N$, state transitions no longer obey the Markov property, since a state of the modulator transmitting a symbol does not only depend on its adjacent peer, but also depend on more previous ones. Therefore, in the following discussions on deriving the transmit probabilities of modulated symbols, we consider two cases of $K - 1 < N$ and $K - 1 \geq N$, respectively.

We firstly define $\delta_{m,n}$ as the backward run-length of "1" starting from the binary sign $\zeta_{m,n}$, which can be expressed as

$$\delta_{m,n} = \begin{cases} \zeta_{m,n} * (\delta_{m',N} + 1) = \zeta_{m,n} * (\lambda_{m'}^{(b)} + 1), n = 1, \\ \zeta_{m,n} * (\delta_{m,n-1} + 1), 1 < n \leq N, \end{cases} \tag{2.78}$$

where $\boldsymbol{\zeta}_{m'}$ is the previous transmitted binary sequence before $\boldsymbol{\zeta}_m$.

Theorem 2.5 *When we have $K - 1 < N$, the Markov chain of all the states is portrayed in Fig. 2.16, while the transition probability from the state $s_{m'}$ to the state s_m is expressed as*

$$Pr\_s(s_m|s_{m'}) = \begin{cases} 0, & \lambda_{m'}^{(b)} + \lambda_m^{(f)} > K - 1, \\ Pr\_c(\delta_{m,1}|\lambda_{m'}^{(b)}) \prod_{n=1}^{N-1} Pr\_c(\delta_{m,n+1}|\delta_{m,n}), \\ \qquad \lambda_{m'}^{(b)} + \lambda_m^{(f)} \leq K - 1. \end{cases} \tag{2.79}$$

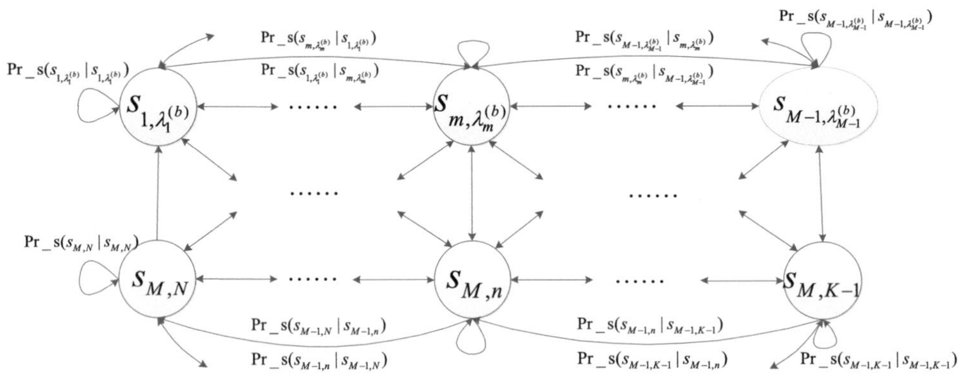

Fig. 2.17 Markov chain for the case of $K-1 \geq N$

Proof Please refer to Sect. 2.3.2.7 for detailed proof.

The transition probability matrix of this Markov chain is denoted as **Pr_s** having the size of $M \times M$, whose element on the m'-th row and the m-th column represents the state transition probability $\mathrm{Pr\_s}(s_m|s_{m'})$. Then, the stationary probability vector $\boldsymbol{\pi}_s$ of all possible states is obtained by solving the following equations

$$\begin{cases} \boldsymbol{\pi}_s = \boldsymbol{\pi}_s \cdot \mathbf{Pr\_s}, \\ \sum\limits_{m=1}^{M} \pi_{s,m} = 1. \end{cases} \tag{2.80}$$

The stationary probability τ_m of the symbol s_m being transmitted is expressed as $\tau_m = \pi_{s,m}$, where $\pi_{s,m}$ represents the m-th element in $\boldsymbol{\pi}_s$.

When we have $K-1 \geq N$, we cannot use a similar Markov chain as we did in Fig. 2.16, where the Markov property cannot be satisfied, as exemplified in Fig. 2.15. We have to redefine the states for retrieving the Markov property. Different from the Markov chain of the case $K-1 < N$, we denote each state as $s_{m,q}$, where $q = \lambda_m^{(b)}$. As discussed before, when we have $m < M$, the binary sequence $\boldsymbol{\zeta}_m$ has a definite value of $\lambda_m^{(b)}$. Therefore, a modulated symbol s_m may correspond to a specific state $s_{m,q}$. However, if we have $m = M$, the backward run-length $\lambda_M^{(b)}$ of "1" can be $\{N, N+1, \cdots, K-1\}$, which depends on the previous transmitted binary sequence. Therefore, the modulated symbol s_M may correspond to $K-N$ different states of $s_{M,q}$ $(q = N, N+1, \cdots, K-1)$. Given a state $s_{m',q'}$, the backward run-length of "1" of the binary sequence $\boldsymbol{\zeta}_{m'}$ is already known as q'. Then, the transition probability from the state $s_{m',q'}$ to $s_{m,q}$ is readily obtained according to the Markov analysis of the unary code. Therefore, the transition between two adjacent states $s_{m',q'}$ and $s_{m,q}$ obeys the Markov property.

Theorem 2.6 *When we have $K-1 \geq N$, the Markov chain of all the states is portrayed in Fig. 2.17, while the state transition probability $Pr\_s(s_{m,q}|s_{m',q'})$ is derived as*

$$Pr\_s(s_{m,q}|s_{m',q'}) = \begin{cases} 0, & q' + \lambda_m^{(f)} > K-1,\ m \neq M, \\ Pr\_c(\delta_{m,1}|q') \prod_{n=1}^{N-1} Pr\_c(\delta_{m,n+1}|\delta_{m,n}), & \\ \qquad q' + \lambda_m^{(f)} \leq K-1,\ m \neq M, & \\ 0, & q \neq q' + N,\ m = M, \\ \prod\limits_{k=q'}^{q} Pr\_c(k+1|k),\ q = q' + N,\ m = M. & \end{cases} \tag{2.81}$$

Proof Please refer to Sect. 2.3.2.8 for detailed proof.

The resultant state transition probability matrix of this Markov chain is denoted as **Pr_s** having the size of $(M+K-N-1) \times (M+K-N-1)$. Then, the stationary probabilities $\boldsymbol{\pi}_s$ of all $M+K-N-1$ states can be obtained by solving similar equations as Eq. (2.80). For the cases of $m = 1 \cdots, M-1$, the probability that s_m being transmitted is just the stationary probability of $s_{m,q}$, where $m = 1 \cdots, M-1$ and $q = \lambda_m^{(b)}$. For the case of $m = M$, the probability that s_M being transmitted is the sum of the stationary probabilities $s_{M,q}$, where $q = N, N+1, \cdots, K-1$. Finally, the resultant probability τ_m of the symbol s_m being transmitted is then expressed as

$$\tau_m = \begin{cases} \pi_{s,m}, & m < M, \\ \sum_{i=M}^{M+K-N-1} \pi_{s,i}, & m = M. \end{cases} \tag{2.82}$$

2.3.2.3 SWIPT Performance Analysis

When M-QAM modulator is conceived, the minimum Euclidean distance in its constellation is expressed as $\xi = \sqrt{\frac{4P_{\max}}{(\sqrt{M}-1)^2}}$, where $P_{\max}$ is the maximum transmit power of the modulated signal. For a complex modulated symbol $s_m = A_{m,I} + jA_{m,Q}$, we have $A_{m,I}, A_{m,Q} \in \{\frac{1-\sqrt{M}}{2}\xi, \frac{3-\sqrt{M}}{2}\xi, \cdots, \frac{\sqrt{M}-3}{2}\xi, \frac{\sqrt{M}-1}{2}\xi\}$. After obtaining the stationary transmission probabilities $\{\tau_m | m = 1, \cdots, M\}$ of all possible symbols, the average transmit power P_t of the SWIPT transmitter is then formulated as:

$$P_t = \sum_{m=1}^{M} \tau_m \kappa_{t,m}, \tag{2.83}$$

where $\kappa_{t,m} = ((A_{m,I})^2 + (A_{m,Q})^2)/2$ represents the transmit power of the symbol s_m. When s_m is transmitted, the received power for WPT is obtained as:

$$\kappa_{r,m} = \rho\left((\Omega||\mathbf{g}||A_{m,I} + z_{a,I})^2 + (\Omega||\mathbf{g}||A_{m,Q} + z_{a,Q})^2\right)/2, \tag{2.84}$$

by exploiting Eq. (2.74) and by reformulating the complex antenna noise as $z_a = z_{a,I} + jz_{a,Q}$.

The envelope $\{|g_i||i = 1, \cdots, N_t\}$ of every element in the multipath fading vector $\mathbf{g}$ follows a Rician distribution, where the probability density function (PDF) for a instantaneous value of $|g_i| = \gamma$ is expressed as:

$$f_{|g_i|}(\gamma) = \frac{\gamma}{b_0} e^{-\frac{\gamma^2+b_i^2}{2b_0}} I_0\left(\frac{\gamma b_i}{b_0}\right), \forall i = 1, \cdots, N_t, \tag{2.85}$$

where $2b_0$ is the average power of the non-line-of-sight (NLOS) component, b_i^2 is the average power of the line-of-sight (LOS) component of $|g_i|$, and $I_n(\cdot)$ represents the n-th order modified Bessel function of the first kind. According to [45], the PDF of $||\mathbf{g}|| = \gamma$ is then formulated as:

$$f_{||\mathbf{g}||}(\gamma) = \frac{\gamma^{N_t}}{b_0 b^{N_t-1}} e^{-\frac{\gamma^2+b^2}{2b_0}} I_{N_t-1}\left(\frac{\gamma b}{b_0}\right), \tag{2.86}$$

where $b^2 = \sum_{i=1}^{N_t} b_i^2$.

By conceiving the nonlinear energy harvesting model [46] in our system, the average energy harvesting amount when the symbol s_m is transmitted is derived as:

$$P_{r,m} = \frac{\Psi - P_{\text{sat}}\Xi}{1-\Xi}, \tag{2.87}$$

where $\Xi = \frac{1}{1+exp(\alpha_1\alpha_2)}$ and $\Psi = \frac{P_{\text{sat}}}{1+exp(-\alpha_1(\kappa_{r,m}-\alpha_2))}$. In Eq. (2.87), P_{sat} is a constant representing the maximum harvested power at the receiver, while α_1 and α_2 are constants related to the circuit specifications such as the resistance and capacitance.

Corollary 2.1 *According to the symbol transmission probabilities τ_m obtained in Sect. III, the average energy harvesting performance is formulated as:*

$$P_r = \sum_{m=1}^{M} \tau_m \int_0^{+\infty} f_{||\mathbf{g}||}(\gamma) \int_{-\infty}^{+\infty}\int_{-\infty}^{+\infty} \frac{e^{-\frac{z_{a,I}^2+z_{a,Q}^2}{\sigma_a^2}}}{\pi\sigma_a^2} P_{r,m}\, dz_{a,I}\, dz_{a,Q}\, d\gamma. \tag{2.88}$$

Since we cannot obtain the exact expressions for τ_m, P_r of (2.88) is in a semi-closed form.

Discussion Based on the transmit pattern of the modulated symbols, the average energy harvesting performance is analyzed by further considering a practical SWIPT transceiver with a M-QAM scheme, a MISO based Rician fading channel and a power splitter. By substituting a specific modulated symbol s_m into Eq. (2.74), when the symbol s_m is transmitted, the received symbol $\hat{s}_m$ for WIT is formulated as:

$$\begin{aligned}\hat{s}_m &= \sqrt{(1-\rho)}(\Omega||\mathbf{g}||s_m + z_a) + z_{\text{cov}} \\ &= \sqrt{(1-\rho)}\Omega||\mathbf{g}||(A_{m,I} + jA_{m,Q}) + (z_{n,I} + jz_{n,Q}),\end{aligned} \tag{2.89}$$

where $z_n = \sqrt{(1-\rho)}z_a + z_{\text{cov}} \sim \mathbb{CN}(0, (1-\rho)\sigma_a^2 + \sigma_{\text{cov}}^2)$ is total AWGN having two orthogonal dimensions $z_{n,I}$ and $z_{n,Q}$. Since the receiver has no knowledge of the input distribution, maximum-likelihood (ML) detector is conceived for demodulating the received signals. Each modulated symbol has a successful demodulation region at the receiver, which is expressed as a square having the length of $\sqrt{(1-\rho)}\Omega||\mathbf{g}||\xi$. Therefore, in order to successfully demodulate the original symbol s_m, the offset $z_{n,I} + jz_{n,Q}$ should be constrained within the range of $[-\sqrt{(1-\rho)}\Omega||\mathbf{g}||\xi/2, \sqrt{(1-\rho)}\Omega||\mathbf{g}||\xi/2]$ for both the in-phase and the quadrature dimension.

Generally, the symbol demodulation error always occurs between two adjacent constellation points, since the probability that one symbol is mis-demodulated as any other non-adjacent symbols can be neglected. According to the Gray mapping rule, there is only one binary difference between the sequences of two adjacent constellation points. Therefore, the BER at the SWIPT receiver could be approximated as $\epsilon_b \approx \frac{\epsilon_s}{N}$, where the SER ϵ_s is calculated as:

$$\epsilon_s = 1 - \int_0^{+\infty} f_{||\mathbf{g}||}(\gamma)\left[1 - 2Q\left(\frac{\sqrt{(1-\rho)}\Omega\gamma\xi/2}{\sqrt{(1-\rho)\sigma_a^2/2 + \sigma_{\text{cov}}^2/2}}\right)\right]^2 d\gamma. \tag{2.90}$$

Since the successful demodulation regions of all the symbols have the same size, the probabilities that any received symbols $\hat{s}_m$ and $\hat{s}_{m'}$ are mis-demodulated to each other are also the same. The BER from "0" to "1" is the same as that from "1" to "0".

Similar to [8] and [42], the mutual information between the encoded information source $\overline{\mathbb{X}}$ and destination $\overline{\mathbb{Y}}$ is exploited for evaluating the WIT performance. In the evaluation of the mutual information, the BER of the modulation schemes in the MISO Rician channel are relied upon for modeling the crossover probability of the resultant binary symmetric channel. The coded source $\overline{\mathbb{X}}$ outputs a binary sequence $\mathbf{X}_L = \{X_1, \cdots, X_L\}$ having a length of L. Then, the encoded destination $\overline{\mathbb{Y}}$ receive a binary sequence $\mathbf{Y}_L = \{Y_1, \cdots, Y_L\}$. Information theoretical channel between the encoded information source and destination can be modeled as a binary symmetric channel having the average BER ϵ_b as the crossover probabilities. The mutual information between $\overline{\mathbb{X}}$ and $\overline{\mathbb{Y}}$ can be formulated as:

$$I(\overline{\mathbb{X}}; \overline{\mathbb{Y}}) = H(\overline{\mathbb{X}}) - H(\overline{\mathbb{X}}|\overline{\mathbb{Y}}). \tag{2.91}$$

Corollary 2.2 *According to the similar analysis in [42], the lower bound of mutual information $I(\overline{\mathbb{X}}; \overline{\mathbb{Y}})$ is finally derived as:*

$$\begin{aligned}I(\overline{\mathbb{X}}; \overline{\mathbb{Y}}) &\geq \sum_{k=0}^{K-1} \pi_{c,k}\left[H(p_k\epsilon_b + (1-p_k)(1-\epsilon_b)) - H(\epsilon_b)\right] \\ &\triangleq I_{\text{inf}}(\overline{\mathbb{X}}; \overline{\mathbb{Y}}),\end{aligned} \tag{2.92}$$

where $H(x) = -x\log_2(x) - (1-x)\log_2(1-x)$, p_k and $\pi_{c,k}$ are the state transition probability and the state stationary probability obtained in Sect. 2.3.2.3, respectively. Since we cannot obtain the exact expression for $\boldsymbol{\pi}_c$, $I_{\text{inf}}(\overline{\mathbb{X}}; \overline{\mathbb{Y}})$ of Eq. (2.92) is in a semi-closed form.

Discussion Based on the Markov information source, the mutual information between the transmitter and the receiver is analyzed by modeling a practical transceiving process as a binary symmetric information channel.

Algorithm 2.4 Bisection method based algorithm for obtaining ρ^*

Require: Codeword distribution $\mathbf{Pr\_x}$; Mutual information threshold I_{th}; Mutual information threshold I_{th}; Convergence threshold ϕ and all the other physical layer parameters.
Ensure: Optimal power splitting ratio ρ^*;
Initialise $\rho_{\text{upper}} \leftarrow 1$, $\rho_{\text{lower}} \leftarrow 0$, $I \leftarrow 0$;
while $|\rho_{\text{upper}} - \rho_{\text{lower}}| > \phi$ **do**
  Update $\rho = (\rho_{\text{upper}} + \rho_{\text{lower}})/2$;
  Obtain lower-bound mutual information $I = I_{\text{inf}}(\overline{\mathbb{X}}; \overline{\mathbb{Y}})$ according to Eq. (2.92);
  if $I > I_{\text{th}}$ **then**
    $\rho_{\text{lower}} = \rho$;
  else
    $\rho_{\text{upper}} = \rho$;
  end if
end while
Return $\rho^* = (\rho_{\text{upper}} + \rho_{\text{lower}})/2$.

2.3.2.4 Unary Coding Design

The unary codeword distribution $\mathbf{Pr\_x}$ as well as the power splitting ratio ρ are jointly optimized, in order to maximize the average power P_r for energy harvesting of the receiver, while guaranteeing the lower bound of mutual information $I_{\text{inf}}(\overline{\mathbb{X}}; \overline{\mathbb{Y}})$ is higher than a threshold. This optimization problem is formulated as:

$$\text{(P2.3.3)} \quad \max_{\mathbf{Pr\_x}, \rho} \ P_r$$
$$\text{s.t.} \quad I_{\text{inf}}(\overline{\mathbb{X}}; \overline{\mathbb{Y}}) \geq I_{\text{th}}. \tag{2.93}$$

Given a threshold I_{th}, it is obvious that the actual mutual information $I(\overline{\mathbb{X}}; \overline{\mathbb{Y}})$ also satisfies the constraint, since we have $I(\overline{\mathbb{X}}; \overline{\mathbb{Y}}) \geq I_{\text{inf}}(\overline{\mathbb{X}}; \overline{\mathbb{Y}}) \geq I_{\text{th}}$. Since both P_r and $I_{\text{inf}}(\overline{\mathbb{X}}; \overline{\mathbb{Y}})$ are expressed in semi-closed form, (P2.3.3) can only be solved by an efficient search based algorithm. Since the genetic algorithm is capable of an avoiding local optimization to a large extent, it is then exploited for solving (P2.3.3).

Theorem 2.7 *Given a specific codeword distribution* $\mathbf{Pr\_x}$*, the optimal power splitting ratio* ρ^* *can be obtained by solving* $I_{\text{inf}}(\overline{\mathbb{X}}; \overline{\mathbb{Y}}) = I_{\text{th}}$ *with the aid of a bisection method, which is summarized in Algorithm 2.4. The complexity of the bisection method is* $\mathcal{O}(\log_2(1/\phi))$*, where* ϕ *is the convergence threshold.*

Proof Please refer to Sect. 2.3.2.9 for detailed proof.

Then, a genetic algorithm is proposed for optimizing the codeword distribution $\mathbf{Pr\_x}$, which is detailed in the pseudocode of Algorithm 2.5. The main steps are summarized as follows:

- *Step 1: Initialization.* Initialize a generation $\mathbb{G}$ having ind_num individuals. Each individual is a legal codeword distribution $\mathbf{Pr\_x}_j = \{p_j(\overline{\chi}_1), \cdots, p_j(\overline{\chi}_K)\}$ satisfying $\sum_{k=1}^{K} p_j(\overline{\chi}_k) = 1$ for $\forall j = 1, \cdots, ind\_num$. This step corresponds to line 1 in Algorithm 2.5.
- *Step 2: Natural selection.* Obtain the fitness of each individual according to the mutual information constraint of (P2.3.3). The fitness of the j-th individual is expressed as:

$$fitness_j = \begin{cases} P_r, & I_{\text{inf}}(\overline{\mathbb{X}}; \overline{\mathbb{Y}}) \geq I_{\text{th}}, \\ 0, & \text{otherwise}. \end{cases} \tag{2.94}$$

The selection probability of each individual is calculated as:

$$prob_j = \frac{fitness_j}{\sum_{j'=1}^{ind\_num} fitness_{j'}}. \tag{2.95}$$

Then, we update the generation $\mathbb{G}$ by selecting new individuals according to the probabilities $\{prob_j | \forall j\}$. This step corresponds to line 6 and line 10 in Algorithm 2.5.

Algorithm 2.5 Genetic algorithm for obtaining ρ^* and $\mathbf{Pr\_x}^*$

Require: Mutual information threshold I_{th}; Mutation probability $mutation\_prob$; Maximum generation number G and all the other physical layer parameters.
Ensure: Optimal power splitting ratio ρ^*; Optimal input distribution $\mathbf{Pr\_x}^*$;
1: Initialise a generation $\mathbb{G}$ according to Step 1; Generation label $i = 1$; Optimal fitness $fitness^*$;
2: **while** $i < G$ **do**
3: $i \leftarrow i + 1$;
4: Obtain the optimal power splitting ratio ρ_j for each individual $\mathbf{Pr\_x}_j$ in $\mathbb{G}$ according to Algorithm 2.4;
5: Obtain the average energy harvesting power P_r and lower-bound mutual information $I_{\text{inf}}(\overline{\mathbb{X}}; \overline{\mathbb{Y}})$ for each individual;
6: Obtain the fitness of $\mathbb{G}$ according to (2.94);
7: **if** $\max\{fitness_j\} > fitness^*$ **then**
8: Update $fitness^* \leftarrow \max\{fitness_j\}$. Update ρ^* and $\mathbf{Pr\_x}^*$ according to the individual having the maximum fitness;
9: **end if**
10: Update the generation $\mathbb{G}$ by selecting individuals according to the probabilities (2.95);
11: Obtain offsprings after crossover according to Step 3 and update the generation $\mathbb{G}$;
12: As for each individual, generate a value φ following uniformly distribution $\mathcal{U}(0, 1)$. If $\varphi > mutation\_prob$, operate mutation according to Step 4. Then, update the generation $\mathbb{G}$ after traversing all the individuals;
13: **end while**

- *Step 3: Crossover.* Crossover is operated after a natural selection. The generation $\mathbb{G}$ is updated once more by generating offsprings of two randomly selected parental individuals one after another. Given a specific pair of parental individuals $\mathbf{Pr\_x}_{(father)}$ and $\mathbf{Pr\_x}_{(mother)}$, their offsprings $\mathbf{Pr\_x}_{(offspring)}$ is expressed as:

$$\begin{aligned} p_{(offspring)}(\mathcal{X}_k) =& a \cdot p_{(father)}(\mathcal{X}_k) + (1-a) \cdot p_{(mother)}(\mathcal{X}_k), \\ & \forall j = 1, \cdots, ind\_num, k = 1, \cdots, K. \end{aligned} \tag{2.96}$$

Note that the offspring satisfies $\sum_{k=1}^{K} p_{(offspring)}(\mathcal{X}_k) = 1$, since we have $\sum_{k=1}^{K} p_{(father)}(\mathcal{X}_k) = 1$ and $\sum_{k=1}^{K} p_{(mother)}(\mathcal{X}_k) = 1$. The crossover factor a determines the fraction of characteristics that the offspring inherits from the parental individuals. In our algorithm, we choose $a = 0.5$. This step corresponds to line 11 in Algorithm 2.5.
- *Step 4: Mutation.* The last step of genetic algorithm is mutation, in order to prevent the solution from local optimum. For each individual $\mathbf{Pr\_x}_j$, the mutation is operated by adding a random Δp following the normal distribution $\mathcal{N}(0, \nu)$, which is expressed as:

$$p_j(\mathcal{X}_k) = p_j(\mathcal{X}_k) + \Delta p, k = 1, \cdots, K. \tag{2.97}$$

Then, the individual is updated as:

$$p_j(\mathcal{X}_k) = \frac{p_j(\mathcal{X}_k) - \min(0, \min(\mathbf{Pr\_x}_j))}{\sum \mathbf{Pr\_x}_j - K \min(0, \min(\mathbf{Pr\_x}_j))}, \tag{2.98}$$

for $\forall k = 1, \ldots, K$. A larger variance ν results in a faster convergence of the genetic algorithm, but a lower optimality. This step corresponds to line 12 in Algorithm 2.5.

The complexity of Algorithm 2.5 is $\mathcal{O}(G^2)$ [42], where G is the maximum number of generations.

2.3.2.5 Simulation Results

Our optimal coding design and impact of M-QAM on SWIPT performance are validated by both Monte-Carlo simulation and theoretical analysis. We consider $N_t = 4$ antennas at the transmitter and one antenna at the receiver. The square root of the path-loss[6] is set to $\Omega = 10^{-1.5}$, while the noise power is $\sigma_a^2 = \sigma_{\text{cov}}^2 = -50$ dBm. The transmitter has a maximum transmit power constraint of $P_{\max} = 23$ dBm [47]. The Gray mapping from binary sequence to modulated symbols is

[6] According to the path-loss equation $G_t G_r (\frac{c}{4\pi d f_c})^2$ in [42], if we set the antennas gain as $G_t = G_r = 10$ dBi and the carrier frequency as $f_c = 800$ MHz, the transmission distance is about $d = 9.43$ m, where c is the speed of the light.

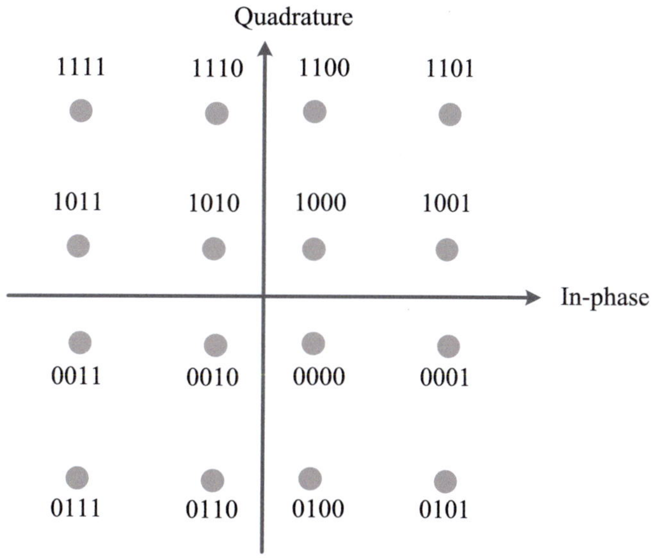

Fig. 2.18 Gray mapping for 16-QAM

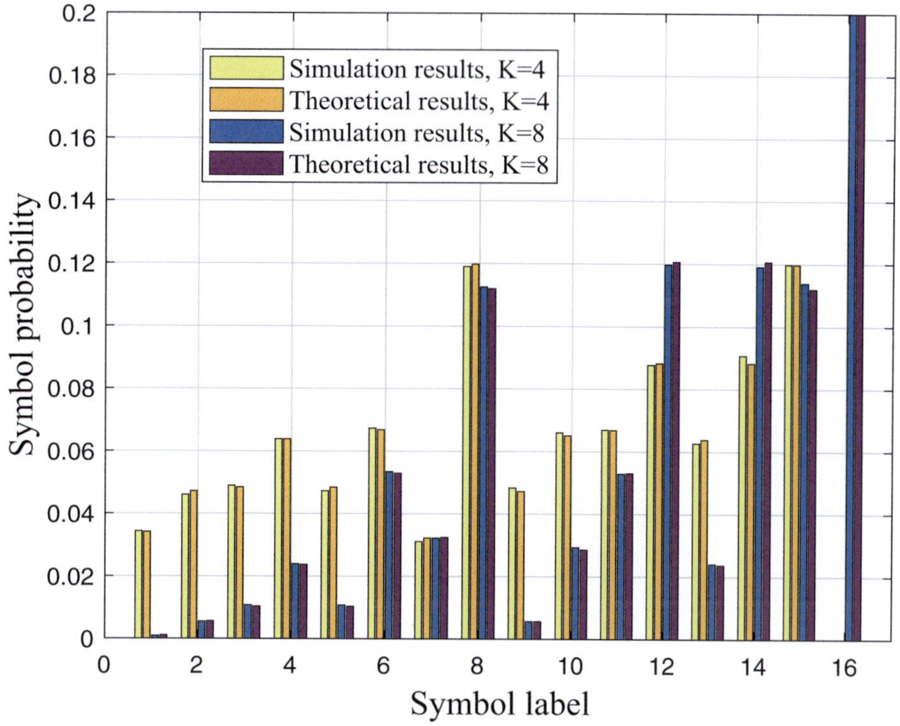

Fig. 2.19 Validation of symbol probability Markov analysis

conceived, where the constellation of 16-QAM is exemplified in Fig. 2.18. For the nonlinear energy harvesting circuits, we set $P_{\text{sat}} = 10.71mW$, $\alpha_1 = 0.8963$ and $\alpha_2 = 0.6614$ [46]. For the Rician fading channel in Eq. (2.85), we set $b_1^2 = \cdots = b_{N_t}^2 = 2b_0 = 0.5$ [45].

Firstly, we validate our theoretical analysis, when 16-QAM is adopted. In Fig. 2.19, we compare the theoretical transmit probabilities of all the modulated symbols to the simulation counterparts, given an arbitrary codeword distribution $\mathbf{Pr\_x} = \{0.42, 0.25, 0.07, 0.26\}$ for 4-level unary code and 8-level unary code with $\mathbf{Pr\_x} = \{0.16, 0.25, 0.13, 0.04, 0.05, 0.05, 0.23, 0.09\}$, respectively. The Monte-Carlo simulation is adopted by generating 10^6 codewords for obtaining those probabilities. Observe from Fig. 2.19 that the simulation and theoretical results perfectly match with each other, which confirms the accuracy of our Markov modeling. Note that when 4-level unary code is

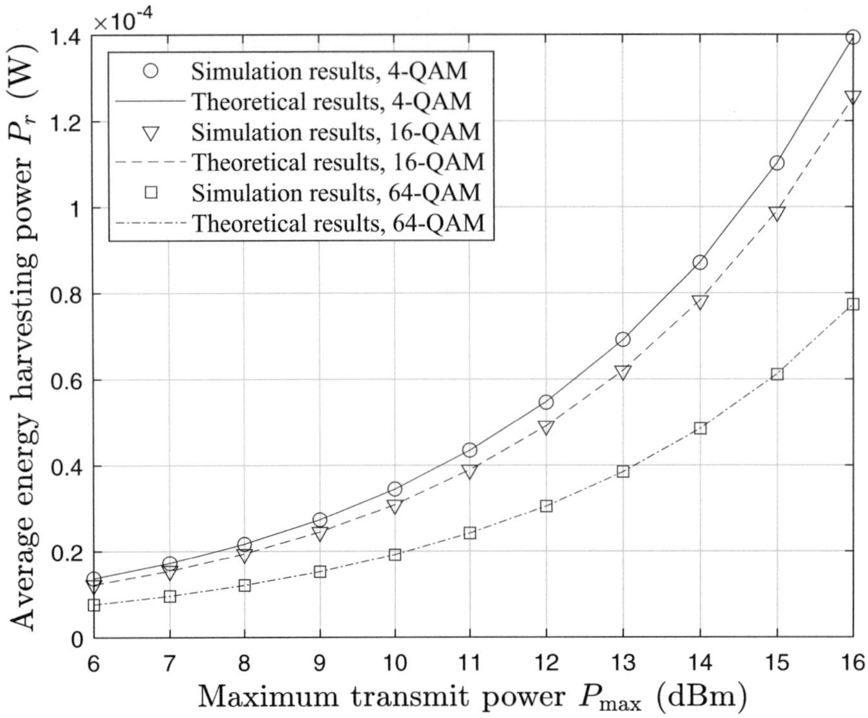

Fig. 2.20 Validation of WPT analysis

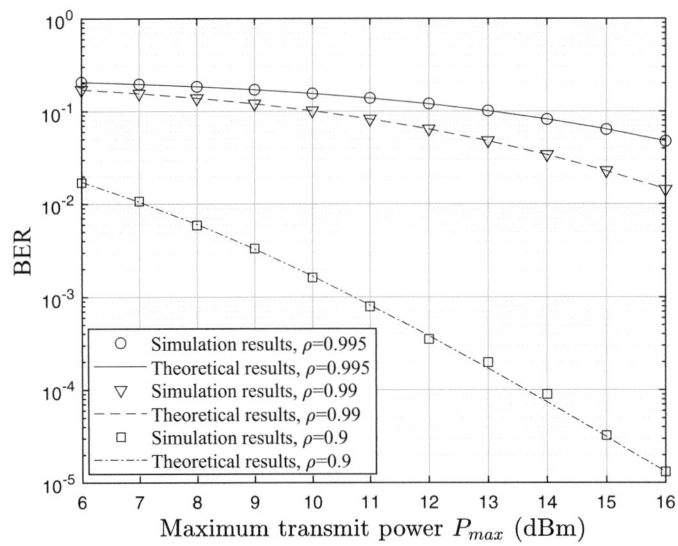

Fig. 2.21 Validation of WIT analysis

conceived, the probability of the 16-th symbol being sent is zero, since the maximum run-length of binary sign "1" is $K - 1 = 3$. In Figs. 2.20 and 2.21, we depict the average energy harvesting power as well as the receive BER versus maximum transmit power $P_{\max}$. Observe from Fig. 2.20 that a lower order M results in a higher energy harvesting power, since it has a higher average power. Moreover, observe from Fig. 2.21 that a higher power splitting ratio ρ results in a higher BER, since the signal strength for the demodulation is weakened.

We then evaluate the SWIPT performance of our genetic algorithm aided coding design in Figs. 2.22, 2.23, and 2.24. Observe from Fig. 2.22 that when we increase the minimum mutual information constraint I_{th}, the average energy harvesting power P_r is reduced. Given a K-level unary code and a fixed modulator, any coding design cannot achieve a mutual information higher than the corresponding channel capacity. Therefore, observe from Fig. 2.22 that, every coding and modulation scheme can only satisfy the mutual information constraint lower than a certain value. For example, when 16-

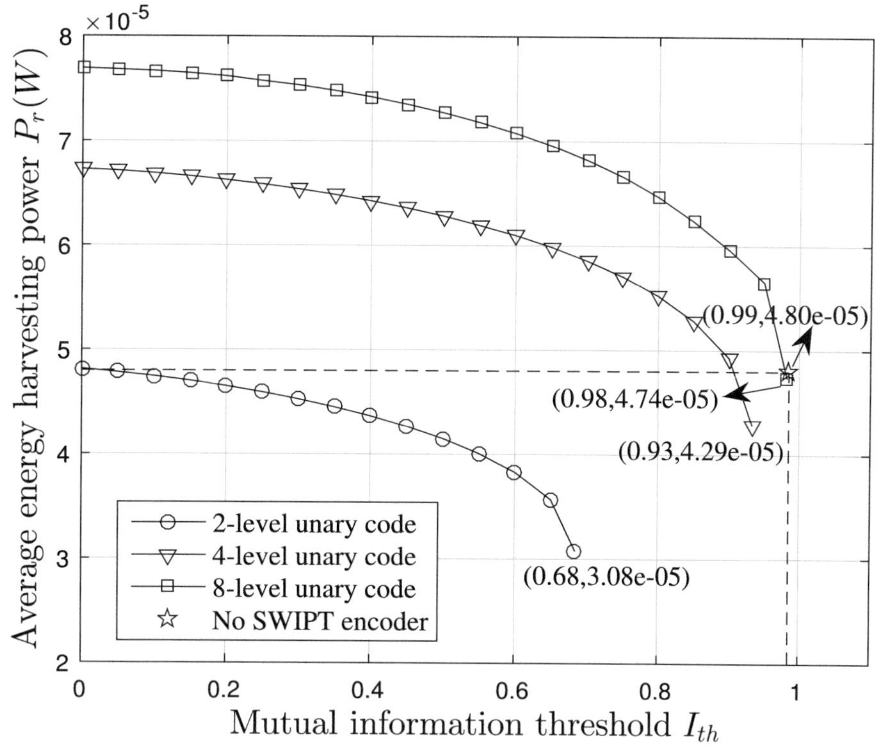

Fig. 2.22 Average energy harvesting power versus mutual information threshold, when 16-QAM modulator is conceived

QAM modulator is adopted, the upper bound mutual information of 2-level unary code is 0.68, while that of 8-level unary code is 0.98. Observe from Fig. 2.22 that 8-level unary code has the best WPT performance. Moreover, we may observe from Fig. 2.23 that when 4-QAM modulator is adopted, the average energy harvesting power P_r does not change as we increase I_{th}. This is because that all the symbols in 4-QAM's constellation have the same transmit power. Therefore, the codeword distribution does not have any impact on the WPT performance. Moreover, a higher order modulator results in a lower WPT performance. This is because when the maximum transmit power $P_{\max}$ is fixed, a higher order modulator achieves a lower average transmit power. Furthermore, a higher order modulator also results in a higher BER, which indicates that the WPT performance has to be sacrificed in order to compensate for the degradation of the WIT performance. We also compare the WPT and WIT performance of the system having unary coding aided SWIPT encoder and that without SWIPT encoder, where the source and channel encoder generates equally distributed binary signs "1" and "0".[7] Observe from Figs. 2.22 and 2.23 that the system without SWIPT encoder has a fixed mutual information and a fixed average energy harvesting power. Its attainable mutual information is always higher than that of the system having SWIPT encoder. This is because, according to the classic information theory, additional coding process (SWIPT encoder in our system) may always reduces the mutual information between the source and the destination. However, we may arbitrarily adjust the WIT and WPT performance by the coding design of the SWIPT encoder. For instance, when 16-QAM is conceived, the WPT performance attained by the SWIPT encoder is 40% higher than that without a SWIPT encoder. Moreover, the performance improvement of the SWIPT encoder is as high as 70%, when 256-QAM is adopted.

Observe from Fig. 2.24 that the average energy harvesting power P_r increases as we increase the level of the unary code. This is because when K becomes higher, the SWIPT encoder becomes more flexible to control the attainable SWIPT performance. Furthermore, a higher order modulator results in a more rapidly increasing P_r. This is because a modulated symbol may represent more binary signs, which requires a higher K to achieve the optimal WPT performance.

We also compare the energy harvesting performance of the unary code to that of the Huffman code in Fig. 2.25. We assume the same number of codewords for the Huffman code. Its optimal codeword distribution is also obtained by solving (P2.3.3). Then, the average energy harvesting power of the Huffman code is obtained by the Monte-Carlo simulation. Observe from Fig. 2.25 that our unary code outperforms the Huffman counterpart in terms of the WPT performance. Moreover, a larger number K of the codewords results in a higher WPT performance, since it brings in a higher flexibility.

[7] A typical channel encoder, such as LDPC and polar code, always output equally distributed binary signs.

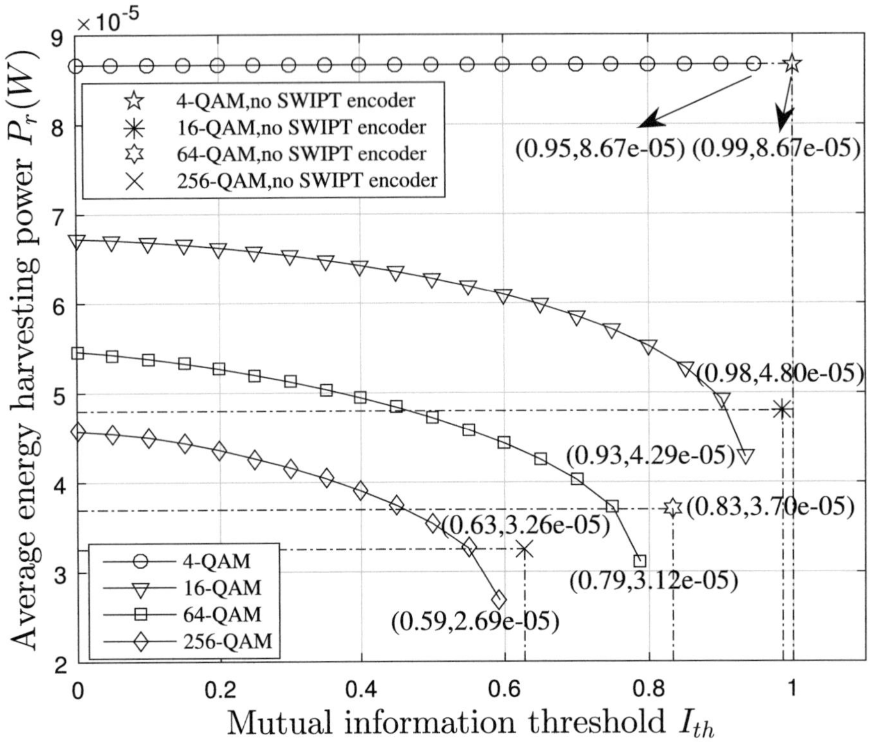

Fig. 2.23 Average energy harvesting power versus modulation order, when 4-level unary code is conceived

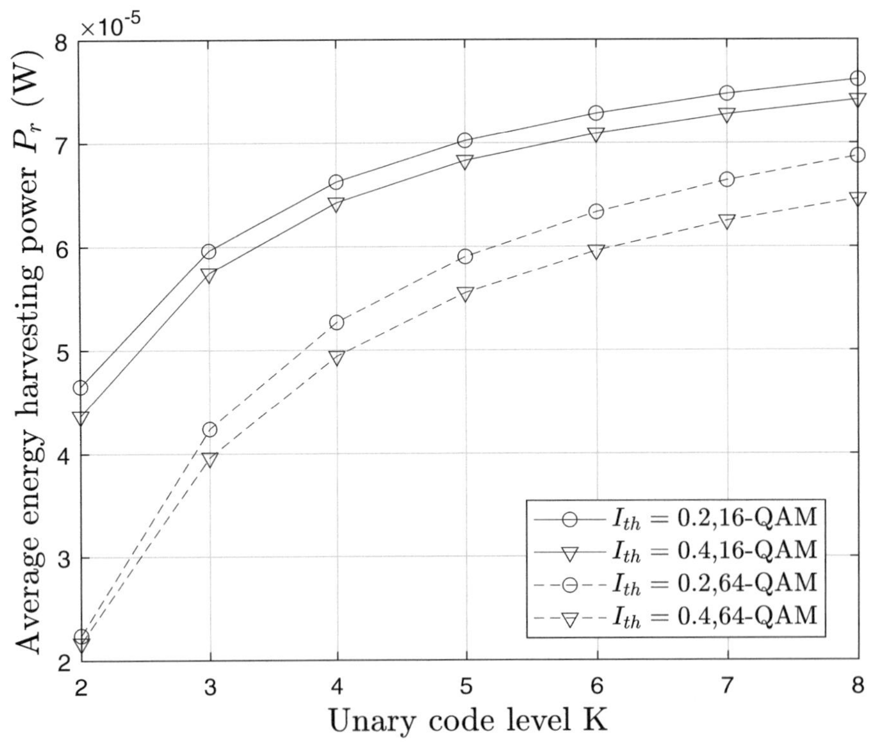

Fig. 2.24 Average energy harvesting power versus unary code level *K*

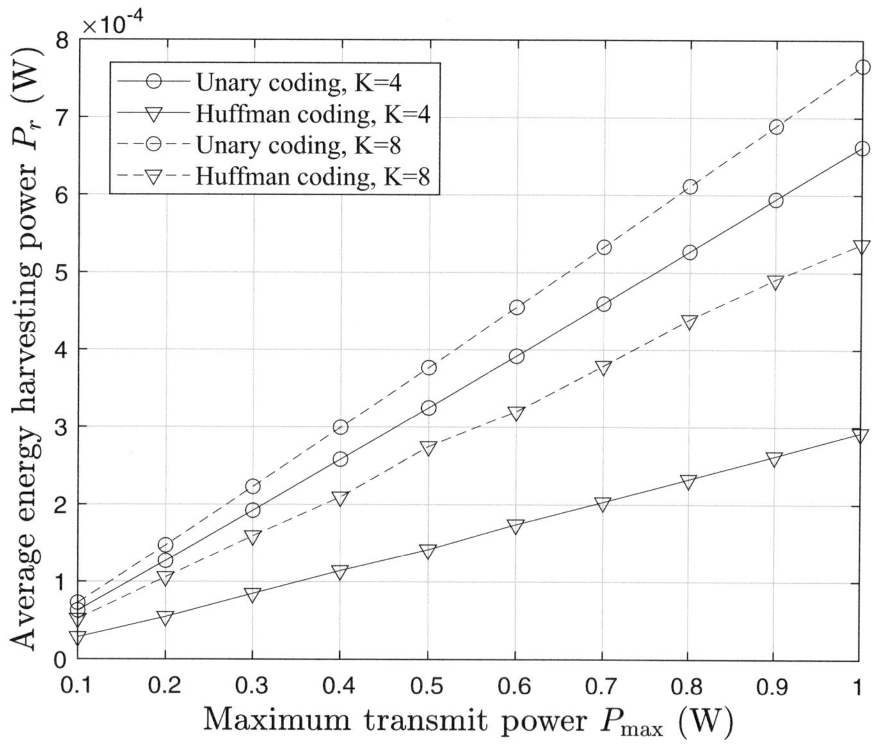

Fig. 2.25 Average energy harvesting power versus codeword number K, where 16-QAM modulator is conceived

2.3.2.6 Proof of Theorem 2.4

We define $p_k = \Pr\{C_{l+1} = k+1|C_l = k\}$ $(k = 0, \cdots, K-1)$ as the transition probability from the state $C_l = k$ to the state $C_{l+1} = k+1$. Then, p_k is formulated as Eq. (2.76). If the current state is $C_l = k$ $(k = 0, \cdots, K-2)$, the coded information source $\overline{\mathbb{X}}$ outputs a binary sign "1" with a probability of p_k, and it then transits to a new state of $C_{l+1} = k+1$. $\overline{\mathbb{X}}$ outputs a binary sign "0" with a probability of $1 - p_k$, and it then transits to the state $C_{l+1} = 0$. Specifically, when the current state is $C_l = K-1$, $\overline{\mathbb{X}}$ must output a binary sign "0", since the run-length of "1" reaches its maximum. Therefore, the state transition probabilities are resultant obtained as Eq. (2.75).

2.3.2.7 Proof of Theorem 2.5

In a K-level unary coding system, the maximum run-length of "1" is $K-1$. Therefore, if the sum of $\lambda_{m'}^{(b)}$ and $\lambda_m^{(f)}$ is larger than $K-1$, it is impossible to transit from the state $s_{m'}$ to the state s_m. When we have $\lambda_{m'}^{(b)} + \lambda_m^{(f)} \leq K-1$, the transition probability $\Pr\_s(s_m|s_{m'})$ is easily obtained according to the state transition probabilities of coded information source.

2.3.2.8 Proof of Theorem 2.6

When we have $m < M$, since the backward run-length of sign "1" is determined for every state, the state transition probability $\Pr\_s(s_{m,q}|s_{m',q'})$ can be obtained similar to the case of $K-1 < N$, which is formulated as

$$\Pr\_s(s_{m,q}|s_{m',q'}) = \begin{cases} 0, & q' + \lambda_m^{(f)} > K-1, \\ \Pr\_c(\delta_{m,1}|q') \prod_{n=1}^{N-1} \Pr\_c(\delta_{m,n+1}|\delta_{m,n}), & \\ & q' + \lambda_m^{(f)} \leq K-1. \end{cases} \tag{2.99}$$

When we have $m = M$, since there is no sign "0" in the binary sequence ζ_M, the backward run-length $q = \lambda_M^{(b)}$ depends on the symbol $s_{m'}$ transmitted in the previous time-slot, which is expressed as $q = \lambda_M^{(b)} = \lambda_{m'}^{(b)} + N = q' + N$. Therefore, the probability of the state $s_{m',q'}$ transiting to $s_{M,q}$ is expressed as

$$\text{Pr\_s}(s_{M,q}|s_{m',q'}) = \begin{cases} 0, & q \neq q' + N, \\ \prod_{k=q'}^{q} \text{Pr\_c}(k+1|k), & \text{otherwise}. \end{cases} \tag{2.100}$$

Finally, the transition probabilities for the case of $K - 1 \geq N$ is summarized in Eq. (2.81).

2.3.2.9 Proof of Theorem 2.7

We firstly discuss the monotonicity of P_r w.r.t. ρ. If we increase ρ, $\kappa_{r,m}$ also increases. According to Eq. (2.88), P_r monotonously increases w.r.t. ρ.

Since function $Q\left(\frac{\sqrt{(1-\rho)}\Omega||\mathbf{g}||\xi/2}{\sqrt{(1-\rho)\sigma_a^2/2+\sigma_{\text{cov}}^2/2}}\right)$ monotonically increase w.r.t. ρ, observe from Eq. (2.90) that ϵ_s also increases when we increase ρ. So does ϵ_b. We then discuss the monotonicity of $I_{\text{inf}}(\overline{\mathbb{X}}; \overline{\mathbb{Y}})$ w.r.t. ϵ_b. The lower bound mutual information of Eq. (2.92) can be reformulated as:

$$I_{\text{inf}}(\overline{\mathbb{X}}; \overline{\mathbb{Y}}) = \sum_{k=0}^{K-1} \pi_{c,k} \Delta_k, \tag{2.101}$$

where $\Delta_k = H(p_k\epsilon_b + (1 - p_k)(1 - \epsilon_b)) - H(\epsilon_b)$. The first-order derivative of Δ_k w.r.t. ϵ_b is derived as

$$\frac{d\Delta_k}{d\epsilon_b} = (1 - 2p_k)\log_2 \frac{(2p_k - 1)\epsilon_b + 1 - p_k}{p_k - (2p_k - 1)\epsilon_b} + \log_2 \frac{\epsilon_b}{1 - \epsilon_b}. \tag{2.102}$$

We have $\frac{d\Delta_k}{d\epsilon_b} = 0$ when $\epsilon_b = 0.5$. The second-order derivative of Δ_k w.r.t. ϵ_b is expressed as

$$\frac{d^2\Delta_k}{d\epsilon_b^2} = \frac{1}{\ln 2} \frac{p_k(1 - p_k)}{\epsilon_b(1 - \epsilon_b)\left((2p_k - 1)\epsilon_b + 1 - p_k\right)\left(p_k - (2p_k - 1)\epsilon_b\right)}. \tag{2.103}$$

When the BER ϵ_b satisfies $0 < \epsilon_b < 0.5$, we have $\min(0.5, 1 - p_k) < (2p_k - 1)\epsilon_b + 1 - p_k < \max(0.5, 1 - p_k)$ and $\min(0.5, 1 - p_k) < p_k - (2p_k - 1)\epsilon_b < \max(0.5, 1 - p_k)$. Therefore, the second-order derivative satisfies $\frac{d^2\Delta_k}{d\epsilon_b^2} > 0$ subject to constraint of $0 < \epsilon_b < 0.5$. As a result, $\frac{d\Delta_k}{d\epsilon_b}$ monotonically increase w.r.t. ϵ_b. Since we have $\frac{d\Delta_k}{d\epsilon_b}|_{\epsilon_b=0.5} = 0$, it is easy to obtain $\frac{d\Delta_k}{d\epsilon_b}|_{\epsilon_b<0.5} < 0$, which indicates that Δ_k monotonically decreases w.r.t. ϵ_b. Furthermore, since ϵ_b is a monotonously increasing function of ρ, the lower bound mutual information $I_{\text{inf}}(\overline{\mathbb{X}}; \overline{\mathbb{Y}})$ monotonically decreases w.r.t. ρ, when we have $\epsilon_b < 0.5$.

Since the BER ϵ_b is normally lower than 0.5, the optimal power splitting ratio ρ^* could be obtained by solving $I_{\text{inf}}(\overline{\mathbb{X}}; \overline{\mathbb{Y}}) = I_{\text{th}}$.

2.4 Exercises

1. Assuming a fully shuffled deck of playing cards (containing 52 cards), question
   (a) What is the amount of information given by any particular arrangement?
   (b) What is the amount of information that can be obtained if 13 cards are taken from the deck, all of which are given different numbers of points?
2. Prove that $H(X_1X_2\ldots H_n) < H(X_1) + H(X_2) + \ldots + H(X_N)$.

3. There is a binary smooth martensitic chain with known $p(0/0) = 0.8$, $p(1/1) = 0.7$, find its symbolic entropy. Compose a binary Huffman code by combining three symbols into one and find the average code word length and coding efficiency of the new symbol.
4. Each frame of a TV image can be considered to be composed of 3×10^5 pixels, all of which change independently, and each pixel takes 128 different luminance levels and sets the probability of occurrence of luminance levels, how much information does each frame contain? If a broadcaster chooses 1000 Chinese characters out of about 10,000 to describe a television image, how many Chinese characters does the broadcaster need to describe the image appropriately?
5. Let the binary symmetric channel be a memoryless channel with channel matrix $\begin{bmatrix} \bar{p} & p \\ p & \bar{p} \end{bmatrix}$ where $p > 0$, $\bar{p} < 1$, $\bar{p} + p = 1$, $\bar{p} >> p$ Try to write the channel matrix P for $N = 3$ times extended memoryless channel.
6. Assuming source $\begin{pmatrix} X \\ P(X) \end{pmatrix} = \begin{Bmatrix} a_1 & a_2 \\ p & 1-p \end{Bmatrix}$ $(p < 0.5)$, its distortion is Hamming distortion. When the allowable average distortion $D = 0.5p$, how many binary data symbols does each source symbol need to be represented on average?

References

1. Cano, C., Pittolo, A., Malone, D., Lampe, L., Tonello, A. M., & Dabak, A. G. (2016). State of the art in power line communications: From the applications to the medium. *IEEE Journal on Selected Areas in Communications. 34*(7), 1935–1952. ISSN: 0733-8716.
2. Wu, J., Wu, H., Li, C., Li, W., He, X., & Xia, C. (2012). Advanced four-pair architecture with input current balance function for power over ethernet (PoE) system. *IEEE Transactions on Power Electronics, 28*(5), 2343–2355. ISSN: 0885-8993.
3. Gastpar, M. (2007). On capacity under receive and spatial spectrum-sharing constraints. *IEEE Transactions on Information Theory, 53*(2), 471–487. ISSN: 0018-9448.
4. Varshney, L. R. (2008). Transporting information and energy simultaneously. In *2008 IEEE International Symposium on Information Theory* (pp. 1612–1616). ISBN: 978-1-4244-2256-2.
5. Pimentel, C., & Alajaji, F. (2009). Packet-based modeling of Reed-Solomon block-coded correlated fading channels via a Markov finite queue model. *IEEE Transactions on Vehicular Technology, 58*(7), 3124–3136. ISSN: 0018-9545.
6. Abadi, M. N., & Cardeno, L. (2015). Renyi entropies and large deviations for the first match function. *IEEE Transactions on Information Theory, 61*(4), 1629–1639. ISSN: 0018-9448.
7. Barbero, A. I., Rosnes, E., Yang, G., & Ytrehus, O. (2011). Constrained codes for passive RFID communication. In *2011 Information Theory and Applications Workshop* (pp. 1–9). IEEE. ISBN: 978-1-4577-0360-7.
8. Fouladgar, A. M., Simeone, O., & Erkip, E. (2014). Constrained codes for joint energy and information transfer. *IEEE Transactions on Communications, 62*(6), 2121–2131. ISSN: 1558-0857.
9. Tandon, A., Motani, M., & Varshney, L. R. (2014). Constant subblock composition codes for simultaneous energy and information transfer. In *2014 Eleventh Annual IEEE International Conference on Sensing, Communication, and Networking Workshops (SECON Workshops)* (pp. 45–50).
10. Tandon, A., Motani, M., & Varshney, L. R. (2015). Real-time simultaneous energy and information transfer. In *2015 IEEE International Symposium on Information Theory (ISIT)* (pp. 1124–1128).
11. Tandon, A., Motani, M., & Varshney, L. R. (2016). Subblock-constrained codes for real-time simultaneous energy and information transfer. *IEEE Transactions on Information Theory, 62*(7), 4212–4227. ISSN: 0018-9448.
12. Dawy, Z., Saad, W., Ghosh, A., Andrews, J. G., & Yaacoub, E. (2016). Toward massive machine type cellular communications. *IEEE Wireless Communications, 24*(1), 120–128. ISSN: 1536-1284.
13. Krikidis, I., et al. (2014). Simultaneous wireless information and power transfer in modern communication systems. *IEEE Communications Magazine, 52*(11), 104–110. ISSN: 0163-6804.
14. Zhou, X., Zhang, R., & Ho, C. K. (2013). Wireless information and power transfer: Architecture design and rate-energy tradeoff. *IEEE Transactions on Communications, 61*(11), 4754–4767. ISSN: 0090-6778.
15. Clerckx, B., & Bayguzina, E. (2016). Waveform design for wireless power transfer. *IEEE Transactions on Signal Processing, 64*(23), 6313–6328. ISSN: 1053-587X.
16. Zeng, Y., Clerckx, B., & Zhang, R. (2017). Communications and signals design for wireless power transmission. *IEEE Transactions on Communications, 65*(5), 2264–2290. ISSN: 0090-6778.
17. Zhang, R., & Ho, C. K. (2013). MIMO broadcasting for simultaneous wireless information and power transfer. *IEEE Transactions on Wireless Communications, 12*(5), 1989–2001. ISSN: 1536-1276.
18. Lv, K., Hu, J., Yu, Q., Yang, K. (2017). Throughput maximization and fairness assurance in data and energy integrated communication networks. *IEEE Internet of Things Journal, 5*(2), 636–644. ISSN: 2372-2541.
19. Yang, J., Hu, J., Lv, K., Yu, Q., & Yang, K. (2018). Multi-dimensional resource allocation for uplink throughput maximisation in integrated data and energy communication networks. *IEEE Access, 6*, 47163–47180. ISSN: 2169-3536.
20. Gautam, S., Lagunas, E., Chatzinotas, S., & Ottersten, B. (2019). Relay selection and resource allocation for SWIPT in multi-user OFDMA systems. *IEEE Transactions on Wireless Communications, 18*(5), 2493–2508. ISSN: 1536-1276.
21. Zhao, Y., Hu, J., Diao, Y., Yu, Q., & Yang, K. (2018). Modelling and performance analysis of wireless LAN enabled by RF energy transfer. *IEEE Transactions on Communications, 66*(11), 5756–5772. ISSN: 0090-6778.

22. Zhao, Y., Hu, J., Guo, R., Yang, K., & Leng, S. (2018). Enhanced CSMA/CA protocol design for integrated data and energy transfer in WLANs. In *Proceedings of IEEE Globecom 2018* (pp. 1–6).
23. Zhao, Y., Wang, D., Hu, J., & Yang, K. (2018). H-ap deployment for joint wireless information and energy transfer in smart cities. *IEEE Transactions on Vehicular Technology, 67*(8), 7485–7496. ISSN: 1939-9359.
24. Fouladgar, A. M., & Simeone, O. (2012). On the transfer of information and energy in multi-user systems. *IEEE Communications Letters, 16*(11), 1733–1736. ISSN: 1089-7798.
25. Hu, J., Zhao, Y., & Yang, K. (2019). Modulation and coding design for simultaneous wireless information and power transfer. *IEEE Communications Magazine, 57*(5), 124–130. ISSN: 1558-1896.
26. Varshney, L. R. (2012). On energy/information cross-layer architectures. In *2012 IEEE International Symposium on Information Theory Proceedings* (pp. 1356–1360). IEEE. ISBN: 978-1-4673-2579-0.
27. Popovski, P., Fouladgar, A. M., & Simeone, O. (2013). Interactive joint transfer of energy and information. *IEEE Transactions on Communications, 61*(5), 2086–2097. ISSN: 0090-6778.
28. Yin, S., & Qu, Z. (2015). Rate-optimal coding design in joint transfer of energy and information. *IEEE Communications Letters, 19*(5), 715–718. ISSN: 1089-7798.
29. Tandon, A., Motani, M., & Varshney, L. R. (2014). On code design for simultaneous energy and information transfer. In *2014 Information Theory and Applications Workshop (ITA)* (pp. 1–6). IEEE. ISBN: 978-1-4799-3589-5.
30. Babar, Z., Izhar, M. A. M., Nguyen, H. V., Botsinis, P., Alanis, D., Chandra, D., Ng, S. X., Maunder, R. G., & Hanzo, L. (2017). Unary-coded dimming control improves ON-OFF keying visible light communication. *IEEE Transactions on Communications, 66*(1), 255–264. ISSN: 0090-6778.
31. Zhang, W., Song, Z., Brejza, M. F., Wang, T., Maunder, R. G., & Hanzo, L. (2016). Learning-aided unary error correction codes for non-stationary and unknown sources. *IEEE Access, 4*, 2408–2428. ISSN: 2169-3536.
32. Zhang, W., Brejza, M. F., Wang, T., Maunder, R. G., & Hanzo, L. (2015). Irregular trellis for the near-capacity unary error correction coding of symbol values from an infinite set. *IEEE Transactions on Communications, 63*(12), 5073–5088. ISSN: 0090-6778
33. Cover, T. M., & Thomas, J. A. (2006). *Elements of information theory* (2nd ed.). John Wiley & Sons.
34. Yu, X., Chu, J., Yu, K., Teng, T., & Li, N. (2019). Energy-efficiency optimization for IoT-distributed antenna systems with SWIPT over composite fading channels. *IEEE Internet of Things Journal, 7*(1), 197–207. ISSN: 2372-2541.
35. Chen, F., Fu, J., Wang, Z., Zhou, Y., & Qiu, W. (2019). Joint communication and computation resource optimization in FD-MEC cellular networks. *IEEE Access, 7*, 168444–168454. ISSN: 2169-3536.
36. Huang, F., Chen, J., Wang, H., Ding, G., Gong, Y., & Yang, Y. (2019). Multiple-UAV-assisted SWIPT in internet of things: User association and power allocation. *IEEE Access, 7*, 124244–124255. ISSN: 2169-3536.
37. Zhao, Y., Hu, J., Ding, Z., & Yang, K. (2019). Joint interleaver and modulation design for multi-user SWIPT-NOMA. *IEEE Transactions on Communications, 67*(10), 7288–7301. ISSN: 1558-0857.
38. Im, C., Lee, J., & Lee, C. (2020). A multi-tone amplitude modulation scheme for wireless information and power transfer. *IEEE Transactions on Vehicular Technology, 69*(1), 1147–1151. ISSN: 1939-9359.
39. Rajaram, A., Jayakody, D. N. K., Chen, B., Dinis, R., & Affes, S. (2020). Modulation-based simultaneous wireless information and power transfer. *IEEE Communications Letters, 24*(1), 136–140. ISSN: 2373-7891.
40. Claessens, S., Pan, N., Schreurs, D., & Pollin, S. (2019). Multitone FSK modulation for SWIPT. *IEEE Transactions on Microwave Theory and Techniques, 67*(5), 1665–1674. ISSN: 1557-9670.
41. Krikidis, I., & Psomas, C. (2019). Tone-index multisine modulation for SWIPT. *IEEE Signal Processing Letters, 26*(8), 1252–1256. issn: 1558–2361.
42. Hu, J., Li, M., Yang, K., Ng, S. X., & Wong, K. K. (2019). Unary coding controlled simultaneous wireless information and power transfer. *IEEE Transactions on Wireless Communications, 19*(1), 637–649.
43. Mitani, T., Kawashima, S., & Nishimura, T. (2017) Analysis of voltage doubler behavior of 2.45-GHz voltage doubler-type rectenna. *IEEE Transactions on Microwave Theory and Techniques, 65*(4), 1051–1057. ISSN: 0018-9480.
44. Yuan, F., Jin, S., Wong, K. K., Zhao, J., & Zhu, H. (2017). Wireless information and power transfer design for energy cooperation distributed antenna systems. *IEEE Access, 5*, 8094–8105. ISSN: 2169-3536.
45. Alfano, G., & De Maio, A. (2007). Sum of squared shadowed-rice random variables and its application to communication systems performance prediction. *IEEE Transactions on Wireless Communications, 6*(10), 3540–3545.
46. Clerckx, B., Zhang, R., Schober, R., Ng, D. W. K., Kim, D. I., & Poor, H. V. (2018). Fundamentals of wireless information and power transfer: From RF energy harvester models to signal and system designs. *IEEE Journal on Selected Areas in Communications, 37*(1), 4–33. ISSN: 1558-0008.
47. Zhang, H., Feng, M., Long, K., Karagiannidis, G. K., Leung, V. C., & Poor, H. V. (2019) Energy efficient resource management in SWIPT enabled heterogeneous networks with NOMA. *IEEE Transactions on Wireless Communications, 19*(2), 835–845.

Chapter 3
Waveform Design for Integrated Data and Energy Transfer

3.1 Overview of Classic Modulation

Digital data is usually in the form of binary data streams, namely the sequence of "0" and "1". The aim of communication system is to transmit those data to the destination reliably with specific channel. However, the natural properties of the channel may cause severe damage to the transmitted digital data, including noise, attenuation, distortion, and interference. In order to minimize the damage as much as possible, some specific signals that can simultaneously represent the binary stream and match the channel characteristic must be found. The process of mapping a digital sequence into a signal for transmission over a communication channel is called digital modulation.

The modulator of digital communication system will map k bits to the corresponding signal waveform set $s_m(t)$, where $1 \leq m \leq M$ and $M = 2^k$. Assume that the modulated signals are transmitted every T_s seconds, where the T_s is the signal transmission period, and the symbol transmission rate or symbol rate is defined as

$$R_s = \frac{1}{T_s}. \tag{3.1}$$

Due to k bits are carried by every signal, the bit transmission period is defined as

$$T_b = \frac{T_s}{k} = \frac{T}{\log_2 M}. \tag{3.2}$$

Furthermore, the bit transmission rate is then defined as

$$R_b = kR_s = R_s \log_2 M. \tag{3.3}$$

The mapping between the binary sequence and the modulated signal sequence can be with memory and without memory. In memoryless modulation, the binary sequence is divided into sub-sequences of which the length is k. Every sub-sequence is mapped to a specific signal $s_m(t)$ in the waveform set without considering the previous modulated signals. In memory modulation, the current modulated signal is generated according to the current k bits and the previous $(L-1)k$ bits. The finite state machine with $2^{(L-1)k}$ states is defined in this modulation method. If the modulator is in the state $S_{t-1} \in \{1, 2, \cdots, 2^k\}$ at time slot $t - 1$ and the input sub-sequence is $I_t \in \{1, 2, \cdots 2^k\}$, the output signal is $s_{m_t}(t)$, which follows:

$$m_t = f_m(S_{t-1}, I_t), \tag{3.4}$$

$$S_t = f_s(S_{t-1}, T_t), \tag{3.5}$$

where S_t is the state at time slot t and $f_m(\cdot)$ and $f_s(\cdot)$ are the designed transition functions.

J. Hu, K. Yang, *Integrated Data and Energy Transfer in Wireless Networks*, Textbooks in Telecommunication Engineering,
https://doi.org/10.1007/978-3-031-60514-7_3

3.1.1 Memoryless Modulation

3.1.1.1 Pulse Amplitude Modulation

Pulse Amplitude Modulation (PAM) is a modulation method in which the amplitude of a pulse carrier varies with the baseband signal. If the pulse carrier is an impulse pulse sequence, the sampling theorem is the principle of PAM. As a matter of fact, due to the inability to implement real impulse pulse trains, narrow pulse trains are usually adopted. There are two types of PAM modulation signals that can actually be achieved, one is PAM with natural sampling, and the other is that with flat top sampling. These two classifications are distinguished by the variation of carrier pulse amplitude.

3.1.1.2 Phase-Shift Keying

Phase-shift keying (PSK) is a digital modulation scheme that transmits data by changing or modulating the phase of a reference signal. The carrier phase varies based on binary input (1 or 0). The phase-shift difference of $\frac{360}{M}$ degrees is used in MPSK. For example, phase-shift difference of 180° is used between binary 1 and binary 0 in BPSK.

3.1.1.3 Frequency Shift Keying

Frequency shift keying (FSK) is a digital modulation technique that allows data transmission based on changing the carrier frequency of a reference signal. The simplest method to generate an FSK signal is to switch between two independent oscillators based on whether the input data bits are 0 or 1. The waveform generated by this method has discontinuous phase at the time of switching, so this FSK signal is called a discontinuous FSK signal. Due to the discontinuity of the phase, spectrum expansion may occur. With the development of digital processing technology, continuous phase FSK modulation technology is increasingly exploited.

3.1.1.4 Quadrature Amplitude Modulation

Quadrature amplitude keying (QAM) is a method combining PAM and PSK, thereby doubling the effective bandwidth. QAM signal has two carriers of the same frequency, but the phase difference is 90° (quarter cycle). One signal is called the I signal, and the other signal is called the Q signal. From a mathematical perspective, one signal is represented as a sine and the other as a cosine. The two modulated carriers have been mixed during transmission. After reaching the destination, the carrier is separated, and the data is extracted separately and mixed with the original modulation information.

3.1.2 Memory Modulation

3.1.2.1 Continuous Phase Modulation

Continuous phase modulation (CPM) is a non-linear modulation scheme that carries information through the carrier phase. The transmitted signal has the characteristics of constant envelope and continuous phase variation. The general expression for CPM transmission signals can be defined as

$$s_c(t, \mathbf{a}) = \sqrt{\frac{2E_s}{T}} \cos\left(2\pi f_0 t + \phi(t, \mathbf{a})\right), kT \le t \le (k+1)T, \tag{3.6}$$

where E_s is the average energy carried by each symbol, T is the symbol transmission period, f_0 is the carrier frequency, $\mathbf{a} = (a_0, a_1, \cdots)$ is the transmitted information sequence, and $\phi(t, \mathbf{a})$ is the carrier phase function at time slot t, which can be expressed as

$$\phi(t, \mathbf{a}) = 2\pi h \sum_{i=0}^{k} a_i q(t - iT), \tag{3.7}$$

where h is the modulation index and $q(t)$ is the phase pulse function. The $q(t)$ is a continuous function that can be expressed as

$$q(t) = \int_{-\infty}^{t} g(\tau)d\tau = \begin{cases} 0 & \mathrm{t} \leq 0, \\ 0.5 & \mathrm{t} > 0, \end{cases} \tag{3.8}$$

where $g(t)$ is the frequency pulse function and L is the memory length that determines that the current transmission signal is affected by the previous L transmission symbol.

Specifically, the CPM is called continuous phase frequency shift keying (CPFSK) when $L = 1$. Furthermore, the CPM is called minimum shift keying (MSK) when $M = 2$.

3.2 Multiuser Modulation for Integrated Data and Energy Transfer

Radio frequency (RF) signals can be relied upon for conventional wireless information transfer (WIT) and for challenging wireless power transfer (WPT), which triggers the significant research interest in the topic of simultaneous wireless information and power transfer (SWIPT). By further exploiting the advanced non-orthogonal multiple access (NOMA) technique, we are capable of improving the spectrum efficiency of the resource-limited SWIPT system. In our SWIPT system, a hybrid access point (H-AP) superimposes the modulated symbols destined to multiple WIT users by exploiting the power-domain NOMA, while WPT users are capable of harvesting the energy carried by the superposition symbols. In order to maximize the amount of energy transferred to the WPT users, we propose a joint design of the energy interleaver and the constellation rotation based modulator in the symbol-block level by constructively superimposing the symbols destined to the WIT users in the power-domain. Furthermore, a transmit power allocation scheme is proposed to guarantee the symbol error ratio (SER) of all the WIT users. By considering the sensitivity of practical energy harvesters, the simulation results demonstrate that our scheme is capable of substantially increasing the WPT performance without any remarkable degradation of the WIT performance.

3.2.1 Introduction

3.2.1.1 Background

In the upcoming era of 5G and Internet of Things (IoT), massive machine-type communications are enabled by the deployment of low-power IoT devices, which triggers more difficulties on the spectrum efficiency [1], energy efficiency [2], or security [3]. In order to accommodate the explosive growth of the machine-type tele-traffic, non-orthogonal multiple access (NOMA) [4, 5] emerges as a critical technique for substantially increasing the spectrum efficiency. The NOMA aided transmitter differentiates the symbols destined to the different users by allocating various transmit power to these symbols superimposed in the same time–frequency resource block. The successive interference cancellation (SIC) technique [6] is adopted by the users for recovering their requested symbols from the superposition one. For a specific user, the symbols requested by the other users are sequentially demodulated and removed from the superposition symbol until its own requested one is successfully recovered. In order to further increase the spectrum efficiency, the multiple-input multiple-output (MIMO) technique [7] is combined with the NOMA by providing additional multiplexing gain in the spatial domain. By allowing the downlink and uplink transmissions in the same time–frequency resource block, the full-duplex technique may further double the spectrum efficiency of the NOMA [8], if the self-interference can be appropriately mitigated.

Furthermore, in order to alleviate their energy shortage, radio frequency (RF) signal based wireless power transfer (WPT) can be relied upon for remotely charging these battery-powered IoT devices [9]. However, coordinating the conventional wireless information transfer (WIT) and the wireless power transfer (WPT) in the same RF band is a challenging task, which thus stimulates substantial research interest in the topic of simultaneous wireless information and power transfer (SWIPT) [10]. In SWIPT systems, the simultaneous reception of both the information and energy is realized by invoking a signal splitter at the receiver, which is capable of splitting the received RF signal either in the time-domain [11] or in the power-domain [12] or in the spatial domain [13]. Specifically, the WIT users recover the requested information from the received RF signals, while the WPT users harvest energy from their received RF signals.

3.2.1.2 Related Works

Recently, some pioneering works have studied the multiuser SWIPT-NOMA system. For instance, Zheng et al. [14] considered a NOMA aided cooperative NOMA network, where a transmitter communicates with a pair of receivers via a single wireless powered relay. The optimal power allocation scheme was found for maximizing the total downlink throughput. Moreover, Liu et al. [15] proposed a novel protocol for the NOMA aided cooperative network, in which the users close to the transmitters operate as the wireless powered relays for the sake of forwarding the information to the distant users. Xu et al. [16] proposed a joint design of the transmit beamforming and receive power splitting in a similar NOMA aided cooperative network, which aims for maximizing the achievable throughput of the so-called strong users, while ensuring the minimum throughput requirements of other "weak users". Diamantoulakis et al. [17] studied a wireless powered communication network, in which the uplink transmission of multiuser is supported by the NOMA. Furthermore, Alsaba et al. [18] studied a NOMA aided cooperative system, where the "strong user" adopts the full-duplex technique for simultaneously receiving the RF signal from the base station and for relaying information to the "weak user". The sum-throughput was then maximized by invoking a two-step convex optimization method. Mohammadali et al. [19] studied a NOMA-SWIPT system having a battery-powered access point and a pair of energy harvesting users. They investigated the information–energy trade-off, while studying the performance of cognitive radio aided NOMA. In their scheme, the transmit power is allocated to the user having poor channel conditions in order to satisfy certain quality of service (QoS) requirements. Dai et al. [20] investigated a millimetre-wave aided NOMA-SWIPT system on order to further enhance the spectrum efficiency. Moreover, Wei et al. [21] proposed a transceiver design in a cooperative NOMA-SWIPT system, where the power allocation, the power splitting (PS) ratio, and the receiver filter, as well as the transmit beamformer are jointly optimized for maximizing the SWIPT performance.

In the most of the existing works concerning the NOMA aided WIT and SWIPT systems, the WIT performance was always characterized by the classic Shannon–Hartley channel capacity, which inherently assumes infinite Gaussian distributed continuous channel input. However, the attainable performance is always overestimated, since the channel input of any practical WIT or SWIPT system is normally finite and discrete [22]. As a result, we may attain more accurate performance when a practical modulator is considered. Cheng et al. [23] studied the receive spatial modulation (RSM) aided SWIPT, where the optimal power allocation scheme between the information signal and the energy signal is optimized for achieving a balance between the information decoding and the energy harvesting requirements of the receiver. Mohjazi et al. [24] analyzed the end-to-end bit error ratio (BER) performance of the differential modulation in SWIPT aided cooperation networks, where the wireless powered relay adopts the amplify-and-forward approach for relaying the information to the destination. Unfortunately, few works really consider the modulation design of improving the SWIPT performance. There have been a number of works studying the modulation design in NOMA aided WIT systems [25, 26]. Furthermore, Liu et al. [27] proposed a beneficial amalgamation of the spatial modulation and the sparse code-division multiple access (SCDMA), which is capable of supporting a very high user load by exploiting the multiplexing gains provided by the multiple antennas and the non-orthogonality of the SCDMA.

In order to alleviate the impact of the cascaded errors imposed by bursty interference and deep fading, the symbols have to be shuffled and reordered by the interleaver before transmission. For instance, Li et al. [28] studied a bit-interleaving scheme in space–time trellis coding in block-fading channels. Jun et al. [29] proposed a serially concatenated continuous phase modulation by adopting symbol-level interleaving. On the other hand, constellation rotation was proposed for increasing the constrained capacity by rotating the constellations of different users with certain angles. In [30], Neng et al. studied constellation rotation of NOMA with SIC receiver, while the bit error rate was reduced and the capacity was enlarged compared with conventional NOMA. Lin et al. [31] proposed an optimal inter-constellation rotation based on minimum distance criterion for uplink NOMA, while the robustness and fairness among users are both guaranteed.

3.2.1.3 Contributions

Introducing the WPT into the same RF band may make the precious spectrum even more congested. As a result, we have to exploit the high spectrum efficiency of the NOMA by satisfying both the WPT and WIT users' various requirements [14]. However, superimposing the modulated RF signals destined to multiple users in the same time–frequency resource block may inevitably degrade the actual energy carried by the resultant superposition signal, since these modulated signals having diverse phases may be destructively superimposed. Therefore, the resultant WPT performance is impaired. Recently, most existing works [19–23] for NOMA-SWIPT system mainly focus on the power allocation and the beamforming design for maximizing the SWIPT performance, where the classic Shannon capacity is relied upon for characterizing the upper bound WIT performance. However, in a practical NOMA-SWIPT system with a practical modulation scheme having finite alphabet, this overestimated WIT performance may result in unfair resource allocation among the WIT and WPT performance. Apart from the front-end transceiver and air-interface design, the potential of modulation design on improving the SWIPT performance has not been well understood. Unfortunately, few works studied the impact of modulation design in the NOMA-SWIPT system, except for our first attempt [32]. We proposed a constellation rotation modulation scheme in [32] for a three-user NOMA-SWIPT system, which includes a pair of WIT users and a single WPT user. The modulated signal destined to the WIT user pair is constructively superimposed in order to maximize its WPT performance. However, our design in [32] is only suitable for supporting a pair of WIT users, whose requested symbols are superimposed on a single subcarrier (or a single resource block). In practice, we have to map symbols of different WIT users to the multiple subcarriers available. Therefore, the energy interleaving is originally proposed for the sake of mitigating the destructive superposition of multiple symbols scheduled on each subcarrier.

In order to improve the WPT performance of the NOMA aided SWIPT system, we aim for maximizing the amount of energy carried by the superposition signals by jointly designing the energy interleaver and the constellation rotation aided modulator of the transmitter in the symbol-block level, which may not remarkably degrade the WIT performance. To the best of our knowledge, this is the first attempt to the interleaver and modulator design of the multiuser NOMA aided SWIPT system. Our novel contributions are summarized as below:

- A novel *transceiver architecture* of the NOMA aided SWIPT system is proposed for simultaneously transferring energy to multiple WPT users and transferring information to multiple WIT users.
- The transmitter design is obtained by jointly optimizing the *constellation rotation aided modulator* and by optimizing the *energy interleaver* for the sake of maximizing the attainable WPT performance.
- A feasible transmit *power allocation scheme* among the WIT users is obtained without any remarkable degradation of the WIT performance in terms of symbol error ratio (SER).
- The impact of the *practical energy harvester's sensitivity* on the attainable WPT performance is also evaluated.

3.2.2 System Model and Transceiver Architecture

Our NOMA aided SWIPT system consists of K_I WIT users denoted as $\{u_1^I, \cdots, u_{K_I}^I\}$, K_E WPT users denoted as $\{u_1^E, \cdots, u_{K_E}^E\}$, and a single hybrid access point (H-AP), which is capable of simultaneously delivering information to the WIT users and transferring energy to the WPT users. All the users and the H-AP are equipped with a single antenna. The H-AP transmits the superposition signal to the WIT users by adopting the power-domain NOMA technique, while the WPT users harvest energy from this superposition signal by exploiting the broadcast nature of the wireless channels. The transmitter architecture of the H-AP, the receiver architecture of the WIT users, and that of the WPT users are all portrayed in Fig. 3.1. The definitions of the key mathematical notations appearing in this chapter are provided in Table 3.1.

3.2.2.1 Transmitter Architecture of the H-AP

We illustrate the transmitter architecture of the H-AP in the middle of Fig. 3.1, where the H-AP transmits all requested symbols on N different subcarriers $\{\cos(2\pi f_n t)|n = 1, \cdots, N\}$. The transmitter architecture includes the power allocator, the modulator, the energy interleaver, and the constellation rotator.

As illustrated in the middle of Fig. 3.1, the symbols requested by the WIT user $\{u_k^I|k = 1, \cdots, K_I\}$ are assigned allocated with different power $\{P_{n,k}|k = 1, \cdots, K_I\}$ on the n-th subcarrier, respectively. An appropriate power allocation scheme is capable of alleviating the mutual interference induced by the superposition symbol.

Since the WIT receivers require the knowledge of the energy interleaver and the constellation rotators for the sake of successful demodulation, this knowledge also needs to be transmitted to the receivers as the control overhead. In order to reduce the control overhead, both the energy interleaving and the constellation rotation operate in a symbol-block level. Each symbol-block consists of L modulated symbols. The transmitter generates $N * L$ modulated symbols for every WIT user,

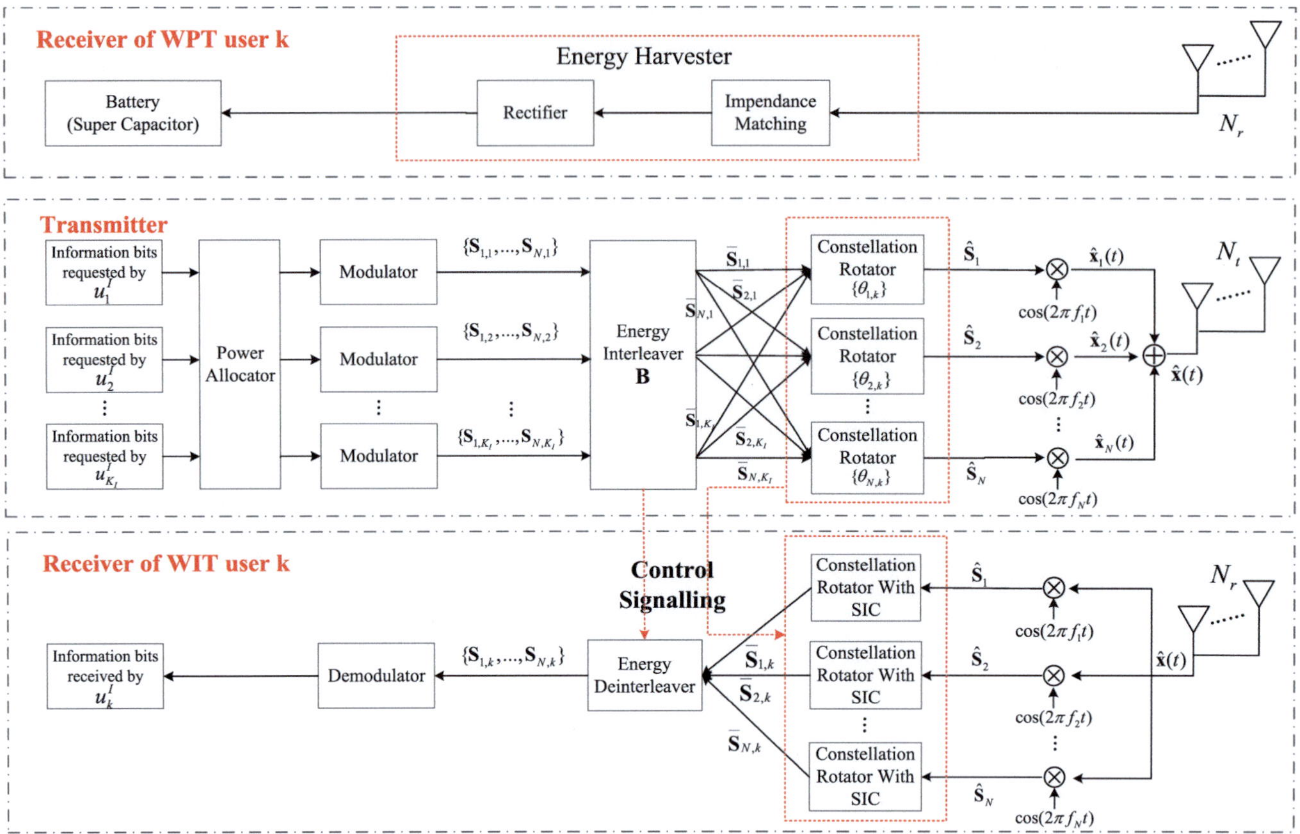

Fig. 3.1 Transceiver architecture

Table 3.1 Definition of the key mathematical notations

Symbols	The meaning of these symbols	Symbols	The meaning of these symbols
P_n	Total transmit power on the n-th subcarrier	$P_{n,k}$	Transmit power of u_k^I on the n-th subcarrier
$\mathbf{S}_{n,k}$	The n-th symbol-block of u_k^I	$s_{n,k,l}$	The l-th symbol in $\mathbf{S}_{n,k}$
$A_{n,k,l}$	Normalized amplitude of $s_{n,k,l}$	$\phi_{n,k,l}$	Phase of $s_{n,k,l}$
$\overline{\mathbf{S}}_{n,k}$	The symbol-block of u_k^I on the n-th subcarrier	$\overline{s}_{n,k,l}$	The l-th symbol in $\overline{\mathbf{S}}_{n,k}$
$\overline{A}_{n,k,l}$	Normalized amplitude of $\overline{s}_{n,k,l}$	$\overline{\phi}_{n,k,l}$	Phase of $\overline{s}_{n,k,l}$
$\widehat{\mathbf{S}}_{n,k}$	$\overline{\mathbf{S}}_{n,k}$ after constellation rotation	$\widehat{s}_{n,k,l}$	The l-th symbol in $\widehat{\mathbf{S}}_{n,k}$
$\widehat{A}_{n,k,l}$	Normalized amplitude of $\widehat{s}_{n,k,l}$	$\widehat{\phi}_{n,k,l}$	Phase of $\widehat{s}_{n,k,l}$
$\widehat{\mathbf{S}}_n$	Superimposed $\widehat{\mathbf{S}}_{n,k}$ of K_I WIT users	$\widehat{s}_{n,l}$	The l-th symbol in $\widehat{\mathbf{S}}_n$
$\widehat{x}_{n,l}$	Modulated signal of $\widehat{s}_{n,l}$	$\widehat{E}_{n,l}$	Energy carried by $\widehat{x}_{n,l}$
$\mathbf{B}$	Energy interleaving matrix	$\boldsymbol{\Theta}$	Constellation rotation angle matrix
$\xi_{n,k,l}$	Constellation distortion of $\widehat{s}_{n,k,l}$	η_n	Utility function on the n-th subcarrier

which are divided into N symbol-blocks.[1] The symbol-blocks requested by WIT user u_k^I are denoted as $\{\mathbf{S}_{n,k}|n = 1, \cdots, N\}$, which are originally scheduled to be transmitted on the n-th carrier.[2]

For the modulated symbols transmitted on the n-th subcarrier to the WIT user u_k^I, the minimum Euclidean distance in the constellation is $2d_{n,k}$. Without loss of generality, M-QAM modulator is adopted as a case study in this manuscript, since it has been widely adopted in modern wireless communications, e.g., the Long-Term Evolution (LTE) and 802.11-based

[1] The number of the symbol-blocks requested by a WIT user is equal to the number of subcarriers.

[2] This scheduling may be altered by the energy interleaver.

system. Specifically, when the M-QAM is adopted, we have $d_{n,k} = \sqrt{\frac{3P_{n,k}}{M-1}}$ [33]. If the symbol-block $\mathbf{S}_{n,k}$ is transmitted to the WIT user u_k^I on the n-th subcarrier, the l-th symbol $s_{n,k,l}$ within it can be expressed as

$$\begin{aligned} s_{n,k,l} &= d_{n,k}\sqrt{\left(A_{n,k,l}^I\right)^2 + \left(A_{n,k,l}^Q\right)^2}e^{j\phi_{n,k,l}} \\ &= d_{n,k}A_{n,k,l}e^{j\phi_{n,k,l}}, \end{aligned} \tag{3.9}$$

for $l = 1, \cdots, L$, where $\phi_{n,k,l}$ represents the phase and $A_{n,k,l}$ represents the amplitude normalized by $d_{n,k}$. Moreover, $A_{n,k,l}^I$ and $A_{n,k,l}^Q$ are the normalized amplitudes of in-phase and quadrature in the constellation of M-QAM, respectively. Specifically, we have $A_{n,k,l}^I, A_{n,k,l}^Q \in \{1-\sqrt{M}, 3-\sqrt{M}, \ldots, \sqrt{M}-3, \sqrt{M}-1\}$.

The symbol-blocks requested by all the WIT users flow into an energy interleaver for the following pair of purposes:

- Converting the serial symbol-blocks requested by every WIT user into the parallel ones.
- Scheduling the transmissions of all the symbol-blocks on the N subcarriers. The scheduling scheme should avoid the superposition of the symbols having opposite phases, since it may degrade the actual energy carried by the resultant superposition symbols.

The energy interleaver can be represented by a three-dimensional binary tensor $\mathbf{B} = \{b_{k,m,n}\}$ having the size of $K_I \times N \times N$, where $b_{k,m,n} = 1$ indicates that the symbol-block $\mathbf{S}_{m,k}$ is transmitted on the n-th subcarrier, while $b_{k,m,n} = 0$ indicates that $\mathbf{S}_{m,k}$ is not transmitted on the n-th subcarrier. After the energy interleaving, all the symbol-blocks are scheduled on the N subcarriers. The symbol-block requested by WIT user u_k^I is denoted as $\overline{\mathbf{S}}_{n,k}$, which is transmitted on the n-th subcarrier.

After the energy interleaving, the modulated symbol-blocks transmitted on every subcarrier then flow into the respective constellation rotator, as portrayed in the middle of Fig. 3.1. The symbol-blocks requested by different WIT users are then rotated with some certain angles, respectively, in order to counteract the energy degradation induced by superimposing the symbols having opposite phases. After the constellation rotation, the symbol-block $\overline{\mathbf{S}}_{n,k}$ is then converted into $\widehat{\mathbf{S}}_{n,k}$. The constellation rotation angles are also represented by a matrix $\boldsymbol{\Theta} = \{\theta_{n,k}\}$ having the size of $N \times K_I$, in which $\theta_{n,k}$ indicates the constellation rotation angle of the symbol-block $\overline{\mathbf{S}}_{n,k}$.

Finally, all the rotated symbol-blocks $\{\widehat{\mathbf{S}}_{n,k} | k = 1, \cdots, K_I\}$ are superimposed together. The resultant superposition symbol-block is represented by $\widehat{\mathbf{S}}_n$. After being modulated onto the n-th subcarrier, the corresponding signal $\widehat{x}_n(t)$ is then broadcast to all the WIT and WPT users.

3.2.2.2 Receiver Architecture of WIT Users

The receiver architecture of the WIT users is illustrated in the bottom of Fig. 3.1. At the receiver of WIT user u_k^I, the superposition symbol-blocks $\{\widehat{\mathbf{S}}_1, \cdots, \widehat{\mathbf{S}}_N\}$ transmitted on N subcarriers are firstly recovered from the received analog signal. Then, these symbol-blocks are further processed by the constellation rotation based SIC in order to extract the symbols requested by u_k^I, as presented in the bottom of Fig. 3.1.

According to the classic SIC, the symbols having the higher power should be firstly demodulated. Without loss of generality, we assume $P_{n,1} < \cdots < P_{n,K_I}$ on each subcarrier. Accordingly, the minimum Euclidean distances of the WIT users' constellations satisfy the inequality of $d_{n,1} < \cdots < d_{n,K_I}$. Then, the WIT user u_k^I demodulates its requested symbol-block $\overline{\mathbf{S}}_{n,k}$ on the n-th subcarrier by obeying the following steps:

- *Step 1:* Initialize the label of the firstly demodulated WIT user as $i = K_I$.
- *Step 2:* Rotate the superposition symbol-block $\widehat{\mathbf{S}}_n$ with an angle of $-\theta_{n,i}$.
- *Step 3:* Recover the symbol-block $\overline{\mathbf{S}}_{n,i}$ according to the maximum-likelihood (ML) algorithm [34], by regarding the symbol-blocks of all the other WIT users in $\widehat{\mathbf{S}}_n$ as the interference. If $i = k$, then $\overline{\mathbf{S}}_{n,i}$ is the symbol-block requested by u_k^I on the n-th subcarrier. Otherwise, remove $\overline{\mathbf{S}}_{n,i}$ from $\widehat{\mathbf{S}}_n$ and come back to Step 2 by setting $i = i - 1$.

After the constellation rotation based SIC, the recovered symbol-blocks $\overline{\mathbf{S}}_{n,k}$ of u_k^I on all the subcarriers are then processed by an energy deinterleaver for the following purposes:

- Convert the parallel symbol-blocks to the serial ones.
- Sort the symbol-blocks to their original order according to $\mathbf{B}$ generated at the transmitter.

The sorted symbol-blocks after the energy deinterleaver are represented by $\{\mathbf{S}_{1,k}, \cdots, \mathbf{S}_{N,k}\}$. After the demodulation, these symbols are finally decoded into information bits.

The receiver architecture of WPT users is illustrated in the top of Fig. 3.1. The WPT users only harvest energy from the superposition signal destined to the WIT users by exploiting the broadcast nature of wireless channel. A typical energy harvester consists of an impedance matching circuit and a rectifier [10]. The impedance matching circuit guarantees an efficient energy transfer from the receive antennas to the electronic loads, while the rectifier converts the RF signals into the direct current (DC), whose energy can be finally stored in the battery or in the super capacitor.

The energy harvesters of the WPT users can only be activated, if the power of the received superposition signal is higher than a threshold P_{th}, which is also regarded as the sensitivity of the energy harvester [35]. For simplicity, we assume an identical sensitivity for the energy harvesters of all the WPT. Specifically, the actual energy harvested by the WPT user u_k^E within a symbol duration T is $E_k^{\text{H}} = P_k^{\text{H}} T \cdot \mathbf{1}(P_k^{\text{H}} \geq P_{\text{th}})$, where P_k^{H} is the received power of the superposition signal, and $\mathbf{1}(\cdot)$ is an indicator function, which is one only when the bracketed condition is satisfied, but zero, otherwise.

3.2.2.3 Control Signal

For the sake of symbol demodulation and of reordering the interleaved symbol-blocks, the WIT users have to be aware of the energy interleaving schemes and constellation rotation angles adopted by the transmitter. As a result, these control signalling overheads have to be transferred from the transmitter to the WIT users before actual data transmission, which is portrayed in Fig. 3.2. These control signalling overheads can be compressed by adopting some appropriate encoding methods. For example, we have $N!$ symbol-block allocation schemes in total for a specific WIT user, since the transmitter has to establish a one-on-one mapping between N symbol-block of this WIT user and N subcarriers. Therefore, the energy interleaving matrix $\mathbf{B}$ can be encoded into a binary sequence having a length of $K_I \lceil \log_2(N!) \rceil$, where $\lceil x \rceil$ indicates the minimum integer not lower than x. For the constellation rotation angles, we may divide the range of $[-\pi, \pi]$ into D non-overlapping regions, namely $\bigcup_{k=1}^{D} [-\pi + \frac{2(k-1)\pi}{D}, -\pi + \frac{2k\pi}{D}]$. Since constellation rotation angles are relative to one another, we may keep the constellation of an arbitrary WIT user unchanged while rotating the constellation of the other $(K_I - 1)$ WIT users by certain angles. Therefore, all the constellation rotation angles of the WIT users can be encoded as a binary sequence having a length of $N(K_I - 1)\lceil \log_2(D) \rceil$.

3.2.3 Specification of Key Functional Modules

In this section, we will specify the key operations in the transmitter, which includes the energy interleaving, the constellation rotation, and the power allocation.

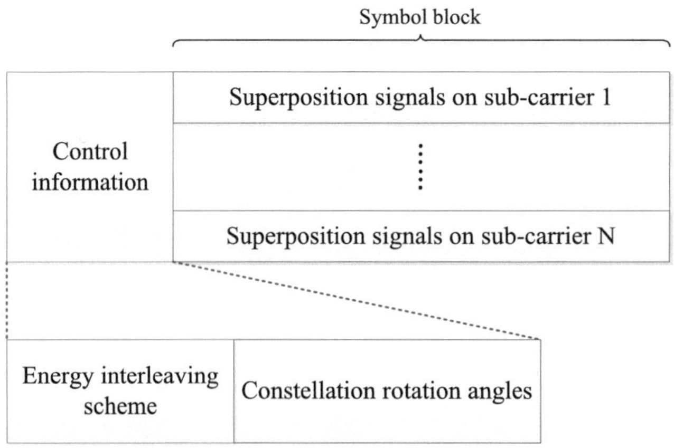

Fig. 3.2 Control signal and data information frame

3.2.3.1 Energy Interleaving

The energy interleaving is adopted at the transmitter in order to prevent the destructive superposition of the symbol-blocks requested by different WIT users on a single subcarrier. Specifically, the WIT user u_k^I requests N symbol-blocks of $\{\mathbf{S}_{1,k}, \cdots, \mathbf{S}_{N,k}\}$ on N subcarriers, as portrayed in Fig. 3.1. The transmitter has to carefully map these N symbol-blocks onto the N subcarriers. Each symbol-block is only allowed to be scheduled on a single subcarrier. The transmitter superimposes K_I symbol-blocks requested by K_I WIT users on every subcarrier by adopting the power-domain NOMA. However, superimposing the symbols having various amplitudes and phases sometimes results in substantial degradation of the energy carried by the resultant superposition symbol, particularly when the symbols requested by the WIT users have opposite phases. As a result, we have to design an efficient interleaving scheme in order to appropriately map the symbol-blocks onto the subcarriers and to avoid the degradation of the energy carried by the resultant superposition signal, which is regarded as the energy interleaving.

Observe from Fig. 3.3a that when we do not adopt the energy interleaving, the energy carried by the superposition symbol-blocks $\mathbf{S}_1$ and $\mathbf{S}_2$ is substantially reduced. This is because some of the symbols in the blocks of $\mathbf{S}_{1,1}$ and $\mathbf{S}_{1,2}$ and some in $\mathbf{S}_{2,1}$ and $\mathbf{S}_{2,2}$ have opposite phases. By contrast, observe from Fig. 3.3b that when the energy interleaving is adopted, the transmitter may map the symbol-block $\mathbf{S}_{2,2}$ onto the first subcarrier and map the symbol-block $\mathbf{S}_{1,2}$ onto the second subcarrier. Therefore, the resultant superposition symbol-blocks $\overline{\mathbf{S}}_1$ and $\overline{\mathbf{S}}_2$ still carry considerable energy. As a result, the energy interleaving substantially increases the WPT performance of the superposition symbols.

According to the energy interleaving tensor $\mathbf{B}$, we have

$$\overline{\mathbf{S}}_{n,k} = \sum_{m=1}^{N} b_{k,m,n} \mathbf{S}_{m,k}. \tag{3.10}$$

Furthermore, the l-th symbol $\overline{s}_{n,k,l}$ in the block $\overline{\mathbf{S}}_{n,k}$ can be expressed as

$$\begin{aligned} \overline{s}_{n,k,l} &= \sum_{m=1}^{N} b_{k,m,n} s_{m,k,l} = \sum_{m=1}^{N} b_{k,m,n} d_{n,k} A_{m,k,l} e^{j\phi_{m,k,l}} \\ &= d_{n,k} \overline{A}_{n,k,l} e^{j\overline{\phi}_{n,k,l}}. \end{aligned} \tag{3.11}$$

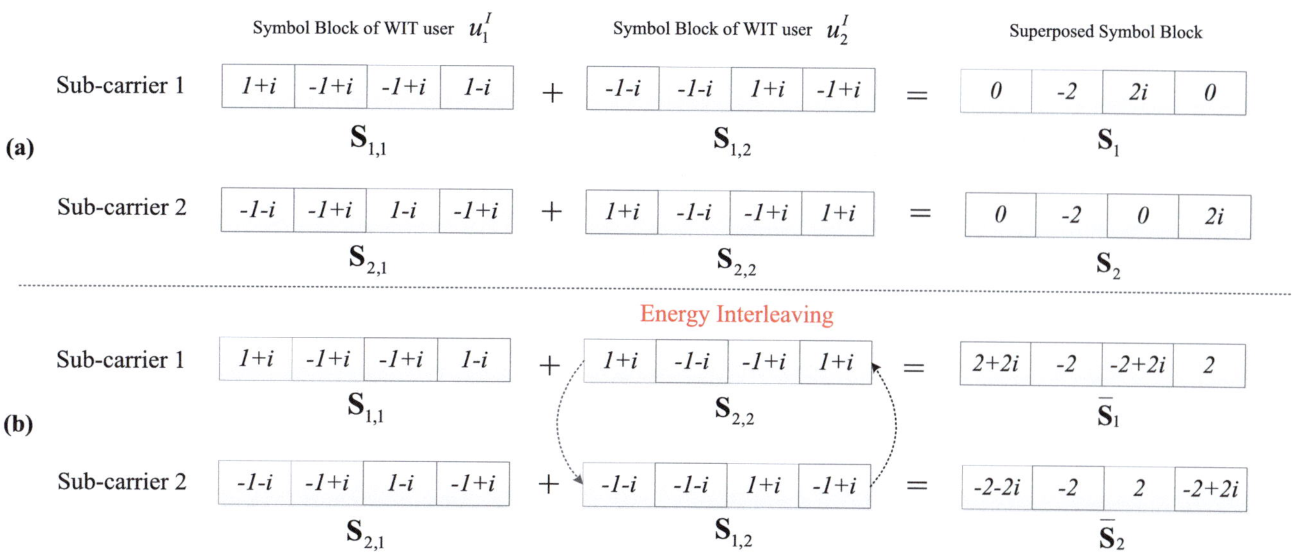

Fig. 3.3 An example of energy interleaving having a pair of WIT users and a pair of subcarriers. Each WIT user requests for a pair of symbol-blocks, where the symbols are modulated by 4-QAM. The transmit power allocated to the symbols requested by the WIT user pair is identical

In (3.11), $\overline{A}_{n,k,l}$ and $\overline{\phi}_{n,k,l}$ are the normalized amplitude and phase of the symbol $\overline{s}_{n,k,l}$, which are formulated as

$$\begin{cases} \overline{\phi}_{n,k,l} = & \arctan\left(\dfrac{\sum\limits_{m=1}^{N} A_{m,k,l} b_{k,m,n} \sin(\phi_{m,k,l})}{\sum\limits_{m=1}^{N} A_{m,k,l} b_{k,m,n} \cos(\phi_{m,k,l})}\right), \\ \overline{A}_{n,k,l} = & \left[\left(\sum\limits_{m=1}^{N} A_{m,k,l} b_{k,m,n} \cos(\phi_{m,k,l})\right)^2 \right. \\ & \left. + \left(\sum\limits_{m=1}^{N} A_{m,k,l} b_{k,m,n} \sin(\phi_{m,k,l})\right)^2\right]^{\frac{1}{2}}. \end{cases} \tag{3.12}$$

3.2.3.2 Constellation Rotation

Constellation rotation aims for avoiding the destructive superposition among the symbol-blocks transmitted on a single subcarrier by rotating the symbol-blocks requested by different WIT users with different angles. As a result, the WPT performance of the resultant superposition symbol-block can be substantially improved.

The symbol superposition between a pair of WIT users with and without the constellation rotation has been exemplified in Fig. 3.4. Specifically, observe from Fig. 3.4a that the actual power carried by the superposition symbol $\overline{s}_{n,l} = \overline{s}_{n,1,l} + \overline{s}_{n,2,l}$ is lower than the sum power of the individual symbols $\overline{s}_{n,1,l}$ and $\overline{s}_{n,2,l}$. The opposite phases of $\overline{s}_{n,1,l}$ and $\overline{s}_{n,2,l}$ substantially impair the attainable WPT performance. By contrast, if we rotate u_1^I's constellation with a certain angle $\theta_{n,1}$, while keeping the constellation of the WIT user u_2^I unchanged, we may observe from Fig. 3.4 that the resultant superposition symbol $\widehat{s}_{n,l} = \widehat{s}_{n,1,l} + \widehat{s}_{n,2,l}$ after the constellation rotation carries a higher power than its counterpart $\overline{s}_{n,l}$.

After the constellation rotation, the symbol-block $\overline{\mathbf{S}}_{n,k}$ is converted into

$$\widehat{\mathbf{S}}_{n,k} = \overline{\mathbf{S}}_{n,k} e^{j\theta_{n,k}}, \tag{3.13}$$

where $\theta_{n,k}$ represents the rotation angle for all the symbols in the symbol-block $\overline{\mathbf{S}}_{n,k}$. The l-th symbol $\widehat{s}_{n,k,l}$ in $\widehat{\mathbf{S}}_{n,k}$ is thus formulated as

$$\widehat{s}_{n,k,l} = \overline{s}_{n,k,l} e^{j\theta_{n,k}} = d_{n,k} \overline{A}_{n,k,l} e^{j(\overline{\phi}_{n,k,l} + \theta_{n,k})}. \tag{3.14}$$

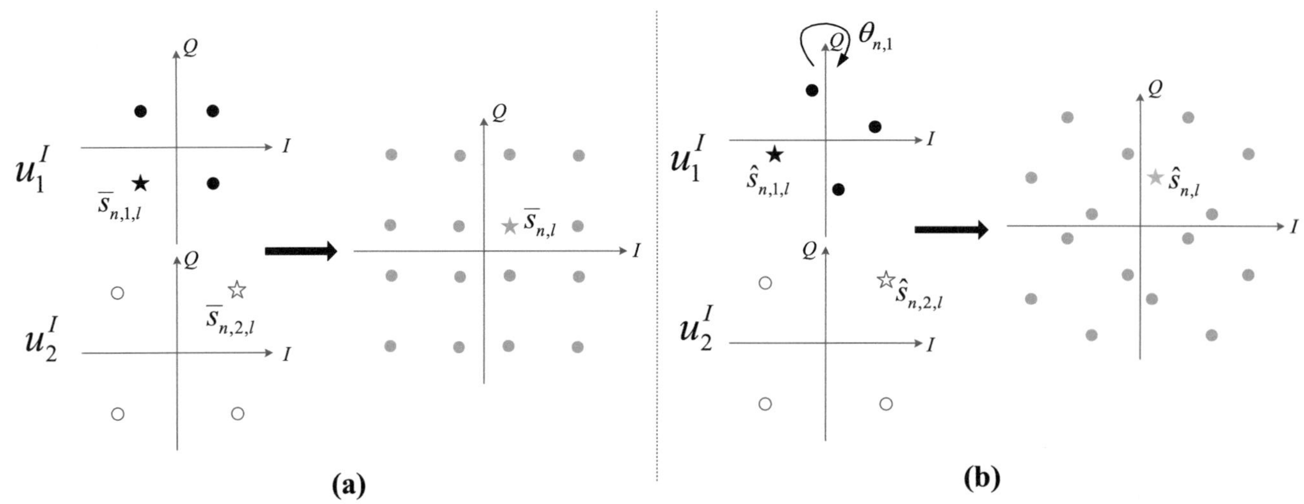

Fig. 3.4 Examples of the symbol superposition between a pair of WIT users: (**a**) Without constellation rotation; (**b**) With constellation rotation. The modulated symbols are obtained by invoking 4-QAM. The symbols requested by the WIT user u_1^I have a lower transmit power than those requested by the WIT user u_2^I. The symbol requested by the WIT user u_1^I is denoted as the filled star, while the symbol requested by u_2^I is denoted as the unfilled star. The resultant superposition symbol is denoted as the gray star

Then, after the superposition of the K_I WIT users' symbol-blocks, the resultant superposition symbol-block $\widehat{\mathbf{S}}_n$ transmitted on the n-th subcarrier is formulated as

$$\widehat{\mathbf{S}}_n = \sum_{k=1}^{K_I} \widehat{\mathbf{S}}_{n,k}. \tag{3.15}$$

Moreover, the l-th superposition symbol $\widehat{s}_{n,l}$ in $\overline{\mathbf{S}}_n$ is derived as

$$\widehat{s}_{n,l} = \sum_{k=1}^{K_I} \widehat{s}_{n,k,l} = \sum_{k=1}^{K_I} d_{n,k} \widehat{A}_{n,k,l} e^{j\widehat{\phi}_{n,k,l}}, \tag{3.16}$$

where we have $\widehat{A}_{n,k,l} = \overline{A}_{n,k,l}$ and $\widehat{\phi}_{n,k,l} = \overline{\phi}_{n,k,l} + \theta_{n,k}$.

Finally, the modulated RF signal $\widehat{x}_{n,l}(t)$ corresponding to the superposition symbol $\widehat{s}_{n,l}$ can be formulated as

$$\begin{aligned} \widehat{x}_{n,l}(t) &= \mathrm{Re}\left[\widehat{s}_{n,l} e^{j2\pi f_n t}\right] \\ &= \sum_{k=1}^{K_I} d_{n,k} \widehat{A}_{n,k,l} \cos(2\pi f_n t + \widehat{\phi}_{n,k,l}), \\ &\qquad (l-1)T \le t \le lT. \end{aligned} \tag{3.17}$$

3.2.3.3 Power Allocation for WIT Assurance

Although the attainable WPT performance can be increased by the constellation rotation, it may incur the adverse symbol offset, which may degrade the associated WIT performance. According to the SIC, when demodulating the symbol $\overline{s}_{n,2,l}$ (or $\widehat{s}_{n,2,l}$), its counterpart $\overline{s}_{n,1,l}$ (or $\widehat{s}_{n,1,l}$) is regarded as interference, as exemplified in Fig. 3.5. Observe from Fig. 3.5a that without the constellation rotation, the superposition symbol $\overline{s}_{n,l}$ is still within the successful demodulation region of $\overline{s}_{n,2,l}$. By contrast, in the case of the constellation rotation, the superposition symbol $\widehat{s}_{n,l}$ is drifted beyond the successful demodulation region of $\widehat{s}_{n,2,l}$, as presented in Fig. 3.5b, which substantially degrades the demodulation performance of the symbol $\widehat{s}_{n,2,l}$. Since we adopt the SIC for sequentially demodulating the symbols requested by different WIT users, we may suffer from severe error propagation when we demodulate the symbols of the rest of WIT users.

According to the SIC, after demodulating the symbol-blocks $\{\widehat{\mathbf{S}}_{n,k+1}, \cdots, \widehat{\mathbf{S}}_{n,K_I}\}$ requested by other WIT users and removing them from the superposition symbol, the WIT user u_k^I then rotates the residual superposition symbol with an angle

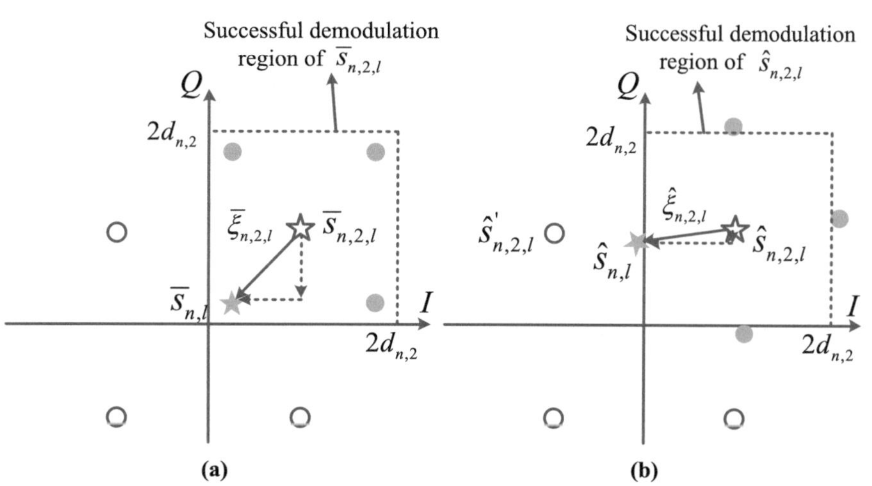

Fig. 3.5 The successful demodulation region of the WIT user u_2^I's symbol: (**a**) without constellation rotation; (**b**) with constellation rotation. The assumptions are in line with those made in Fig.3.4

of $-\theta_{n,k}$ for demodulating its own requested symbol-block $\widehat{\mathbf{S}}_{n,k}$, according to the SIC specified in Sect. 3.2.2.2. As a result, the distortion induced by the l-th symbols belonging to the blocks $\{\widehat{\mathbf{S}}_{n,1}, \cdots, \widehat{\mathbf{S}}_{n,k-1}\}$ on the l-th symbol belonging to the block $\widehat{\mathbf{S}}_{n,k}$ can be formulated as

$$\begin{aligned}\widehat{\xi}_{n,k,l} &= \sum_{i=1}^{k-1} \widehat{s}_{n,i,l} e^{-j\theta_{n,k}} \\ &= \sum_{i=1}^{k-1} d_{n,i} \widehat{A}_{n,i,l} \cos(\widehat{\phi}_{n,i,l} - \theta_{n,k}) + j \sum_{i=1}^{k-1} d_{n,i} \widehat{A}_{n,i,l} \sin(\widehat{\phi}_{n,i,l} - \theta_{n,k}) \\ &\triangleq \widehat{\xi}^{I}_{n,k,l} + j\widehat{\xi}^{Q}_{n,k,l},\end{aligned} \tag{3.18}$$

where $\widehat{\xi}^{I}_{n,k,l}$ represents the in-phase component of the distortion vector $\widehat{\xi}_{n,k,l}$ and $\widehat{\xi}^{Q}_{n,k,l}$ represents the corresponding component in the quadrature phase. In order to reduce the SER of the WIT user u^{I}_{k}, the distortion induced by the symbol superposition should be carefully controlled. Therefore, the residual superposition symbol should be still within the successful demodulation region of the symbol requested by the WIT user u^{I}_{k}, which can be represented by a square having the side length of $2d_{n,k}$, as exemplified in Fig. 3.5. The constraints on the symbol distortion are then formulated as

$$|\widehat{\xi}^{I}_{n,k,l}|, |\widehat{\xi}^{Q}_{n,k,l}| \le d_{n,k}, \tag{3.19}$$

for $l = 1, \cdots, L$, $k = 2, \cdots, K_I$, $n = 1, \cdots, N$, where $|x|$ indicates the absolute value of the complex number x.

In order to satisfy the symbol distortion constraint of (3.19), we opt to control the transmit power $\{P_{n,1}, \cdots, P_{n,K_I}\}$ allocated to the symbols requested by all the K_I WIT users. As a result, the constellation rotation angles may gain the highest degree of freedom for the sake of maximizing the energy carried by the superposition symbol. The following theorem thus provides a feasible power allocation scheme:

Theorem 3.1 *In order to satisfy the symbol distortion constraint of* (3.19)*, the transmit power* $\{P_{n,1}, \cdots, P_{n,K_I}\}$ *of the symbols requested by all the WIT users should satisfy the following inequality:*

$$\sqrt{P_{n,k}} \ge A_{\max} \sum_{i=1}^{k-1} \sqrt{P_{n,i}}, \tag{3.20}$$

for $k = 2, \cdots, K_I$*, where* $A_{\max}$ *represents the normalized maximum amplitude, when the M-QAM is adopted for generating the modulated symbols.*

In order to guarantee the fairness among users but to satisfy the symbol distortion constraint (3.19), we take the equality of (3.20). Therefore, the transmit power of the symbols requested by the WIT user u^{I}_{k} can be expressed as

$$\sqrt{P_{n,k}} = A_{\max} \sum_{i=1}^{k-1} \sqrt{P_{n,i}},\ k = 2, \cdots, K_I. \tag{3.21}$$

Given the total transmit power $P_n = \sum_{k=1}^{K_I} P_{n,k}$ on the n-th subcarrier, the transmit power of the symbols requested by the WIT users can be thus formulated as

$$P_{n,k} = \begin{cases} \left(1 + \sum\limits_{k=2}^{K_I} \left(\frac{A_{\max} - A^{k}_{\max}}{1 - A_{\max}}\right)^2\right)^{-1} P_n, & k = 1 \\ A_{\max} \sum\limits_{i=1}^{k-1} \sqrt{P_{n,i}}, & k = 2, \cdots, K_I \end{cases}. \tag{3.22}$$

3.2.4 *Optimal Design of Energy Interleaver and Constellation Rotator*

3.2.4.1 Problem Formulation

The actual energy $\widehat{E}_{n,l}$ carried by the l-th superposition signal $\widehat{x}_{n,l}(t)$ transmitted to all the WIT users on the n-th subcarrier can be expressed as (3.23).

$$\begin{aligned}\widehat{E}_{n,l} &= Tf_n \int_{-\frac{1}{2f_n}}^{\frac{1}{2f_n}} \left(\widehat{x}_{n,l}(t)\right)^2 dt \\ &= \frac{T}{2}\left(\sum_{k_1=1}^{K_I}\sum_{k_2=1}^{K_I} d_{n,k_1} d_{n,k_2} \widehat{A}_{n,k_1,l} \widehat{A}_{n,k_2,l} \cos\left(\widehat{\phi}_{n,k_1,l} - \widehat{\phi}_{n,k_2,l}\right)\right). \end{aligned} \tag{3.23}$$

Given the specific power allocation scheme (3.22) among the symbols requested by different WIT users, the joint design of the energy interleaver **B** and the constellation rotator $\boldsymbol{\Theta}$ for maximizing the energy carried by the superposition symbols on the multi-carriers can be formulated as

$$\text{(P3.3.1): } \max_{\boldsymbol{\Theta},\mathbf{B}} \quad \widehat{E} = \sum_{n=1}^{N}\sum_{l=1}^{L} \widehat{E}_{n,l}, \tag{3.24}$$

$$\text{s. t.} \quad -\pi \le \theta_{n,k} < \pi, \ 1 \le k \le K_I, \ 1 \le n \le N, \tag{3.24a}$$

$$b_{k,m,n} = 0 \text{ or } 1, \ 1 \le k \le K_I, \ 1 \le n \le N, \ 1 \le m \le N, \tag{3.24b}$$

$$\sum_{m=1}^{N} b_{k,m,n} = 1, \ 1 \le k \le K_I, \ 1 \le n \le N, \tag{3.24c}$$

$$\sum_{n=1}^{N} b_{k,m,n} = 1, \ 1 \le k \le K_I, \ 1 \le m \le N, \tag{3.24d}$$

where (3.24a) and (3.24b) constrain the optional range of **B** and $\boldsymbol{\Theta}$, and (3.24c) and (3.24d) indicate that for a single WIT user u_k^I, only a single symbol-block is allowed to be transmitted on a single subcarrier.

The optimization problem (P3.3.1) is a mixture of integer and real number programming. It is impossible to obtain a closed-form solution. This problem can be addressed by sequentially solving subproblem of the constellation rotator design and that of the energy interleaver design. Note that our design actually maximizes the total energy carried by the multi-carrier signal $\widehat{\mathbf{x}}(t)$ before it goes to the radio front, as illustrated in Fig. 3.1. Therefore, our design may naturally adapt to any arbitrary number of antennas.

3.2.4.2 Constellation Rotator Design

Given a specific energy interleaver **B**, the original optimization problem (P3.3.1) can be simplified as

$$\text{(P3.3.2): } \max_{\boldsymbol{\Theta}} \quad \widehat{E} = \sum_{n=1}^{N}\sum_{l=1}^{L} \widehat{E}_{n,l}, \quad \text{s. t.} \quad \text{(3.24a)}. \tag{3.25}$$

As portrayed in Fig. 3.1, every subcarrier has a constellation rotator. Therefore, the constellation rotator design on a specific subcarrier is independent with others. As a result, the optimization problem (P3.3.2) can be further decomposed into N subproblems in order to obtain the optimal constellation rotation angles for every WIT user on the N subcarriers, respectively.

Algorithm 3.1 Iterative algorithm for obtaining constellation rotation angles

INPUT: User number K_I; Symbol blocks $\overline{\mathbf{S}}_{n,1}, \cdots, \overline{\mathbf{S}}_{n,K_I}$; Convergence accuracy Λ;
OUTPUT: Constellation rotation angles $\theta^*_{n,1}, \cdots, \theta^*_{n,K_I}$.

```
Initialise Iteration counter q ← 1; θ^(0)_{n,1} = ··· = θ^(0)_{n,K_I} ← 0; Randomly choose θ^(1)_{n,1}, ··· , θ^(1)_{n,K_I} within the range of [−π, π); Iteration user k ← 1;
while Not all |θ^(q)_{n,k} − θ^(q−1)_{n,k}| < Λ do
    while k ≤ K_I do
        Obtain θ_{n,k} according to (3.27), θ^(q+1)_{n,k} ← θ_{n,k};
        k ← k + 1;
    end while
    k ← 1, q ← q + 1;
end while
Obtain θ*_{n,k} ← θ^(q)_{n,k} for k = 1, ··· , K_I;
```

Specifically, the optimal constellation rotator design on the n-th subcarrier can be formulated as

$$\text{(P3.3.3):} \quad \max_{\{\theta_{n,1},\cdots,\theta_{n,K_I}\}} \quad \widehat{E}_n = \sum_{l=1}^{L} \widehat{E}_{n,l}, \quad \text{s. t.} \quad (3.24\text{a}). \tag{3.26}$$

Given the symbol-blocks $\{\overline{\mathbf{S}}_{n,1}, \ldots, \overline{\mathbf{S}}_{n,K_I}\}$ transmitted on the n-th subcarrier, the utility function of this subcarrier can be defined as $\eta_n(\overline{\mathbf{S}}_{n,1}, \ldots, \overline{\mathbf{S}}_{n,K_I}) = \widehat{E}_{n,\max}$, where $\widehat{E}_{n,\max}$ is the resultant maximum by solving the problem (P3.3.3).

Observe from (P3.3.3) and (3.23) that the non-convex cosine function exists in the objective of (3.26). As a result, the optimization problem (P3.3.3) is not convex, and we cannot solve it by exploiting the convex optimization. However, we may obtain its sub-optimal solution by exploiting the method of alternating optimization [36].

The iterative algorithm for solving (P3.3.3) has the following steps:

- *Step 1*: Randomly initialize the constellation rotation angles $\{\theta_{n,1}, \cdots, \theta_{n,K_I}\}$ according to the constraint (3.24a).
- *Step 2*: Optimize the constellation rotation angle $\theta_{n,k}$ for the symbols requested by the WIT user u_k^I for $k = 1, \cdots, K_I$ by solving (P3.3.3), when the other angles $\{\theta_{n,1}, \cdots, \theta_{n,K_I}\}/\{\theta_{n,k}\}$ are regarded as constants. According to the trigonometry, the optimal rotation angle $\theta_{n,k}$ is derived as

$$\theta_{n,k} = -\arctan\left(\frac{\sum\limits_{l=1}^{L}\sum\limits_{i=1,i\neq k}^{K_I} \alpha_{n,i,k,l}\sin(\beta_{n,i,k,l})}{\sum\limits_{l=1}^{L}\sum\limits_{i=1,i\neq k}^{K_I} \alpha_{n,i,k,l}\cos(\beta_{n,i,k,l})}\right), \tag{3.27}$$

  where $\alpha_{n,i,k,l} = d_{n,k}d_{n,i}\overline{A}_{n,k,l}\overline{A}_{n,i,l}$, and $\beta_{n,i,k,l} = \overline{\phi}_{n,k,l} - \overline{\phi}_{n,i,l} - \theta_{n,i}$.
- *Step 3*: Sequentially optimize the rotation angles for all the WIT users by using the same method of *Step 2*.
- *Step 4*: Repeat Steps 3 and 4 until the constellation rotation angles converge to their optimums $\{\theta^*_{n,1}, \cdots, \theta^*_{n,K_I}\}$.

The pseudocode of the iterative algorithm is then provided in Algorithm 3.1.

In Step 2 of every iteration, we may obtain the optimal solution of $\theta_{n,k}$, although the problem is non-convex. We let $E_n(\theta^{(q)}_{n,k})$ represent the energy carried by the superposition symbol on the n-th subcarrier, when $\theta^{(q)}_{n,k}$ is updated at the q-th iteration according to Algorithm 3.1. The following inequalities can be readily obtained:

$$\begin{aligned} E_n\left(\theta^{(q)}_{n,k}\right) &\leq E_n\left(\theta^{(q)}_{n,k+1}\right) \leq \cdots \leq E_n\left(\theta^{(q)}_{n,K_I}\right) \\ &\leq E_n\left(\theta^{(q+1)}_{n,1}\right) \leq \cdots \leq E_n\left(\theta^{(q+1)}_{n,k}\right). \end{aligned} \tag{3.28}$$

This is because in every round of optimization, the objective of (P3.3.3) is a nondecreasing function. Furthermore, according to (3.23), we also have

$$E_n\left(\theta_{n,k}^{(q)}\right) \leq \sum_{l=1}^{L} \frac{T}{2}\left(\sum_{k_1=1}^{K_I}\sum_{k_2=1}^{K_I} d_{n,k_1} d_{n,k_2} \overline{A}_{n,k_1,l} \overline{A}_{n,k_2,l}\right),$$
$$\forall 1 \leq k \leq K_I, q = 1, 2, \cdots, \quad (3.29)$$

which indicates that $\{E_n(\theta_{n,k}^{(q)})\}$ is an increasing function with respect to the iteration round q, while $\{E_n(\theta_{n,k}^{(q)})\}$ is also upper-bounded. Therefore, Algorithm 3.1 finally converges as the iteration round q increases. When the algorithm converges, it returns us the sub-optimal solution.

3.2.4.3 Energy Interleaver Design

The utility of a specific symbol-blocks' transmission arrangement $\{\overline{\mathbf{S}}_{n,1}, \ldots, \overline{\mathbf{S}}_{n,K_I}\}$ on the n-th subcarrier can be characterized by the function of $\eta_n(\overline{\mathbf{S}}_{n,1}, \ldots, \overline{\mathbf{S}}_{n,K_I}) = \widehat{E}_{n,\max}$, which is obtained by solving problem (P3.3.3). Since the transmission arrangement of the symbol-blocks on all the N subcarriers is determined by the energy interleaving tensor $\mathbf{B}$, the utility function of the n-th subcarrier can also be expressed as $\eta_n(\mathbf{B})$. Furthermore, the optimal constellation rotation angles $\boldsymbol{\Theta}$ are also functions of the energy interleaving tensor $\mathbf{B}$. Therefore, the original optimization problem (P3.3.1) can be simplified as

$$\text{(P3.3.4):} \max_{\mathbf{B}} \ \widehat{E} = \sum_{n=1}^{N} \eta_n(\mathbf{B}), \quad \text{s.t.} \quad (3.24\text{b}), (3.24\text{c}), (3.24\text{d}), \quad (3.30)$$

in which the energy interleaver tensor $\mathbf{B}$ is the only parameter needs to be optimized for maximizing the total energy transferred by the transmitter.

Since $\mathbf{B}$ is a binary tensor having all its elements either 0 or 1, the optimization problem (P3.3.4) is an integer programming problem. The exhaustive searching can be relied upon for the sake of obtaining the optimal solution.

A specific WIT user u_k^I requests N symbol-blocks on the N subcarriers in total. A single symbol-block of this WIT user is only allowed to be transmitted on a single subcarrier. Therefore, we have $N!$ schemes for mapping the symbol-blocks of the WIT user u_k^I to the subcarriers. We denote $\mathbb{B} = \{\mathcal{B}_k\}$ having the size of $K_I \times 1$ as the feasible searching space for the energy interleaver tensor $\mathbf{B}$, where the element $\mathcal{B}_k \in \{1, \cdots, N!\}$ represents a symbol-block-to-subcarrier mapping scheme for the WIT user u_k^I. We then traverse all the feasible mapping schemes for every WIT user and choose the optimal one for maximizing the objective of (P3.3.4). The complexity of the exhaustive searching is as high as $\mathcal{O}((N!)^{K_I})$. Although K_I may not be very large in a power-domain NOMA systems, the complexity will be substantially increased, if we have more subcarriers.

In order to reduce the complexity of the exhaustive searching, another greedy algorithm is proposed by sacrificing some optimality of the solution. The main steps of the greedy algorithm are summarized as below:

- *Step 1*: Initialize the optimal utility function of every subcarrier as $\eta_n^* = 0$ for $n = 1, \cdots, N$. A binary matrix $\mathbf{V} = \{v_{k,n}\}$ having the size of $K_I \times N$ is defined. Its element $v_{k,n} = 1$ indicates that the n-th symbol-block of u_k^I has already been scheduled on a specific subcarrier, but $v_{k,n} = 0$ represents that the n-th symbol-block has not been scheduled on any subcarrier. All the elements in $\mathbf{V}$ are initialized as $v_{k,n} = 0$. Only the symbol-blocks having the indicators of $\{v_{k,n} = 0 | k = 1, \cdots, K_I, n = 1, \cdots, N\}$ can be scheduled on the free subcarriers. With the aid of $\mathbf{V}$, it can be thus guaranteed that each symbol-block requested by u_k^I is scheduled on a single subcarrier.
- *Step 2*: When we start to schedule the symbol-blocks of K_I WIT users on the n-th subcarrier, the first $(n-1)$ subcarriers have already accepted the transmissions of K_I symbol-blocks for each. Accordingly, the WIT user u_k^I only has $N-(n-1)$ symbol-blocks to be scheduled, whose binary indicators in the matrix $\mathbf{V}$ are all zeros. Hence, we have $(N-n+1)^{K_I}$ possible schemes in total for scheduling the rest of symbol-blocks of all the WIT users on the n-th subcarrier. We denote all these scheduling schemes as a set $\mathbb{G} = \{\mathcal{G}_p | p = 1, \cdots, (N-n+1)^{K_I}\}$.
- *Step 3*: We then traverse all the scheduling schemes in the set $\mathbb{G}$ and find the optimal one for maximizing the objective of (P4). When a specific scheduling scheme $\mathcal{G}_p \in \mathbb{G}$ is chosen, the symbol-blocks scheduled on the n-th subcarrier are thus determined, which can be denoted as $\{\overline{\mathbf{S}}_{n,1}, \cdots, \overline{\mathbf{S}}_{n,K_I}\}$. Therefore, we are capable of calculating the utility function $\eta_n(\overline{\mathbf{S}}_{n,1}, \cdots, \overline{\mathbf{S}}_{n,K_I})$ of the n-th subcarrier. Finally, the optimal scheduling scheme on the n-th subcarrier can be found for the sake of maximizing its utility function. In order to prevent these symbol-blocks from being scheduled again, we update the corresponding indicators in the matrix $\mathbf{V}$.

Algorithm 3.2 Greedy algorithm

INPUT: User number K_I; Sub-carrier number N; Symbol blocks $\mathbf{S}_{1,1}, \cdots, \mathbf{S}_{N,K_I}$;
OUTPUT: Optimal energy interleaving tensor $\mathbf{B}^*$.

Initialise Optimal utility functions $\eta_n^* \leftarrow 0 (1 \le n \le N)$; Sub-carrier label $n \leftarrow 1$; Symbol block chosen matrix $\mathbf{V} = \{v_{k,n}\} \leftarrow \mathbf{0}$ and $\mathbf{V} \in \mathcal{Z}^{K_I \times N}$;
while $n \le N$ **do**
  Initialise a set $\mathbb{G}$ consists of $(N-n+1)^{K_I}$ different combinations $\mathcal{G}_p (1 \le p \le (N-n+1)^{K_I})$ for K_I users, while each combination consists of K_I different symbol-blocks $\mathbf{S}_{n_k,k} (1 \le k \le K_I,\ 1 \le n_k \le N)$ chosen from K_I different users satisfying $v_{k,n_k} = 0$; Combination chosen label $p \leftarrow 1$; Optimal combination $\mathcal{G}^*$;
  while $p \le (N-n+1)^{K_I}$ **do**
    Choose the combination $\mathcal{G}_p$;
    Obtain $\mathbf{S}_{n_k,k}$ for each u_k^I according to $\mathcal{G}_p$, let $\overline{\mathbf{S}}_{n,k} \leftarrow \mathbf{S}_{n_k,k}$;
    Obtain constellation rotation angles $\{\theta_{n,k}(1 \le k \le K_I)\}$ according to Algorithm 3.1;
    Obtain utility function $\eta_n(\overline{\mathbf{S}}_{n,1}, \cdots, \overline{\mathbf{S}}_{n,K_I})$ by solving P3;
    If $\eta_n(\overline{\mathbf{S}}_{n,1}, \cdots, \overline{\mathbf{S}}_{n,K_I}) > \eta_n^*$, then $\eta_n^* \leftarrow \eta_n(\overline{\mathbf{S}}_{n,1}, \cdots, \overline{\mathbf{S}}_{n,K_I})$ and $\mathcal{G}^* \leftarrow \mathcal{G}_p$; $n_k^* \leftarrow n_k (1 \le k \le K_I)$;
    $p \leftarrow p+1$;
  end while
  $\{b_{k,m,n}^* \leftarrow 1 | m = n_k^*, k = 1, \cdots, K_I\}$ and $\{b_{k,m,n}^* \leftarrow 0 | m \ne n_k^*, k = 1, \cdots, K_I\}$;
  $v_{k,n_k^*} \leftarrow 1$; $n \leftarrow n+1$;
end while
Obtain the optimal energy interleaving tensor $\mathbf{B}^*$.

The pseudocode of the greedy algorithm is provided in Algorithm 3.2. The proposed greedy algorithm concludes after N iterations. Its complexity is $\mathcal{O}(\sum_{i=1}^{N} i^{K_I})$, which is substantially lower than $\mathcal{O}((N!)^{K_I})$ of the exhaustive searching.

3.2.5 *Simulation Results*

Without specific statement, the parameters in our simulation are set as follows: We have $K_I = 3$ WIT users and $K_E = 1$ WPT user in the NOMA-SWIPT system. The classic additive white Gaussian noise (AWGN) channels are conceived between the transmitter and the WIT/WPT users, while the noise power is set to be $\sigma^2 = -80$ dBm. The channel power gain between the WPT user and the H-AP is -30 dB, while those between the WIT users and the H-AP are $\{-53, -60, -70\}$ dB, respectively. The transmitter has 10^6 modulated symbols to be transmitted to every WIT user, which are randomly generated by the QAM based modulator. The symbol duration is set to be $T = 10^{-6}$ s. There are 10 subcarriers in the simulation. The total transmit power on every subcarrier is all $P_n = 1$ W.

3.2.5.1 Convergence and Validity

We firstly investigate the convergence of Algorithm 3.1 in Fig. 3.6, where we maximize the energy carried by a specific subcarrier by solving the optimization problem (P3.3.3). Observe from Fig. 3.6 that Algorithm 3.1 converges within 5 iterations at most, when the 4-QAM and 16-QAM are adopted. Furthermore, we compare the WPT performance of the greedy algorithm and the exhaustive searching-based algorithm for solving the problem (P3.3.4) in Fig. 3.7, where we consider $N = 4$ subcarriers in total. Observe from Fig. 3.7 that the WPT performance of the greedy algorithm is almost the same as that of the exhaustive searching. Therefore, in the rest of simulation, the greedy algorithm is adopted for solving the optimization problem (P3.3.1), since its complexity is far lower than the exhaustive searching-based counterpart.

3.2.5.2 WPT Performance

We evaluate the total energy harvested by the WPT user in Figs. 3.8 and 3.9 by invoking the following different interleaving and modulation schemes:

- *Conventional scheme*: Neither energy interleaving nor constellation rotation is invoked.
- *Energy interleaving only scheme*: Constellation rotation is not invoked.

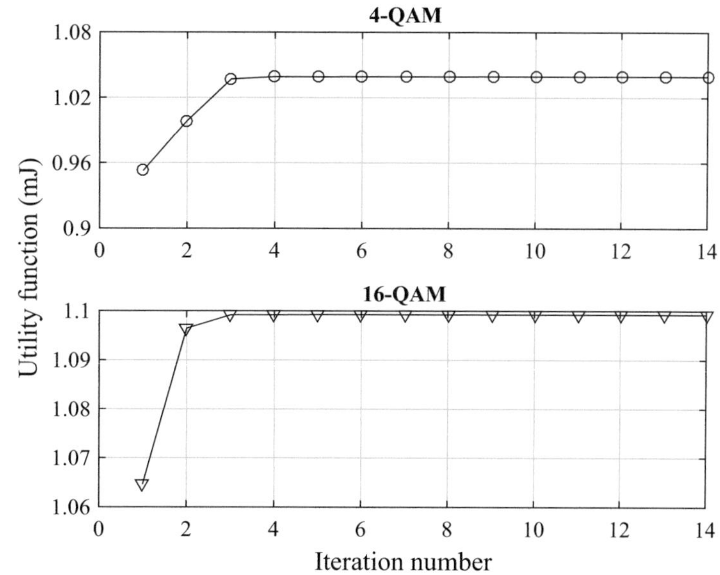

Fig. 3.6 Convergence of Algorithm 3.1

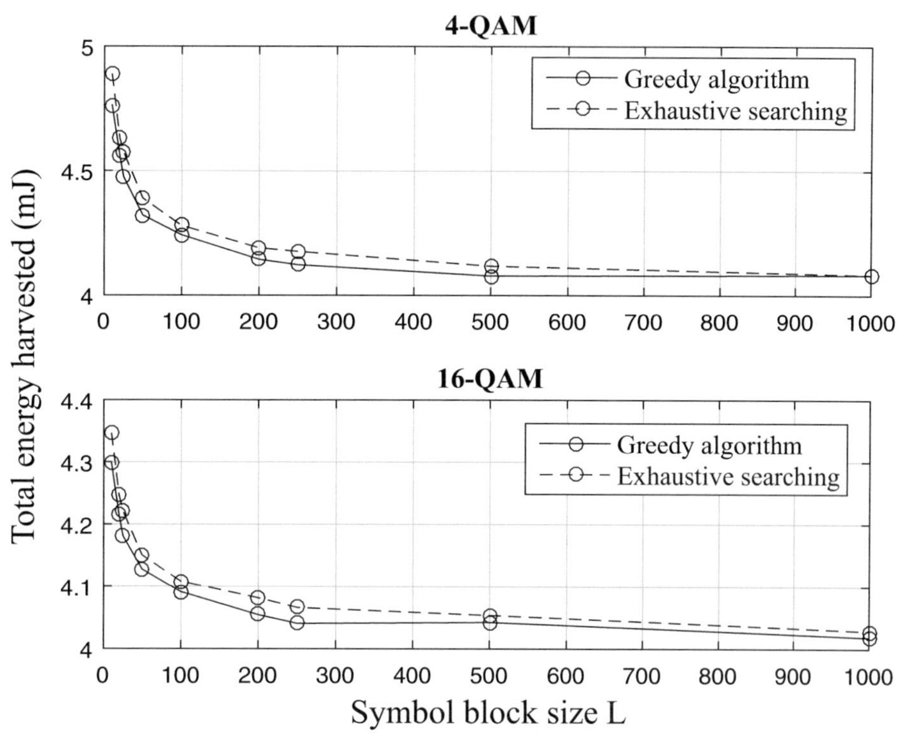

Fig. 3.7 Validity of Algorithm 3.2

- *Constellation rotation only scheme*: Energy interleaving is not invoked.
- *Joint scheme*: Both interleaving and constellation rotation are invoked.

Observe from Figs. 3.8 and 3.9 that the *joint scheme* achieves the highest WPT performance for both the 4-QAM and 16-QAM. Moreover, the *energy interleaving only scheme* and the *constellation rotation only scheme* achieve the moderate WPT performance, which is still much higher than the *conventional scheme*. Furthermore, the WPT performance of the *conventional scheme* keeps constant, but that of all the other three schemes all reduces, as the symbol-block size L increases. This is because a lower value of L represents that the transmitter may update the energy interleaver and the constellation

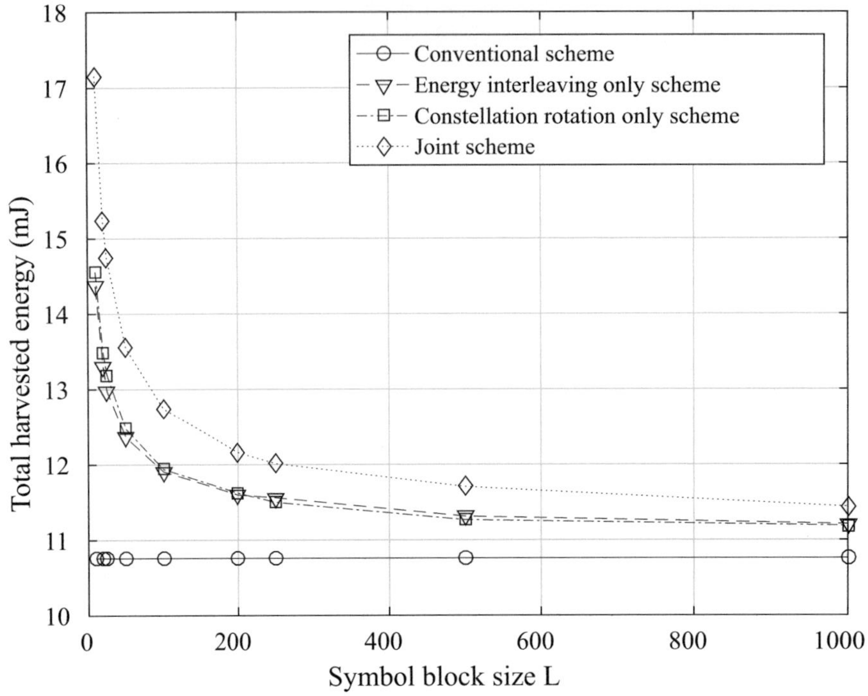

Fig. 3.8 WPT performance with 4-QAM modulation

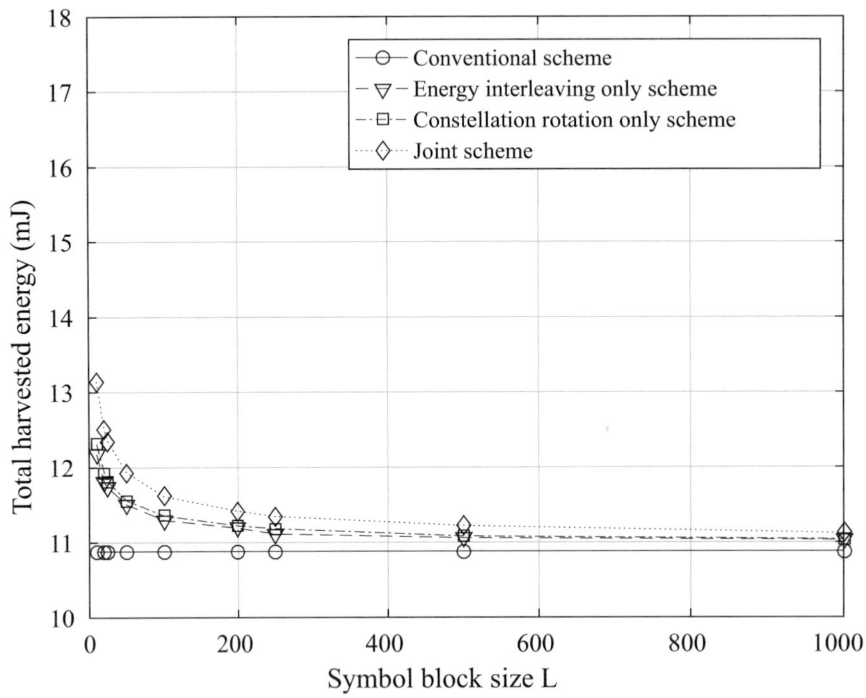

Fig. 3.9 WPT performance with 16-QAM modulation

rotator more frequently, which results in more constructive symbol superposition. However, when we update the energy interleaver and the constellation rotator more frequently, more control signalling has to be transmitted to the WIT receivers, since the knowledge of the energy interleaver and the constellation rotators is indispensable for the demodulation and deinterleaving. Therefore, it is important to design an appropriate symbol-block size L for the sake of compromising between

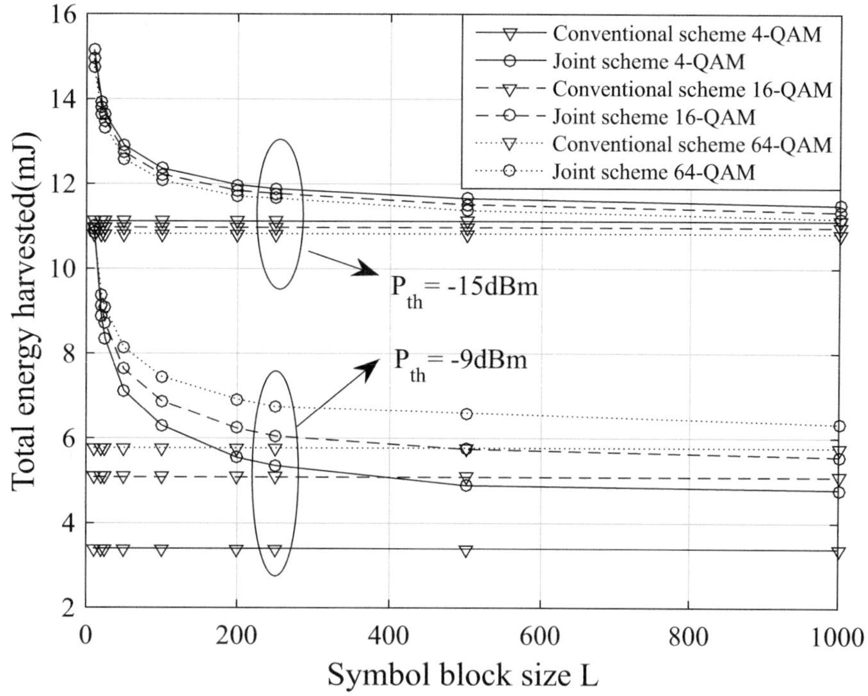

Fig. 3.10 Comparison among modulation schemes with different P_{th}

the WPT performance and the control signalling overhead. By contrast, the WPT performance of the *conventional scheme* is not affected by the symbol-block size L.

Moreover, observe from Figs. 3.8 and 3.9 that the 4-QAM based modulator achieves a higher WPT performance than the 16-QAM based modulators. According to the power allocation scheme of (3.21), when the higher order modulation is adopted, A_{max} also becomes higher, which indicates that the power difference among the symbols requested by different WIT users is enlarged. As a result, when we have a fixed total transmit power, the WPT performance gain incurred by the energy interleaving and the constellation rotation becomes lower, since the energy carried by a symbol having a very high power cannot be substantially degraded by a symbol having a very low power after their superposition. Under this consideration, if we want to achieve a higher energy harvesting gain, a lower rate is a better choice.

3.2.5.3 Sensitivity of Energy Harvester

We investigate the impact of energy harvester's sensitivity P_{th} in Fig. 3.10. The power allocation scheme is derived by substituting $A_{max} = \sqrt{98}$ of the 64-QAM into (3.21). The same power allocation scheme is also invoked for both the 4-QAM and 16-QAM in order to minimize the effect of the power difference on the attainable WPT performance. Observe from Fig. 3.10 that when the energy harvester's sensitivity is low, say $P_{th} = -15$ dBm, the 4-QAM achieves the highest WPT performance, while the 64-QAM achieves the lowest WPT performance. When P_{th} increases to -9 dBm, we observe an inverse trend from Fig. 3.10. This is because a modulation scheme of high order may generate a lot of symbols carrying lower power than the rectifier's sensitivity. The energy carried by these symbols cannot be harvested by the WPT user. By contrast, a modulation scheme of lower order generates symbols carrying almost identical power. If the sensitivity is low, all the symbols can be relied upon for energy harvesting. However, when the sensitivity is high, none of the symbols generated by a low-order modulation scheme is capable of delivering energy to the WPT users. By contrast, since a high-order modulation scheme may generate some symbols carrying very high power, these symbols can still deliver some energy to the WPT users.

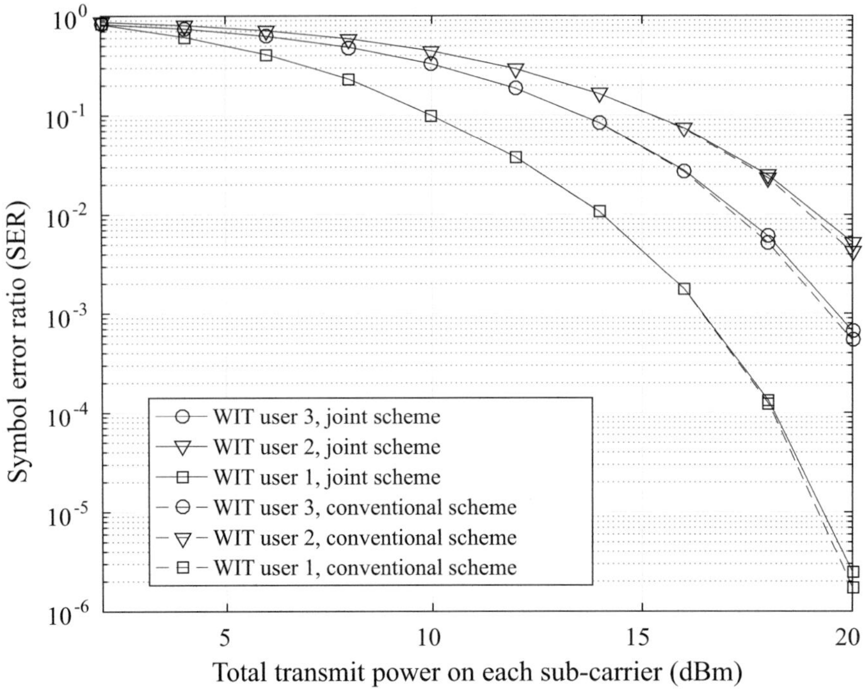

Fig. 3.11 WIT performance with 4-QAM modulation

3.2.5.4 WIT Performance

Finally, we plot the SER performance of the joint and conventional schemes versus the total transmit power on each subcarrier in Figs. 3.11 and 3.12, when 4-QAM and 16-QAM are adopted, respectively. In this simulation, the transmitter has 10^8 modulated symbols to be transmitted to each WIT user in order to achieve more accurate results. Observe from Figs. 3.11 and 3.12 that the joint scheme has almost the same SER performance as the conventional scheme, when the 4-QAM is adopted. When the 16-QAM is adopted, the joint scheme only suffers from a tiny degradation of the SER performance, when compared to the conventional counterpart.

We also compared the SER performance in Fig. 3.13 between our scheme and another fair scheme proposed in [37], where the power allocation scheme is designed by letting the signal to interference and noise ratio (SINR) of all the WIT users be the same. Observe from Fig. 3.13 that when 16-QAM is adopted, for WIT user 3 having the worst channel condition, our scheme achieves a higher SER than the counterpart. However, for WIT user 1 having the best channel condition, our power allocation scheme achieves a much lower SER than the fair scheme. When 64-QAM is adopted, our scheme achieves lower SER differences among WIT users than the counterpart. Moreover, observe from Fig. 3.13c that our power allocation scheme achieves lower average SER among all the WIT users than the fair scheme. In a nutshell, our power allocation scheme outperforms the benchmark of [37] in terms of the SER fairness.

3.2.5.5 Spectrum Efficiency

Let us now discuss the spectrum efficiency of our NOMA-SWIPT system. As for the WIT user u_k^I, the total achievable throughput is formulated as

$$R_k = \sum_{n=1}^{N} B \log\left(1 + \frac{h_{n,k} P_{n,k}}{\sum_{i=1}^{k-1} h_{n,i} P_{n,i} + \sigma^2}\right),$$

where B is the bandwidth of a single subcarrier. Therefore, the spectrum efficiency of the system is derived as

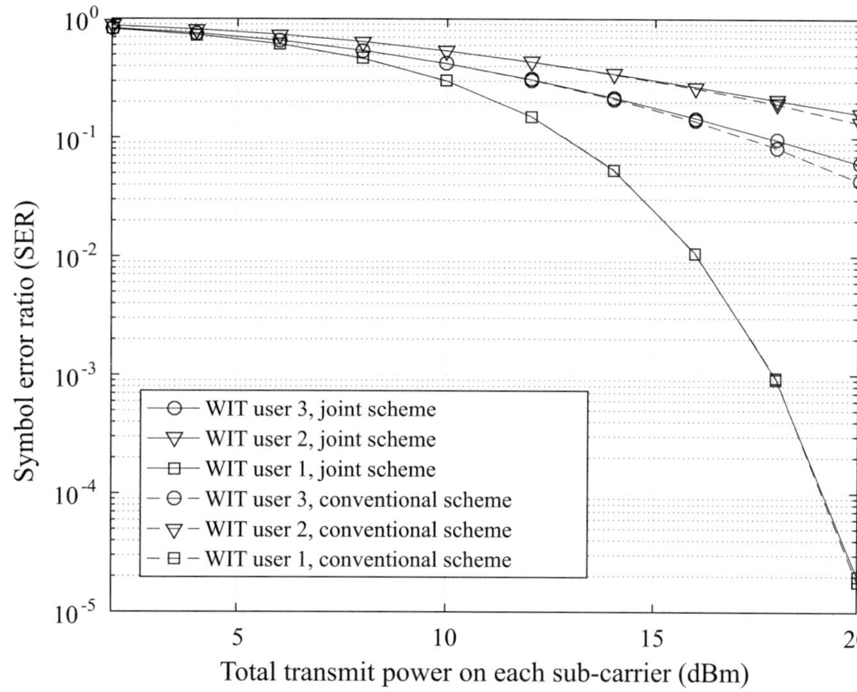

Fig. 3.12 WIT performance with 16-QAM modulation

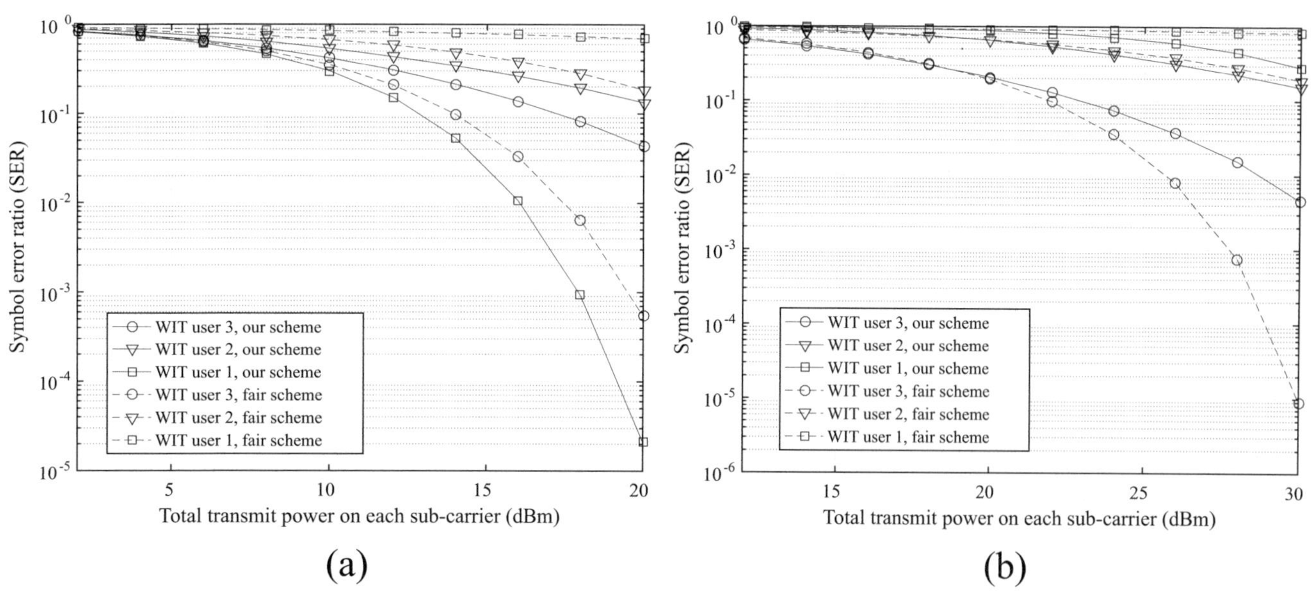

Fig. 3.13 WIT performance with different power allocation schemes. (**a**) 16-QAM. (**b**) 64-QAM

$$\kappa = \frac{\sum_{k=1}^{K_I} R_k}{NB}.$$

In our simulation, the spectrum efficiency is thus calculated as 7.47 bps/Hz, which is around 17% higher than 6.36 bps/Hz of the OFDMA-based counterpart. Meanwhile, observe from Figs. 3.8 and 3.9 that if the symbol-block size is set to be $L = 100$, our joint design is capable of achieving 18% and 6% higher WPT performance than the conventional scheme, when 4-QAM and 16-QAM are adopted, respectively.

3.3 Multi-sinusoidal and OTFS Waveform for Integrated Data and Energy Transfer

3.3.1 Multi-sinusoidal Waveform for Integrated Data and Energy Transfer

Integrated data and energy transfer (IDET) is capable of simultaneously delivering on-demand data and energy to low-power Internet of Everything (IoE) devices. We propose a multi-carrier IDET transceiver relying on superposition waveforms consisting of multi-sinusoidal signals for wireless energy transfer (WET) and orthogonal-frequency-division-multiplexing (OFDM) signals for wireless data transfer (WDT). The outdated channel state information (CSI) in aging channels is employed by the transmitter to shape IDET waveforms. With the constraints of transmission power and WDT requirement, the amplitudes and phases of the IDET waveform at the transmitter and the power splitter at the receiver are jointly optimized for maximizing the average direct current (DC) among a limited number of transmission frames with the existence of carrier frequency offset (CFO). For the amplitude optimization, the original non-convex problem can be transformed into a reversed geometric programming problem, and then it can be effectively solved with existing tools. As for the phase optimization, the artificial bee colony (ABC) algorithm is invoked in order to deal with the non-convexity. Iteration between the amplitude optimization and phase optimization yields our joint design. Numerical results demonstrate the advantage of our joint design for the IDET waveform shaping with the existence of the CFO and the outdated CSI.

3.3.1.1 Introduction

In the era of Internet of Everything (IoE), IoE devices are massively deployed for environment monitoring and data processing. Due to frequent operations, the batteries of the IoE devices can be drained within a short time. Harvesting energy from radio frequency (RF) signals has been envisioned as an effective way for energy supply [38]. Relying on RF signals, integrated data and energy transfer (IDET) technique can provide flexible wireless energy transfer (WET) services, while satisfying the quality of service (QoS) of wireless data transfer (WDT) [10]. However, traditional WDT waveforms have diverse amplitudes for modulating information, which results in unstable WET performance. Therefore, novel waveforms for conveying IDET services should be carefully investigated.

By exploiting multiple transmit antennas, IDET waveforms can be designed in the spatial domain for improving the performance [39–42]. Specifically, Yue et al. [39] designed a hybrid multicast beam with/without perfect channel state information (CSI) for IDET towards multiple data and energy users. Moreover, a robust beamforming design with channel estimation errors was studied in order to minimize the transmit power of a base station delivering IDET services, while satisfying requirements of both data and energy users [40]. In [41], Zhao et al. proposed three transmission schemes based on receive spatial modulation aided IDET system with finite alphabet. Furthermore, Pan et al. [42] firstly introduced an intelligent reflecting surface (IRS) to assist IDET, since incident waveforms towards the IRS can be adjusted to benefit from additional spatial gains.

Coding and modulation can be relied upon for adjusting waveforms for IDET [22, 43–46]. The first design guide for a coding and modulation controlled IDET system was proposed in [22]. Specifically, Zhao et al. [43] proposed a unary coding-based IDET encoder by considering a practical M-QAM. In [44], the codeword distribution in a unary encoder was designed to maximize the WET performance with the required WDT constraint. Moreover, according to [22], different modulation schemes have diverse IDET performance. Therefore, an adaptive modulation-based IDET transceiver was firstly investigated in [45], which significantly increased the attainable rate-energy region. In [46], the non-orthogonal multiple access (NOMA) technique was adopted by a transmitter by superimposing the modulated symbols destined to multiple WDT requesters in the power-domain. Then an energy interleaver and a constellation rotation based modulator were jointly designed for maximizing the attainable WET performance without any degradation of the WDT performance.

Since multi-sinusoidal (multi-tone) waveforms were widely regarded as an efficient way for WET [47], some of the existing works tried to modulate information by exploiting characteristics of these waveforms in order to deliver IDET services [48–50]. Specifically, Kim et al. [48] originally proposed a scheme of modulating information by exploiting different peak-to-average-power ratio (PAPR) of multi-sinusoidal waveforms for delivering IDET services. Krikidis et al. [49] then proposed a tone index of multi-sinusoidal waveforms based modulation (TIM) scheme for IDET. Furthermore, a multi-tone amplitude modulation (MAM) scheme modulates information both on the number of sinusoidal waveforms with different frequencies and on their amplitudes [50].

Furthermore, an orthogonal-frequency-division-multiplexing (OFDM) waveform for WDT is a special type of multi-sinusoidal waveforms with varying amplitudes and phases in its components. Therefore, careful designs on multi-sinusoidal waveforms for IDET have attracted a lot of attentions [51–55]. Specifically, Buckley [51] superimposed the unmodulated multi-carrier signal onto the OFDM signal to improve the WET performance. By varying amplitudes and phases of the superimposed IDET waveform, the WET performance was maximized while satisfying the WDT performance constraint. Moreover, in [52], Kim proposed a realistic IDET system and demonstrated the benefit of jointly designing the OFDM waveform and the multi-sinusoidal waveforms. Buckley et al. [53] analyzed the structure of OFDM waveform, and they exploited cyclic prefix for dedicated WET, which effectively improved the overall IDET performance. In [54], a DC filter is invoked at the receiver to filter the deterministic WET signals from the superposition signals so as to minimize the intra–subcarrier interference. In [55], a joint subcarrier allocation and amplitude adjustment problem was formulated to minimize the outage of WDT in an OFDM assisted IDET system.

However, the drawbacks of above-mentioned works [51–55] can be summarized as follows:

- Perfect carrier frequency synchronization is assumed for IDET waveform design. However, it is not practical in low-power IoE devices with limited signal processing capabilities. The carrier frequency offset (CFO) caused by either the oscillator mismatch between the transmitter and the receiver or the Doppler shift can severely degrade the system performance, since it sabotages the orthogonality among subcarrier. It is common to consider CFO for performance analysis in an OFDM-based system [56].
- Their waveform design requires perfect CSI acquisition in every transmission frame. As a result, numerous energy and communication resources are consumed for CSI acquisition. It is also impractical for an IoE device with simple hardware and energy shortage. Therefore, employing the outdated CSI to shape the IDET signals is of great significance. Aging channels have been widely used for modeling temporal correlation [57–59]. However, they were never considered in the IDET waveform design.

Against this background, our novel contributions are summarized as follows:

- A practical IDET transceiver is proposed. With the aid of inverse fast Fourier transformation (IFFT), the IDET transmitter is capable of generating superposition IDET waveforms with arbitrary amplitudes and phases in the digital domain. The IDET receiver can also adjust its power splitter to balance the WDT and WET performance.
- The carrier asyncrhonization, due to the hardware limitations and the movements, is assumed between the IDET transmitter and receiver. The resultant CFO induces the inter–subcarrier interference, which degrades the IDET performance. Our waveform design is capable of overcoming the adverse effect of the CFO.
- By exploiting the temporal correlation of aging channels, our waveform design only requires the CSI of the first transmission frame within a block to achieve the maximum IDET performance. This significantly reduces the burden of CSI acquisition on the IDET system.
- The average direct current harvested by the IDET receiver among all the transmission frames within a block is maximized by jointly designing the amplitudes and phases for every component of multi-sinusoidal WET waveforms and for every subcarrier of OFDM WDT waveforms. Moreover, the power splitter at the IDET receiver is also optimally designed.
- An iterative algorithm is developed by decomposing the original problem into a subproblem of amplitude optimization and another of phase optimization. Specifically, the amplitude optimization is obtained by exploiting a reversed geometric programming method, while the phase optimization is obtained by an artificial bee colony (ABC) method.

Notation: a is a scalar; **a** is a vector; $||\cdot||_2^2$ refers to the norm of a vector; the operation of $\mathbb{E}$ takes the average of the input random variable.

3.3.1.2 System Model

As portrayed in Fig. 3.14, an IDET transceiver relies upon a superposition waveform constituted by a multi-sinusoidal waveform for WET and an OFDM waveform for WDT. Let us assume that these two kinds of waveforms have N orthogonal components (subcarriers) with frequencies of $\{f_n|n = 1, \cdots, N\}$. We assume that every subcarrier occupies the same bandwidth B and that adjacent subcarriers have a frequency space of Δf. Therefore, the n-th subcarrier has a frequency of $f_n = f_1 + (n-1)\Delta f$. Moreover, the central frequency of the entire IDET waveform is $f_c = (f_1 + f_N)/2$.

The top box of Fig. 3.14 illustrates an IDET transmitter. We define a transmission block consisting of L transmission frames in total. In the l-th transmission frame, the transmit IDET signal $x(t, l)$ consists of the OFDM-based WDT signal $x_I(t, l)$ that carries modulated data symbols $\{x_n(l)|n = 1, \cdots, N\}$ and multi-sinusoidal WET signal $x_E(t, l)$. The n-th symbol can be further expressed as $x_n(l) = s_n(l)e^{j\phi_n(l)}$, which modulates information on its amplitude $s_n(l)$ and phase $\phi_n(l)$. Furthermore, the modulated symbol $x_n(l)$ on the n-th subcarrier obeys a complex Gaussian distribution with zero mean and a unity variance. In order to counteract the adverse channel fading, the phase and amplitude controller shapes the original modulated symbols in both amplitude- and phase-domains. The resultant shaped symbol on the n-th subcarrier is

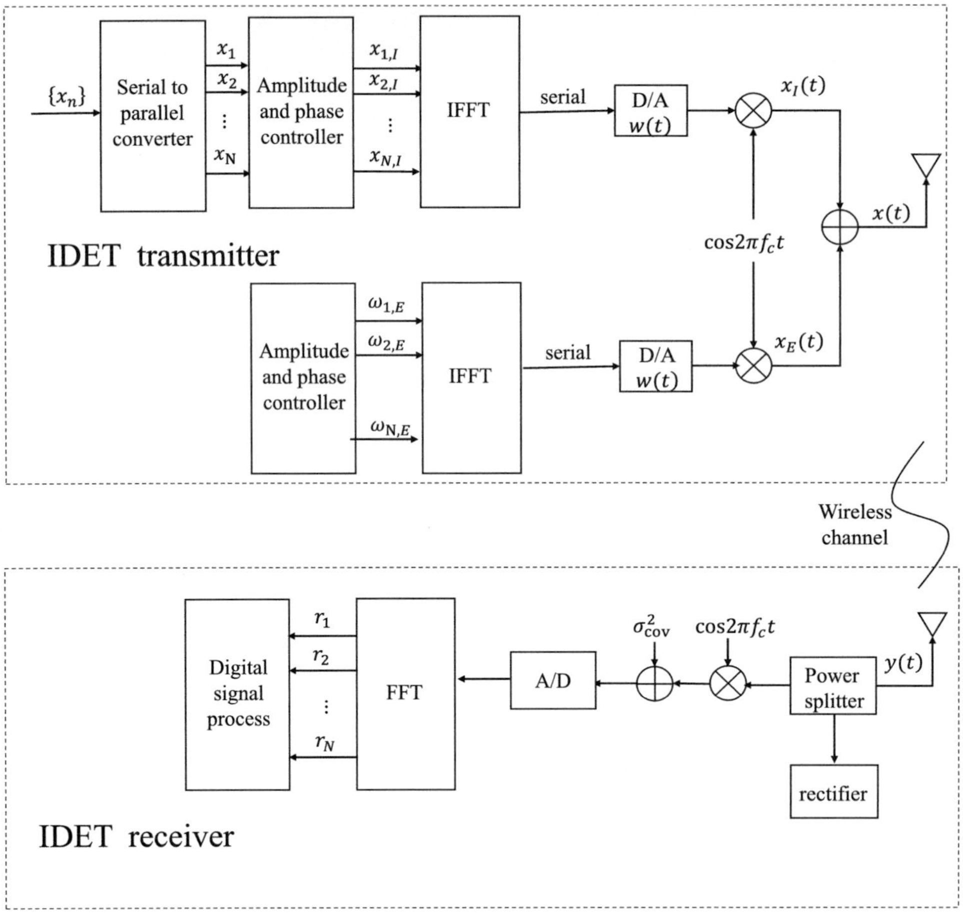

Fig. 3.14 System transmitter and receiver

then expressed as $x_{n,I}(l) = x_n(l)\omega_{n,I}(l) = x_n(l)s_{n,I}(l)e^{j\phi_{n,I}(l)}$, where the complex $\omega_{n,I}(l) = s_{n,I}(l)e^{j\phi_{n,I}(l)}$ represents how we shape the symbol $x_n(l)$ on the n-th subcarrier. By converting the baseband WDT signal into the passband with a central frequency of f_c, the transmit WDT signal during the l-th transmission frame is expressed as

$$x_I(t,l) = \mathrm{Re}\left[\sum_{n=1}^{N} x_n(l)\omega_{n,I}(l)w(t,l)e^{j2\pi f_n t}\right]. \tag{3.31}$$

Note that the transmit rectangular window function $w(t,l)$ can be expressed as

$$w(t,l) = \begin{cases} 1/T, & (l-1)T \le t \le lT, \\ 0, & \text{others}, \end{cases} \tag{3.32}$$

where T is a symbol duration.

Similarly, as shown in the top box of Fig. 3.14, the multi-sinusoidal WET waveforms are also generated by a phase–amplitude controller and subsequent IFFT. We denote baseband WET signals in the l-th transmission frame as $\{\omega_{n,E}(l) = s_{n,E}(l)e^{j\phi_{n,E}(l)} | \forall n = 1, \cdots, N\}$, where the complex $\omega_{n,E}(l) = s_{n,E}(l)e^{j\phi_{n,E}(l)}$ represents how the n-th component (subcarrier) of the multi-sinusoidal waveform is adjusted. Finally, the baseband WET signal is then converted into the passband with a central frequency of f_c. The transmit WET signal during the l-th transmission frame can be finally expressed as

$$x_E(t,l) = \text{Re}\left[\sum_{n=1}^{N} \omega_{n,E}(l) w(t,l) e^{j2\pi f_n t}\right]. \tag{3.33}$$

The passband WET signal $x_E(t,l)$ and WDT signal $x_I(t,l)$ are superimposed in the time-domain, which yields the IDET signal $x(t,l)$ expressed as

$$\begin{aligned} x(t,l) &= x_I(t,l) + x_E(t,l) \\ &= \text{Re}\left[\sum_{n=1}^{N} \left(s_n(l) s_{n,I}(l) e^{j(\phi_n(l)+\phi_{n,I}(l))} + s_{n,E}(l) e^{j(\phi_{n,E}(l))}\right) w(t,l) e^{j2\pi f_n t}\right]. \end{aligned} \tag{3.34}$$

Therefore, a waveform shaper for L transmission frames can be expressed as $\{\boldsymbol{s}_E, \boldsymbol{s}_I, \boldsymbol{\phi}_I, \boldsymbol{\phi}_E\}$, where all the parameters are $N \times L$ matrices.

We assume that the IDET signal propagates through a multipath channel, which has M paths in total. Furthermore, we define $\psi_{n,m}$ as the phase shift in the n-th subcarrier and the m-th signal propagation path, while $a_{n,m}$ represents the corresponding amplitude gain. The channel response in the n-th subcarrier of the m-th path can be denoted as $h_{n,m} = a_{n,m} e^{j\psi_{n,m}}$. Moreover, the complex channel response of the n-th subcarrier can be formulated as

$$h_n = \sum_{m=1}^{M} a_{n,m} e^{j\psi_{n,m}} = a_n e^{j\psi_n}, \tag{3.35}$$

where we have

$$\begin{aligned} a_n &= \left|\sum_{m=1}^{M} a_{n,m} e^{j\psi_{n,m}}\right|, \\ \psi_n &= \arctan \frac{\text{Im}\left(\sum_{m=1}^{M} a_{n,m} e^{j\psi_{n,m}}\right)}{\text{Re}\left(\sum_{m=1}^{M} a_{n,m} e^{j\psi_{n,m}}\right)}. \end{aligned} \tag{3.36}$$

In the aging channel, the channel response in the current transmission frame l is temporally correlated to that of the previous frame $(l-1)$. In the n-th subcarrier, the temporal correlation [58] is modeled as

$$\begin{aligned} h_n(l) &= \alpha h_n(l-1) + \Delta h_n(l) \\ &= \alpha^{l-1} h_n(1) + \sum_{k=1}^{l} \alpha^{l-k} \Delta h_n(k), \end{aligned} \tag{3.37}$$

for $\forall n = 1, \cdots, N$, the correlation factor α satisfies $0 \le \alpha < 1$. We assume the same correlation factor α for all the subcarriers. Moreover, $\{\Delta h_n(l) | \forall n, l\}$ in Eq. (3.37) represents the uncertain components, which obeys an independent and identical Gaussian distribution with a zero mean and a variance of $(1-\alpha^2)\sigma_h^2$. Note that σ_h^2 is a common variance for all the channel response $\{h_n(l) | \forall, n = 1, \cdots, N, l = 1, \cdots, L\}$. Moreover, in a specific subcarrier, $\Delta h_n(l)$ is always independent of $h_n(l-1)$.

In order to avoid CSI acquisition for every transmission frame, the outdated CSI can be invoked by the transmitter to generate an appropriate IDET waveform by exploiting the temporal correlation. As shown in Fig. 3.15, we define a transmission block consisting of L transmission frames. The transmitter only needs to acquire the accurate CSI in the very first transmission frame. This CSI is exploited for designing the IDET waveform for the entire block.

The architecture of an IDET receiver is illustrated at the bottom of Fig. 3.14. We denote the CFO between the receiver and the transmitter as Δf_c. With the existence of the CFO, the received passband signal $y(t,l)$ at the l-th transmission frame can be expressed as

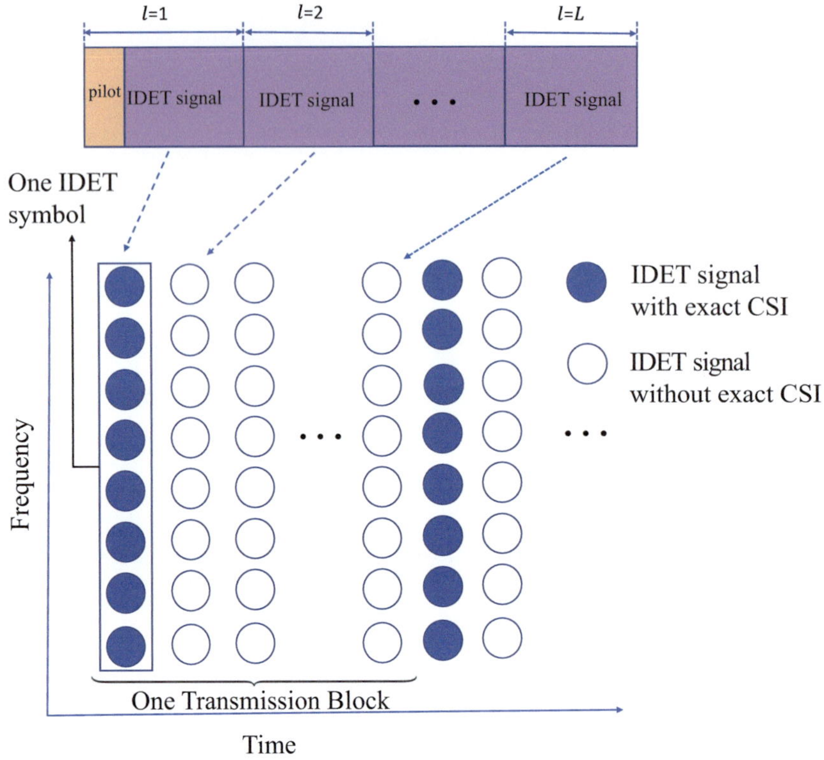

Fig. 3.15 IDET signal in the time-domain and frequency domain

$$y(t,l) = \text{Re}\left[r_I(t,l) + r_E(t,l) + z_a(t)\right], \tag{3.38}$$

where we have

$$r_I(t,l) = \sum_{n=1}^{N} h_n(l) x_{n,I}(l) \omega_{n,I}(l) \hat{w}(t), \tag{3.39}$$

$$r_E(t,l) = \sum_{n=1}^{N} h_n(l) \omega_{n,E}(l) \hat{w}(t). \tag{3.40}$$

Note that $\hat{w}(t) = w(t,l)e^{j2\pi(f_n+\Delta f_c)t}$. To achieve IDET, a power splitter is implemented for splitting $y(t,l)$ into two parts. Note that $\sqrt{\rho(l)}$ of the signal is for energy harvesting. Another portion $\sqrt{1-\rho(l)}$ of the received RF signal is converted into the baseband for the information decoding. The additional noise $z_{cov}(t)$ is imposed by the passband-to-baseband converter on the received signal, which obeys a Gaussian distribution with a zero mean and a variance of σ_{cov}^2. Then, the analog signals pass an analog-to-digital (A/D) converter and a fast Fourier transformer (FFT) to recover the information carried.

For the receiver, $\{\rho(l)|\forall l = 1, \ldots, L\}$ of power splitter for every transmission frame should be designed to balance the WDT and WET performance.

3.3.1.3 Problem Formulation

We conceive a Schottky diode based non-linear energy harvester. After truncating the Taylor series of the diode characteristic equation to the fourth order, the output DC of the l-th transmission frame [60] can be then expressed as

$$i_{\text{out}}(l) = k_2 \mathbb{E}\left\{\mathcal{P}\left[y_{in}^2(t,l)\right]\right\} + k_4 \mathbb{E}\left\{\mathcal{P}\left[y_{in}^4(t,l)\right]\right\}, \tag{3.41}$$

where $\mathcal{P}\{\cdot\}$ calculates the time average of the input stochastic process. It is expressed as

$$\mathcal{P}[y(t,l)] = \frac{1}{T}\lim_{T\to\infty}\int_0^T y(t,l)dt. \tag{3.42}$$

Since the noise power of $z_a(t)$ is much lower than that of the superposition IDET signals, its contribution to the output DC is ignored. By substituting Eq. (3.38) into Eq. (3.43), we then obtain Eq (3.43).

$$\begin{aligned} i_{\text{out}}(l) = k_2\rho\left\{\mathcal{P}\left[r_E^2(t,l)\right]+\mathcal{P}\left[r_I^2(t,l)\right]\right\} + k_4\rho^2\left\{\mathcal{P}\left[r_E^4(t,l)\right]+\mathcal{P}\left[r_I^4(t,l)\right]\right.\\ \left.+6\mathcal{P}\left[r_E^2(t,l)\right]\cdot\mathcal{P}\left[r_I^2(t,l)\right]\right\}. \end{aligned} \tag{3.43}$$

The detailed expressions of the four terms in Eq. (3.43), namely $\mathcal{P}[r_E^2(t,l)]$, $\mathcal{P}[r_I^2(t,l)]$ and $\mathcal{P}[r_E^4(t,l)]$ and $\mathcal{P}[r_I^4(t,l)]$, can be found in Eq. (3.44).

In our system, the transmitter does not acquire the accurate CSI until the next channel estimation. As a result, only the outdated CSI is available at the transmitter.

$$\begin{cases} \mathcal{P}\{r_E^2(t,l)\} = \frac{1}{2}\left[\sum\limits_{n=0}^{N-1}|h_n(1)|^2\alpha^{2(l-1)}s_{n,\text{E}}(l)^2+(1-\alpha^{2(l-1)})s_{n,E}(l)^2\sigma_h^2\right], \\ \mathcal{P}\{r_I^2(t,l)\} = \frac{1}{2}\left[\sum\limits_{n=0}^{N-1}|h_n(1)|^2\alpha^{2(l-1)}s_{n,\text{I}}(l)^2+(1-\alpha^{2(l-1)})s_{n,\text{I}}(l)^2\sigma_h^2\right], \\ \mathcal{P}\{r_E^4(t,l)\} = \frac{3}{8}\left[\sum\limits_{\substack{n_0,n_1,n_2,n_3\\ n_0+n_1=n_2+n_3}}\prod\limits_{j=0}^{3}\alpha^{(l-1)}|h_{n_j}(1)|s_{n_j,E}(l)\right. \\ \qquad\left.+\frac{3\sigma_h^2}{4}(1-\alpha^{2(l-1)})\left[\sum\limits_{n=0}^{N-1}|s_{n,E}(l)|^2\right]^2\right], \\ \mathcal{P}\{r_I^4(t,l)\} = \frac{3}{4}\alpha^{4(l-1)}\left[\sum\limits_{n=0}^{N-1}|h_n(1)s_{n,\text{I}}(l)|^2\right]^2+\frac{3\sigma_h^2}{2}(1-\alpha^{2(l-1)})\left[\sum\limits_{n=0}^{N-1}|s_{n,I}(l)|^2\right]^2. \end{cases} \tag{3.44}$$

In the received RF signal $y(t,l)$ of Eq. (3.38), only the WDT signal $r_I(t)$ carries modulated information, while $r_E(t)$ is dedicated for improving the WET performance. After the passband-to-baseband conversion, the received baseband symbol $r_n(l)$ on the n-th subcarrier during the l-th transmission frame is derived as

$$\begin{aligned} r_n(l) &= \int_{-\infty}^{\infty}\left[r_I(t,l)+r_E(t,l)+\hat{z}(t)\right]e^{-j2\pi f_n t}dt \\ &= \chi_n\int_{-\infty}^{\infty}w(t,l)e^{j2\pi\Delta f_c t}dt+\tilde{z}+\sum_{k=1,k\neq n}^{N}\chi_k\int_{-\infty}^{\infty}w(t,l)e^{j2\pi f_n^k t}dt, \end{aligned} \tag{3.45}$$

where we have $\hat{z}(t) = z_a(t)+z_{\text{cov}}(t)$, $\tilde{z} = \tilde{z}_a+\tilde{z}_{\text{cov}}$, $\chi_n = h_n(l)(x_{n,I}(l)\omega_{n,I}(l)+\omega_{n,E}(l))$, $\chi_k = h_k(l)\left(x_{k,I}(l)\omega_{k,I}(l)\right.$ $\left.+\ \omega_{k,E}(l)\right)$, $f_n^k = f_k - f_n + \Delta f_c$. $\tilde{z}_a$ and $\tilde{z}_{cov}$ are the projection of the baseband noises. Their variances are σ_a^2 and σ_{cov}^2, respectively. Moreover, let $W(f)$ denote the Fourier transformation of the window function $w(t,l)$. Thus, Eq. (3.45) can be formulated as

$$\begin{aligned} r_n(l) &= \chi_n W(\Delta f_c)+\sum_{k=1,k\neq n}^{N}\chi_k W(f_n^k)+\tilde{z} \\ &= \alpha_n\frac{\chi_n}{h_n(l)}W(\Delta f_c)+\sum_{k=1,k\neq n}^{N}\alpha_n^k\frac{\chi_k}{h_k(l)}W(f_n^k)+\tilde{z}, \end{aligned} \tag{3.46}$$

where $\alpha_n = \alpha h_n(l-1) + \Delta h_n(l-1)$, $\alpha_n^k = \alpha h_k(l-1) + \Delta h_n(l-1)$. Since $\Delta h_n(l-1)$ is an uncertain component of the CSI, its impact on the IDET performance could be beneficial or detrimental. In the worst case, the term related to $\Delta h_n(l-1)$ is considered as noise or interference. By contrast, in the best case, the term related to $\Delta h_n(l-1)$ can be considered as an additional useful signal to provide extra IDET performance gain [61]. In order to guarantee the lowest WDT performance, we consider a lower bound of the channel capacity, where the term related to $\Delta h(l-1)$ in Eq. (3.46) is considered to be noise and interference. Therefore, the lower bound of the WDT rate during the l-th transmission frame can be expressed as

$$R_I(l) = \sum_{n=1}^{N} R_n(l) = \sum_{n=1}^{N} B\log_2\left(1 + \left(\frac{(1-\rho(l))(P_{ICI}(l) + P_{I,n,\Delta h}(l))}{(1-\rho(l))\alpha^2 P_{n,I}(l)} + \frac{(1-\rho(l))(P_{E,n,\Delta h}(l) + \sigma_a^2) + \sigma_{cov}^2}{(1-\rho(l))\alpha^2 P_{n,I}(l)}\right)^{-1}\right). \tag{3.47}$$

Since the dedicated WET is deterministic, the receiver can obtain the amplitude and the phase of the WET signal. Therefore, the intra–subcarrier interference imposed by the WET signal can be cancelled. The power $P_{n,I}(l)$ of the desired WDT signal in Eq. (3.47) can be expressed as

$$P_{n,I}(l) = \alpha^{2(l-1)} s_{n,I}(l)^2 |h_n(1)|^2 \times |W(\Delta f_c)|^2 \cos^2(\bar{\psi}_{n,I}(l)), \tag{3.48}$$

where we have $\bar{\psi}_{n,I}(l) = \psi_n(l) + \phi_{n,I}(l)$.While the interference power $P_{n,I,\Delta h}(l)$ and $P_{n,E,\Delta h}(l)$ caused by the outdated CSI during the l-th transmission frame can be expressed as

$$P_{n,E,\Delta h}(l) = (1-\alpha^{2(l-1)}) s_{n,P}(l)^2 \times \sigma_{\mathrm{h}}^2 |W(\Delta f)|^2 \cos^2(\bar{\psi}_{n,E}(l)), \tag{3.49}$$

$$P_{n,I,\Delta h}(l) = (1-\alpha^{2(l-1)}) s_{n,I}(l)^2 \times \sigma_{\mathrm{h}}^2 |W(\Delta f)|^2 \cos^2(\bar{\psi}_{n,I}(l)). \tag{3.50}$$

Similarly, we have $\bar{\psi}_{n,E}(l) = \psi_n(l) + \phi_{n,E}(l)$. Moreover, the power $P_{ICI}(l)$ of the inter-carrier interference in Eq. (3.47) is formulated as Eq. (3.51).

$$\begin{aligned}
P_{n,ICI}(l) &= E[r_n(l) r_n(l)^*] - P_{n,I}(l) - P_{n,E}(l) - P_{n,E,\Delta h}(l) - P_{n,I,\Delta h}(l) \\
&= \sum_{k\neq n}^{N} s_{k,I}(l)^2 \left| \alpha^{(l-1)} h_k(1) + \sum_{k=1}^{l} \alpha^{l-k} \Delta h_n(k) \right|^2 |W(f_k - f_n + \Delta f_c)|^2 \\
&+ \sum_{k\neq n}^{N} \sum_{m\neq n}^{N} s_{k,E}(l) s_{m,E}(l) |\alpha^{(l-1)} h_k(1) + (1-\alpha^{2(l-1)}) \sigma_{\mathrm{h}}^2| \left| \alpha^{(l-1)} h_m(1) \right. \\
&\left. + (1-\alpha^{2(l-1)}) \sigma_{\mathrm{h}}^2 \right| \\
&\times W(f_k - f_n + \Delta f_c) W(f_m - f_n + \Delta f_c) \cos(\bar{\psi}_m(l) - \bar{\psi}_k(l)).
\end{aligned} \tag{3.51}$$

Observe from Eq. (3.51) that both the WET and WDT signal induce the inter-carrier interference. When the perfect carrier synchronization is achieved, namely $\Delta f_c = 0$, any pairs of subcarriers are totally orthogonal with each other. Otherwise, the resultant inter-carrier interference cannot be ignored, and the system performance can be severely degraded.

By acquiring the exact CSI $\boldsymbol{h}(1) = \{h_n(1) = a_n(1) e^{j\psi_n(1)} | \forall n = 1, \cdots, N, \}$ during the first transmission frame of a block having L transmission frames in total, the IDET signal can be optimally shaped by jointly designing the amplitudes $\boldsymbol{s}_E(l)$ of the WET signal, $\boldsymbol{s}_I(l)$ of the WDT signal, the phases $\boldsymbol{\phi}_E(l)$ of the WET signal, the phases $\boldsymbol{\phi}_I(l)$ of the WDT signal, and the power splitting ratio $\rho(l)$ for $\forall l = 1, \cdots, L$. Furthermore, we assume the same transmit power for every transmission frame. We aim to maximize the average output DC for a single transmission block consisting of L transmission frames in total with the maximum transmit power and information rate constraints. As a result, the original optimization problem (P1) can be reformulated as

$$(\text{P3.6.1}): \max_{\substack{\{\mathbf{s}_I(l),\mathbf{s}_E(l),\boldsymbol{\phi}_I(l),\\ \boldsymbol{\phi}_E(l),\rho(l)|\forall l=1,\ldots,L\}}} \frac{1}{L}\sum_{l=1}^{l=L} i_{\text{out}}(l) \tag{3.52}$$

$$\text{s.t.} \quad R_I(l) \geq R_{\min}, \forall l = 1, \cdots, L, \tag{3.52a}$$

$$\frac{1}{2}\left[||\mathbf{s}_I(l)||_2^2 + ||\mathbf{s}_E(l)||_2^2\right] \leq P,$$

$$\forall l = 1, \cdots, L, \tag{3.52b}$$

$$X_a\rho(l) \in [0, 1], \forall l = 1, \cdots, L. \tag{3.52c}$$

According to Eq. (3.43) and Eq. (3.47), $i_{\text{out}}(l)$ and $R_I(l)$ are the functions with respect to variables $\{\mathbf{s}_I(l), \mathbf{s}_E(l), \boldsymbol{\phi}_I(l), \boldsymbol{\phi}_E(l), \rho(l)\}$. Note that (3.52a)–(3.52c) constrain the maximum transmit power P, the minimum information rate $R_{\min}$, and the power splitting ratio $\rho(l)$ for every transmission frame, respectively.

3.3.1.4 Joint Amplitude and Phase Design

Since the transmit power is always the same, (P3.6.1) can be equivalently decoupled into L independent subproblems for maximizing the output DC in every transmission frame. Specifically, the l-th subproblem can be formulated as

$$(\text{P3.6.2}): \max_{\substack{\mathbf{s}_I(l),\mathbf{s}_E(l),\\ \boldsymbol{\phi}_I(l),\boldsymbol{\phi}_E(l),\rho(l)}} i_{\text{out}}(l) \tag{3.53}$$

$$\text{s.t.} \quad R_I(l) \geq R_{\min}, \tag{3.53a}$$

$$\frac{1}{2}\left[||\mathbf{s}_I(l)||_2^2 + ||\mathbf{s}_E(l)||_2^2\right] \leq P, \tag{3.53b}$$

$$\rho(l) \in [0, 1]. \tag{3.53c}$$

Given a specific phase design $\{\phi_{n,E}(l), \phi_{n,I}(l)\}$, (P3.6.2) is transformed into the following amplitude optimization problem:

$$(\text{P3.6.3}): \max_{\mathbf{s}_I(l),\mathbf{s}_E(l),\rho(l)} i_{\text{out}}(l) \tag{3.54}$$

$$\text{s.t.} \quad R_I(l) \geq R_{\min}, \tag{3.54a}$$

$$(3.53\text{b})\text{–}(3.53\text{c}),$$

where $i_{\text{out}}(l) = i_{\text{out}}(\mathbf{s}_I(l), \mathbf{s}_E(l), \rho(l))$ and $R_I(l) = R_I(\mathbf{s}_I(l), \mathbf{s}_E(l), \rho(l))$. Note that the objective of (P3.6.3) is a polynomial function, according to Eq. (3.43). Therefore, (P3.6.3) has a form of Geometric Programming (GP) [62]. By introducing some auxiliary variables, we can transform (P3.6.3) into the following equivalent reversed GP problem:

$$(\text{P3.6.4}): \max_{\mathbf{s}_E(l),\mathbf{s}_I(l),\rho(l)} \tau \tag{3.55}$$

$$\text{s.t.} \quad \tau / \mathrm{i}_{\text{out}}(l) \leq 1, \tag{3.55a}$$

$$R_{th} \Big/ \left(\prod_{n=0}^{N-1} \gamma_n\right) \leq 1, \tag{3.55b}$$

$$\frac{\gamma_n}{v_n}\Big[(1-\rho)\left(P_{I,n,\Delta h}(l) + P_{E,n,\Delta h}(l)\right)$$

$$+(1-\rho)\left(P_{ICI}(l)+\sigma_a^2\right)\Big] \leq 1, \quad \forall n = 1, 2 \ldots N, \tag{3.55c}$$

(3.52b)–(3.52c),

where τ, γ_n, and υ_n are auxiliary variables. In Eq. (3.55c), γ_n satisfies $\log_2 \gamma_n \geq \log_2 R_n(l)$, where $R_n(l)$ is the spectral efficiency in the n-th subcarrier in Eq. (3.47). Moreover, υ_n in Eq. (3.55c) is a function of $\{\mathbf{s}_I(l), \mathbf{s}_E(l), \rho(l)\}$, which is expressed as

$$\begin{aligned}\upsilon_n(\mathbf{s}_I(l), \mathbf{s}_E(l), \rho(l)) &= (1-\rho(l))(P_{n,ICI}(l) \\ &+ P_{n,I}(l) + P_{I,n,\Delta h}(l) + P_{E,n,\Delta h}(l)) + \sigma_a^2.\end{aligned} \tag{3.56}$$

Note that the inverse of υ_n and that of $i_{\text{out}}(l)$ are not polynomials, which prevent the application of GP tools. By employing the classic arithmetic mean and geometric mean (AM–GM) inequality, we approximate a polynomial with a monomial. Specifically, we define that $g_k(\mathbf{s}_E(l), \mathbf{s}_I(l), \rho(l))$ are the monomials in the polynomial $i_{\text{out}}(l) = \sum_{k=1}^{K} g_k(\mathbf{s}_E(l), \mathbf{s}_I(l), \rho(l))$. Similarly, $g_{n,k}(\mathbf{s}_E(l), \mathbf{s}_I(l), \rho(l))$ represents a monomial term of $\upsilon_n(l) = \sum_{k=1}^{K_n} g_{n,k}(\mathbf{s}_E(l), \mathbf{s}_I(l), \rho(l))$. By using AM–GM inequality, we can obtain the following monomials as

$$\hat{i}_{\text{out}}(l) = \prod_{k=1}^{K} \left(\frac{g_k(\mathbf{s}_E(l), \mathbf{s}_I(l), \rho(l))}{t_k}\right)^{-t_k}, \tag{3.57}$$

$$\hat{\upsilon}_n(l) = \prod_{k=1}^{K_n} \left(\frac{g_{n,k}(\mathbf{s}_E(l), \mathbf{s}_I(l), \rho(l))}{t_{n,k}}\right)^{-t_{n,k}}, \forall n = 1, \ldots, N, \tag{3.58}$$

where $\hat{i}_{\text{out}}(l)$ and $\hat{\upsilon}_n(l)$ are the approximations of $i_{\text{out}}(l)$ and $\upsilon_n(l)$, respectively. The value of auxiliary variables $\{t_k, t_{n,k}\}$ should satisfy $\sum_{k=1}^{K} t_k = 1$ and $\{\sum_{k=1}^{K_n} t_{n,k} = 1 | \forall n = 1, \ldots, N\}$. Suppose that we have a set of feasible solutions $\{\mathbf{s}_E^{(i)}(l), \mathbf{s}_I^{(i)}(l), \rho^{(i)}(l)\}$ after the i-th iteration. Then, we define

$$t_k = \frac{g_k(\mathbf{s}_E^{(i)}(l), \mathbf{s}_I^{(i)}(l), \rho^{(i)}(l))}{i_{\text{out}}(\mathbf{s}_E^{(i)}(l), \mathbf{s}_I^{(i)}(l), \rho^{(i)}(l))}, \tag{3.59}$$

$$t_{n,k} = \frac{g_{n,k}(\mathbf{s}_E^{(i)}(l), \mathbf{s}_I^{(i)}(l), \rho^{(i)}(l))}{\upsilon_n(\mathbf{s}_E^{(i)}(l), \mathbf{s}_I^{(i)}(l), \rho^{(i)}(l))} \tag{3.60}$$

As a result, a standard reversed GP can be formulated as

$$(\text{P3.6.5}): \quad \max_{\mathbf{s}_E(\mathbf{l}), \mathbf{s}_I(l), \rho(l)} \quad \tau \tag{3.61}$$

$$\text{s.t.} \quad \tau \Big/ \hat{\mathrm{i}}_{\text{out}}(l) \leq 1, \tag{3.61a}$$

$$R_{th} \Big/ \left(\prod_{n=0}^{N-1} \gamma_n\right) \leq 1, \tag{3.61b}$$

$$\begin{aligned}&\gamma_n \bar{\rho}(l) \left(P_{ICI}(l) + P_{I,n,\Delta h}(l)\right. \\ &\left.+ P_{E,n,\Delta h}(l)\right) + \gamma_n \sigma_a^2 / \hat{\upsilon}_n(l) \leq 1, \forall n = 1, 2, \ldots N,\end{aligned} \tag{3.61c}$$

(3.53b)–(3.53c),

where $\bar{\rho}(l) = 1 - \rho(l)$. Constraints (3.61a)–(3.61c) are all transformed into polynomial forms.

Algorithm 3.3 A reverse GP aided algorithm for amplitude design

Input: The channel state information $h_n(1)$ of every sub-carriers during the first transmission frame of a block; The required accuracy ϵ;
Output: The amplitudes of information signal $\{\mathbf{s}_E(l)\}$ and the amplitudes of power signal $\{\mathbf{s}_I(l)\}$; The power splitting ratio $\{\rho(l)\}$;

1: Initialise $\mathbf{s}_E(l), \mathbf{s}_I(l), \rho(l), \bar{\rho}(l), i = 0, i_{\text{out}}^{(0)} = 0$;
2: **repeat**
3: Update $i \leftarrow i + 1$ and let $\mathbf{s}_E^{(i)}(l), \mathbf{s}_I^{(i)}(l), \rho^{(i)}(l), \bar{\rho}^{(i)}(l) \leftarrow \mathbf{s}_E(l), \mathbf{s}_I(l), \rho(l), \bar{\rho}(l)$
4: Obtain $\hat{v}_n(l) \leftarrow \mathbf{s}_E(l), \mathbf{s}_I(l), \rho(l), \bar{\rho}(l)$ and $\hat{i}_{\text{out}}(l) \leftarrow \mathbf{s}_E^{(i)}(l), \mathbf{s}_I^{(i)}(l), \rho^{(i)}(l), \bar{\rho}^{(i)}(l)$
5: Obtain $\mathbf{s}_E^{(i)}(l), \mathbf{s}_I^{(i)}(l), \rho^{(i)}(l), \bar{\rho}^{(i)}(l)$ by solving (P5) with CVX
6: Let $\mathbf{s}_E(l), \mathbf{s}_I(l), \rho(l), \bar{\rho}(l) \leftarrow \mathbf{s}_E^{(i)}(l), \mathbf{s}_I^{(i)}(l), \rho^{(i)}(l), \bar{\rho}^{(i)}(l)$
7: Let $i_{\text{out}}^{(i)}(l) \leftarrow i_{\text{out}}(\mathbf{s}_E(l), \mathbf{s}_I(l), \rho(l), \bar{\rho}(l))$
8: **until** $|i_{\text{out}}^{(i)}(l) - i_{\text{out}}^{(i-1)}(l)| < \epsilon$

Algorithm 3.4 ABC algorithm for phase design

Input: The channel state information $h_n(1)$ of every sub-carriers during the first transmission frame of a block; The amplitudes of information signal $\{\mathbf{s}_I(l)\}$ and the amplitudes of power signal $\{\mathbf{s}_E(l)\forall l = 1\}$; The size of the transmission block L;
Output: The phases of power signal $\{\boldsymbol{\phi}_E(l)\}$; The phases of information signal $\{\boldsymbol{\phi}_I(l)\}$; The power splitting ratio $\{\rho(l)\}$;

1: Initialise $l = 0, L$;
2: **while** $l < L$ **do**
3: Update $l \leftarrow l + 1$ and initialise $i = 1$
4: Get the M feasible solutions and calculate probability vector $\boldsymbol{P}$ and the fitness vector $\boldsymbol{i}_{\text{out}}$;
5: **while** $i > iter$ **do**
6: The leader bees search and update the nectar sources ;
7: The follower bees search and update the nectar sources;
8: The scout bees search and update the nectar sources;
9: Update the nectar sources;
10: **end while**
11: **end while**

A reversed GP aided iterative amplitude design is detailed in Algorithm 3.3. After initializing variables, calculate the approximated $\hat{i}_{\text{out}}(l)$ and $\hat{v}_n(l)$ of $i_{\text{out}}(l)$ and $v_n(l)$, respectively, as shown in Lines 1–4. (P3.6.5) can be effectively solved by an existing optimization tool, as shown in Line 5 of Algorithm 3.3 [63]. After sufficient iterations, we may achieve the required accuracy, as shown in Line 8 of Algorithm 3.3. With the aid of Algorithm 3.3, we may obtain the amplitude design, when both the phases of the OFDM-based WDT signals and those of the multi-sinusoidal WET signals are given.

Ellipsoid method can be invoked to solve a reversed GP problem. Its complexity is $\mathcal{O}((2N)^2)$ [64]. We assume that Δ iterations are required to accurately approximate a polynomial with a monomial. Thus, the complexity of Algorithm 3.3 is $\mathcal{O}(\Delta(2N)^2)$.

Given a specific amplitude design $\{\mathbf{s}_E(l), \mathbf{s}_I(l)\}$, the original optimization problem (P3.6.2) is reformulated as

$$\text{(P3.6.6)}: \max_{\boldsymbol{\phi}_E(l), \boldsymbol{\phi}_I(l), \rho(l)} \quad i\text{out}(l) \tag{3.62}$$

$$\text{s.t.} \quad R_I(l) \geq R_{\min}, \tag{3.62a}$$

$$\text{(3.53b)–(3.53c)},$$

where $i_{\text{out}}(l) = i_{\text{out}}(\boldsymbol{\phi}_E(l), \rho(l))$, and $R_I(l) = R_I(\boldsymbol{\phi}_E(l), \boldsymbol{\phi}_I(l), \rho(l))$. Since the elements of $\boldsymbol{\phi}_E(l)$ and $\boldsymbol{\phi}_I(l)$ are included in trigonometric functions, we invoke an artificial-intelligence-based algorithm to find the optimal phase design, namely Artificial Bee Colony (ABC), which is suitable for solving a complex non-convex optimization problem [65]. Being combined with both local and global searches, an ABC-based algorithm imitates the behaviors of bees looking for nectar to find a satisfactory solution, which is detailed in Algorithm 3.4. The main steps of Algorithm 3.4 are summarized as below:

- *Step 1*: Randomly generate M feasible solutions as the initial nectar sources in Line 4, where M is the number of nectar sources. Each solution $\boldsymbol{X}_i = \{\boldsymbol{\phi}_E(l), \boldsymbol{\phi}_I(l), \rho(l)\}$ for $\forall i - 1, 2, \ldots, M$ is a $(2N \mid 1)$ dimension vector representing a single nectar source.

  So we have M different nectar sources in total. Calculate the fitness i_{out} of the i-th solution (the i-th nectar source) in Eq. (3.43).

- *Step 2*: Both the leader and the follower bees search the nectar sources in Lines 6 and 7. Specifically, every leader bee randomly searches around a specific nectar source (a feasible solution) by obeying the following strategy expressed as

$$\boldsymbol{X}_i^{j+1} = \boldsymbol{X}_i^j + r\varepsilon\left(\boldsymbol{X}_i^j - \boldsymbol{X}_k^j\right), \tag{3.63}$$

  where j represents the j-th iteration, $\varepsilon \in [0, 1]$ is a constant, $k \in (1, 2, \ldots, M)$ is a random index and different from i, while r is a constant step size. The nectar source is only updated if $\boldsymbol{X}_i^{j+1}$ has a better performance than $\boldsymbol{X}_i^j$. The number of the follower bees is smaller than that of the leader bees. Therefore, we can only assign the follower bees to the nectar sources with a better performance. The probability of a follower bee searching the nectar source $\boldsymbol{X}_i^j$ is obtained as [65]

$$P_i = F_i / \sum_{m=1}^{M} F_i, \tag{3.64}$$

  where $F_i = e^{\frac{i_{\text{out},i}}{\frac{1}{M}\sum_{i=1}^{M} i_{\text{out},i}}}$ is defined as the normalized fitness of this nectar source. Note that a nectar source with a higher normalized fitness has a higher probability to be chosen by a follower bee. Then the follower bee may search around the target nectar source by obeying the same strategy of Eq. (3.63). The nectar source is only updated, if a better one is found.
- *Step 3*: The scout bees search the nectar sources in Line 8. When the fitness of the nectar source cannot be improved within several iterations, the leader bees give up the original nectar source, and they become scout bees searching a new area expressed as

$$\boldsymbol{X}_i^j = \boldsymbol{X}_{\text{min}}^j + \varepsilon\left(\boldsymbol{X}_{\text{max}}^j - \boldsymbol{X}_{\text{min}}^j\right), \tag{3.65}$$

  where $\boldsymbol{X}_{\text{min}}^j$ and $\boldsymbol{X}_{max}^j$ are the lower and upper bounds of the $\boldsymbol{X}^j$, respectively.

The cooperation among the leader bees and the follower bees can find a better local solution. Moreover, with the presence of the scout bees, the situation of being trapped in a local optimum can be avoided [66]. Thus, we can obtain a near-optimal solution $\{\boldsymbol{\phi}_E^*(l), \boldsymbol{\phi}_I^*(l), \rho^*(l)|\forall l = 1, \ldots, L\}$, whose fitness is the highest among all the feasible solutions.

By optimizing the amplitudes and phases sequentially, the optimal solutions to every transmission frame are obtained, as shown in Lines 6–11 in Algorithm 3.5. The original problem (P3.6.1) that maximizes the average output DC among all the L transmission frames is solved after optimizing L subproblems. The general iterative algorithm for the joint design of both amplitudes and phases for the IDET waveforms is detailed in Algorithm 3.5.

Algorithm 3.5 Joint IDET amplitude and phase waveform

Input: The channel state information $h_n(1)$ of every sub-carriers during the first transmission frame of a block; The required accuracy ϵ; The size of the transmission block L;

Output: The phases of information signal $\{\boldsymbol{\phi}_I(l)\forall l = 1, \cdots, L\}$; The phases of power signal $\{\boldsymbol{\phi}_E(l)\forall l = 1, \cdots, L\}$; The amplitudes of information signal $\{\boldsymbol{s}_E(l)\forall l = 1, \ldots, L\}$ and the amplitudes of power signal $\{\mathbf{s}_I(l)\forall l = 1, \ldots, L\}$; The power splitting ratio $\{\rho(l)\forall l = 1, \cdots, L\}$;

1: Initialise $l = 0, L$;
2: **while** $l < L$ **do**
3: Update $l \leftarrow l + 1$
4: Initialise $\mathbf{s}_E(l), \mathbf{s}_I(l), \boldsymbol{\phi}_I(l) = -\boldsymbol{\psi}(1), \boldsymbol{\phi}_E(l) = -\boldsymbol{\psi}(1), \widetilde{i}_{\text{out}} = 0, i_{\text{out}} = 0$
5: **repeat**
6: Update $\widetilde{i}_{\text{out}} = i_{\text{out}}$
7: Obtain $\widetilde{\mathbf{s}}_E(l), \widetilde{\mathbf{s}}_I(l)$ by Algorithm 3.3
8: update $\mathbf{s}_E(l) = \widetilde{\mathbf{s}}_E(l), \mathbf{s}_I(l) = \widetilde{\mathbf{s}}_I(l)$
9: obtain $\widetilde{\boldsymbol{\phi}}_E(l), \widetilde{\boldsymbol{\phi}}_I(l), \widetilde{\rho}(l)$ by Algorithm 3.4
10: Update $\boldsymbol{\phi}_E(l) = \widetilde{\boldsymbol{\phi}}_E(l), \boldsymbol{\phi}_I(l) = \widetilde{\boldsymbol{\phi}}_I(l), \rho(l) = \widetilde{\rho}(l)$
11: Calculate $i_{\text{out}}(\mathbf{s}_P(l), \mathbf{s}_I(l), \boldsymbol{\phi}_I(l), \boldsymbol{\phi}_E(l), \rho(l))$
12: **until** $|i_{\text{out}} - \widetilde{i}_{\text{out}}| < \epsilon$
13: **end while**

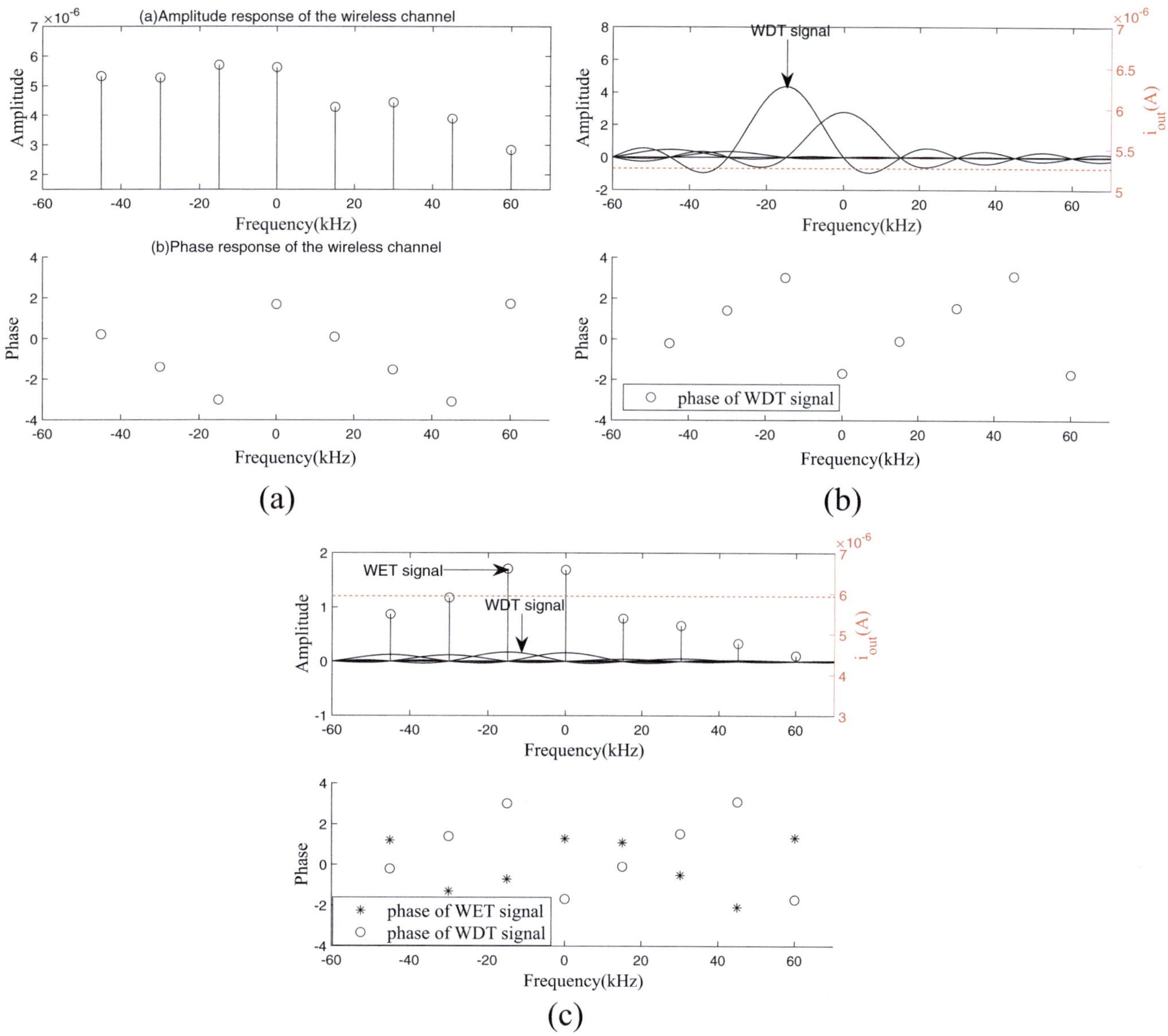

Fig. 3.16 (**a**) Wireless channel response. (**b**) OFDM waveform shaping. (**c**) IDET waveform shaping

3.3.1.5 Simulation Results

For the non-linear energy harvester, we consider the following parameters: $k_2 = 0.0034$, $k_4 = 0.3859$, and $R_{ant} = 50\ \Omega$, which are in line with [51]. The central frequency is $f_c = 2.4$ GHz, and the subcarrier spacing is $\Delta f = 15$ kHz, while $N = 8$ subcarriers are conceived. We define the normalized carrier frequency offset as CFO $= \Delta f_c / \Delta f$. The path-loss from the IDET transmitter to the receiver is modeled as $PL(\text{dB}) = 32.4 + 20\log(d) + 20\log(f_c) + G$, where d is the distance between the IDET transmitter and the receiver [44]. The transmission distance is set to $d = 6.3$ m, the transmit power of the IDET transmitter is $P = 36$ dBm, while the receive antenna gain is $G = 2$ dBi [67]. The small-scale fading of the wireless channel follows the Tapped Delay Line model [68], which models the correlation in the frequency domain.

The channel response during the first transmission frame of a block is portrayed in Fig. 3.16a, where the correlation in the frequency domain is clearly observed. We illustrate the amplitude and phase design for our IDET waveform and those for the conventional OFDM waveform during this transmission frame in Fig. 3.16b, c. Observe from Fig. 3.16b that in the OFDM waveform, a single subcarrier associated with the strongest amplitude response gains most of the transmit power in order to maximize the output DC. However, since our IDET waveform is constituted by an OFDM-based waveform and a multi-sinusoidal waveform, we can observe from Fig. 3.16c that the transmit power is more evenly allocated among all

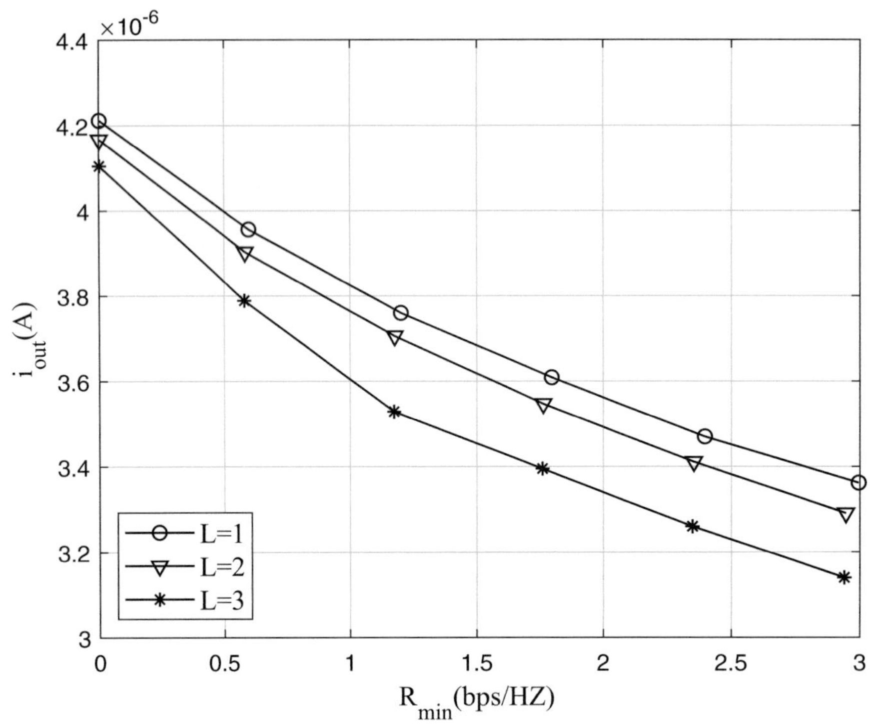

Fig. 3.17 Output DC with different block sizes L for $N = 8$, CFO $= 0.025$, $\alpha = 0.97$, and $\sigma_h^2 = 1$ with $B = 15$ kHz

the subcarriers of the OFDM waveform and among all different frequency components of the multi-sinusoidal waveform. Moreover, our simulation results also show that our IDET waveform is capable of achieving the output DC of 5.93×10^{-6} A, while the OFDM waveform only achieves 5.26×10^{-6} A. Note that in this simulation, the WDT requirement is set to 1 bps/Hz.

We investigate the impact of the block size on the IDET performance. We consider 6 transmission frames in total. For $L = 1$, the transmitter may acquire accurate CSI for every transmission frames. For $L = 2$, we have 3 blocks, while we have 2 blocks for $L = 3$. In these two cases, the transmitter only acquires the accurate CSI during the first transmission frame of a block. Observe from Fig. 3.17 that the output DC becomes lower, when the block size increases. This is because the outdated CSI is used for more transmission frames, as L increases. Moreover, the output DC also reduces, as the WDT requirement increases. This is because in order to satisfy strict WDT requirement, the transmitter allocates more power to the WDT signal, which thus reduces the WET performance.

We study the impact of the aging channel's temporal correlation factor α on the IDET performance. Observe from Fig. 3.18 that the output DC reduces, as we reduce α. This is because a descreasing α results in a lower temporal correlation of the channel. As a result, the outdated CSI becomes even more inaccurate, which substantially reduces the IDET performance. For example, the average output DC is $i_{\text{out}} = 5.18 \times 10^{-6}$ A as $\alpha = 0.99$. The output DC decreases to $i_{\text{out}} = 4.66 \times 10^{-6}$ A as α reduces to 0.95. Note that the common WDT requirement is set to 1 bps/Hz.

We investigate the impact of the CFO on the IDET performance in Fig. 3.19. Observe from Fig. 3.19 that the output DC achieves $i_{\text{out}} = 5.16 \times 10^{-6}$ A with CFO $= 0$. However, i_{out} degrades to 4.43×10^{-6} A, when the CFO increases to 0.05. Note that the common WDT requirement is set to 1.5 bps/Hz. Since the multi-sinusoidal dedicated WET signal is more efficient than the OFDM-based WDT signal in terms of the output DC, the IDET transmitter gradually adapts itself from a pure WET mode by sending multi-sinusoidal WET signals to a pure WDT mode by sending OFDM signals, as the WDT requirement $R_{\min}$ gradually increases from zero to its maximum. The reason is two-fold: (1) The multi-sinusoidal waveform is more efficient than the OFDM waveform for WET. When the WDT requirement is not quite high, the transmitter may allocate more power to the sinusoidal waveform in order to achieve a better output DC. (2) The multi-sinusoidal waveform cannot carry any information. Therefore, when the WDT requirement becomes stricter, the transmitter may reduce the power allocated to the multi-sinusoidal waveform in order to make sure the OFDM waveform satisfy the increasing WDT requirement.

We demonstrate the advantage of our joint design over the existing scheme, when the CFO is considered. Algorithm 3.5 is our joint design on the IDET waveform, while Algorithm 4 was proposed in [51] without considering the presence of the CFO and the outdated CSI. Observe from Fig. 3.20 that when the CFO exists, Algorithm 3.5 outperforms its counterpart.

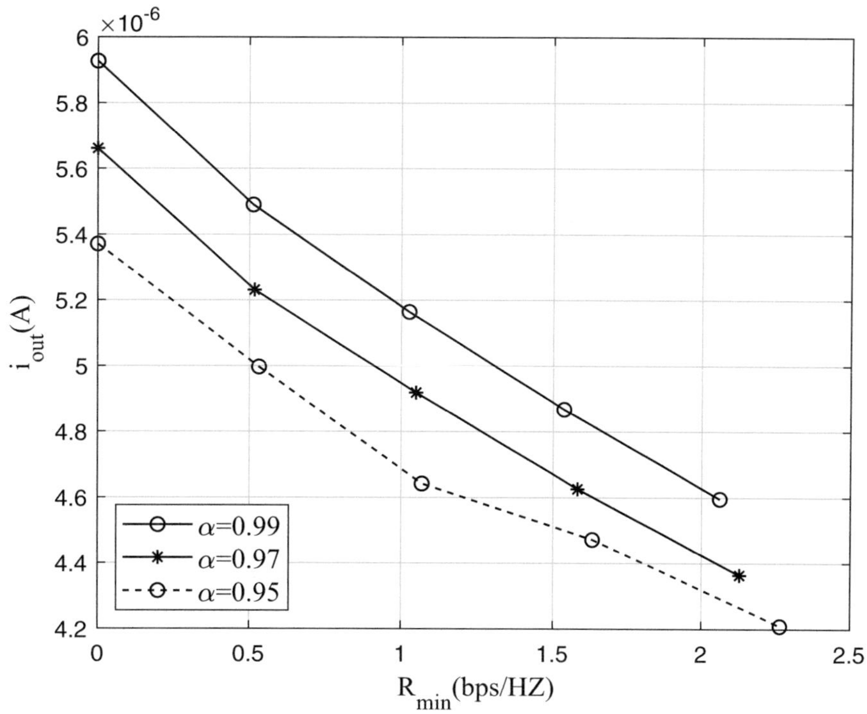

Fig. 3.18 Output DC with different temporal correlation factors α for $N = 8$, CFO $= 0.025$, $L = 2$, and $\sigma_h^2 = 1$ with $B = 15$ kHz

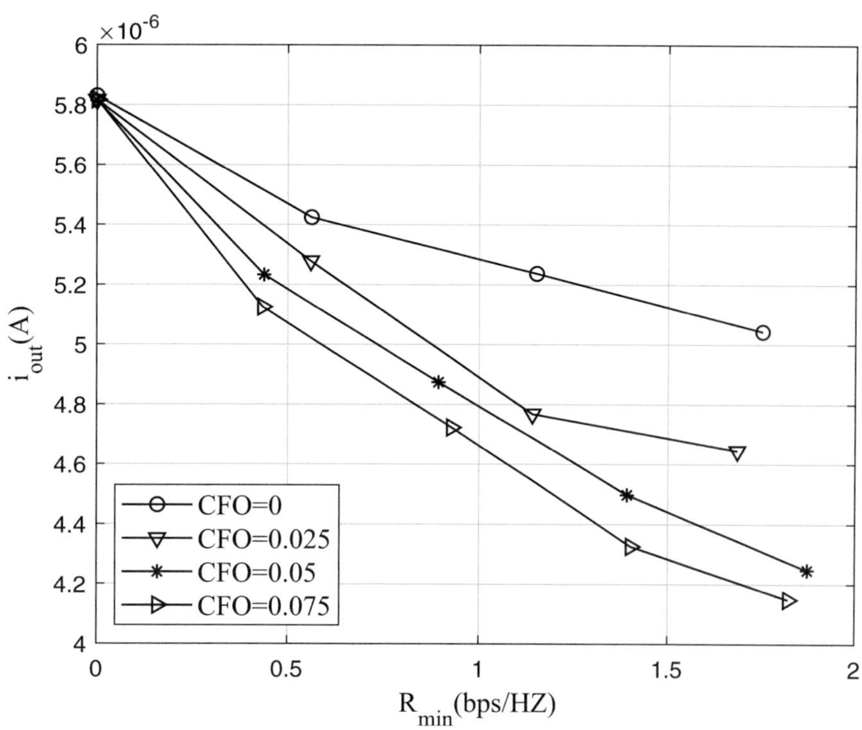

Fig. 3.19 Output DC with different CFOs for $N = 8$, $L = 2$, $\alpha = 0.97$, and $\sigma_h^2 = 1$ with $B = 15$ kHz

For example, Algorithm 3.5 achieves an output DC of $i_{\text{out}} = 6.41 \times 10^{-6}$ A, while Algorithm 3.4 only achieves $i_{\text{out}} = 5.62 \times 10^{-6}$ A, when the WDT requirement is 1 bps/Hz and CFO $= 0.05$. As the CFO increases, the performance gain of Algorithm 3.5 becomes higher.

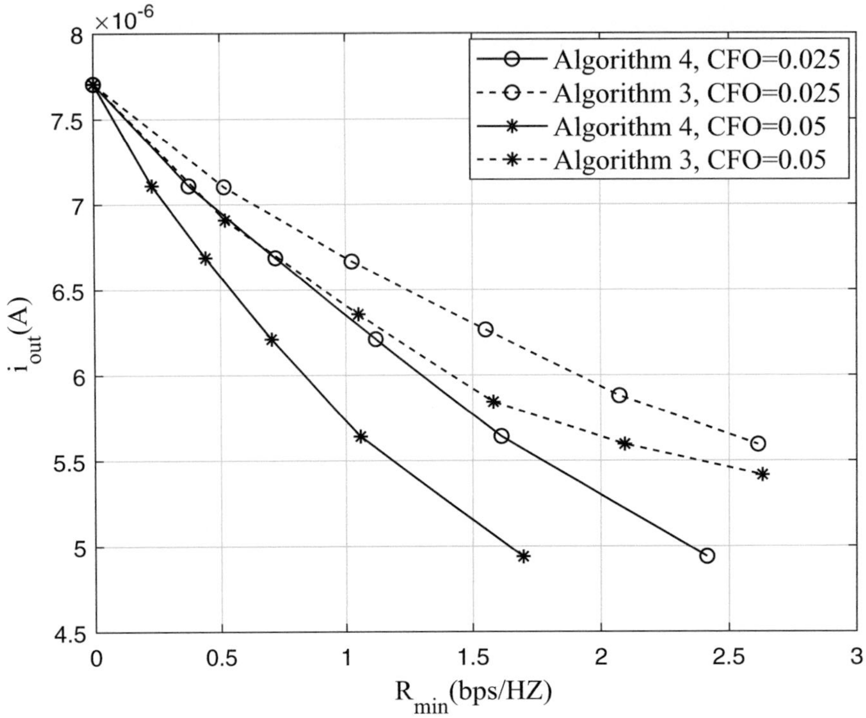

Fig. 3.20 Algorithm compare for $N = 8$, $\alpha = 0.97$, $L = 2$, and $\sigma_h^2 = 1$ with $B = 15$ kHz

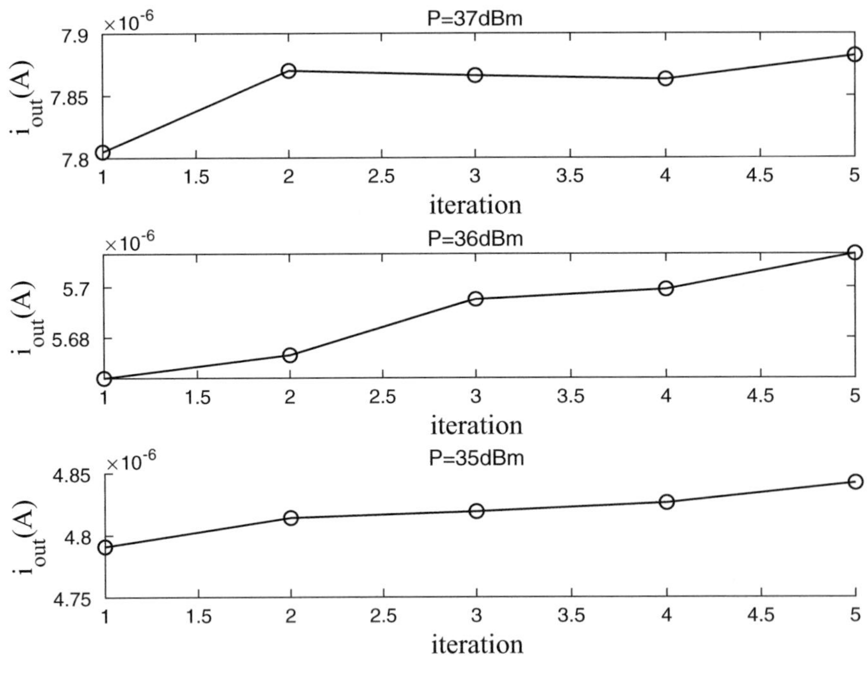

Fig. 3.21 Convergence performance for Algorithm 3.3 for $\alpha = 0.97$ and $\delta_h^2 = 1$

We study the convergence performance of Algorithm 3.5 under different transmit power in Fig. 3.21. Observe from Fig. 3.21 that Algorithm 3.5 has a fast convergence rate, which proves that our optimal design can be solved with a low complexity.

We compare Algorithms 3.3 and 3.5 under different CFO conditions in Fig. 3.22. Algorithm 3.3 only optimizes the amplitudes of the IDET waveform, while Algorithm 3.5 optimizes both their amplitudes and phases. According to Eq. (3.51),

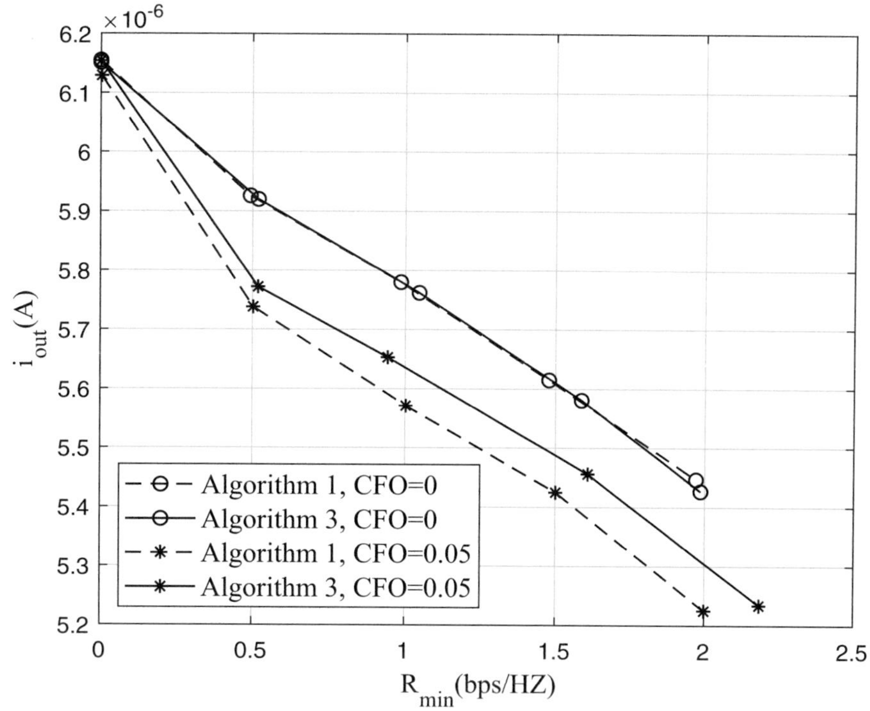

Fig. 3.22 Algorithm compare for $N = 8$, $\alpha = 0.97$, $L = 2$, and $\sigma_h^2 = 1$ with $B = 15\,\text{kHz}$

$\boldsymbol{\phi}_E(l)$ and $\boldsymbol{\phi}_I(l)$ initialized in Line 4 of Algorithm 3.5 are optimal, when perfect carrier synchronization is conceived. Therefore, there is no difference between two algorithms when CFO $= 0$. When the CFO cannot be ignored, by jointly optimizing the amplitudes and phases, Algorithm 3.5 outperforms its counterpart. For example, the average output DC increases from $i_{\text{out}} = 5.21 \times 10^{-6}$ A to $i_{\text{out}} = 5.3 \times 10^{-6}$ A when CFO $= 0.05$ and the WDT requirement is 2 bps/Hz.

3.3.2 *OTFS Waveform for Integrated Data and Energy Transfer*

Integrated data and energy transfer (IDET) is an advanced technology for enabling energy sustainability for massively deployed low-power electronic consumption components. However, the existing work of IDET using the orthogonal-frequency-division-multiplexing (OFDM) waveforms is designed for static scenarios, which would be severely affected by the destructive Doppler offset in high-mobility scenarios. Therefore, we proposed an IDET system based on orthogonal time–frequency space (OTFS) waveforms with the imperfect channel assumption, which is capable of counteracting the Doppler offset in high-mobility scenarios. At the transmitter, the OTFS-IDET system superimposes the random data signals and deterministic energy signals in the delay-Doppler (DD) domain with optimally designed amplitudes. The receiver optimally splits the received signal in the power-domain for achieving the best IDET performance. After formulating a non-convex optimization problem, it is transformed into a geometric programming (GP) problem through inequality relaxations to obtain the optimal solution. The simulation demonstrates that a higher amount of energy can be harvested when employing our proposed OTFS-IDET waveforms than the conventional OFDM-IDET ones in high-mobility scenarios.

3.3.2.1 Introduction

In the upcoming era of Internet of Everything (IoE), it is impossible to recharge massively deployed miniature communication devices by wire. To maintain the energy sustainability of communication devices, a promising solution is composed of wireless data transfer (WDT) and wireless energy transfer (WET), where the rectifier at the receiver converts the radio frequency (RF) signal into the direct current (DC) to be stored in energy buffers or to be invoked for directly driving electronic loads. By splitting the received signal in the time- or power-domain, energy harvesting (EH) and data receiving can be achieved simultaneously.

Over the past decades, numerous works have investigated the signal designs for enabling integrated data and energy transfer (IDET) [10]. For example, the joint power and subcarrier allocation was considered in the IDET system with multiple users [69]. The coding controlled [44] and constellation rotation aided [70] IDET were proposed to improve the EH performance while ensuring the data transmission rate. A practical IDET system based on the phase-shift modulation was studied in [71]. Additionally, the signal design for both the nonlinear and linear energy harvesters was studied in [51].

As discussed above, the existing IDET signal designs were almost based on the classic orthogonal frequency-division multiplexing (OFDM). Moreover, the OFDM-IDET performance was degraded by the frequency offset in the high-mobility scenario had proved in [72]. Therefore, we urgently need an IDET system to combat the adverse effect of the frequency offset induced by high mobility. Fortunately, orthogonal time–frequency space (OTFS) modulation [73], designed in the delay-Doppler (DD) domain rather than in the time–frequency (TF) domain, was proposed to address the frequency offset issue incurred by the Doppler effect.

However, to our best knowledge, research on OTFS-IDET is still a blank in the literature, which motivates our work in this chapter. In summary, our novel contributions are as follows:

- An OTFS-IDET system is proposed. The transmitter sends superimposed random data signals and deterministic energy signals to formulate the optimal IDET waveforms, while the receiver splits the received signal and simultaneously performs the EH and data recovery.
- Considering the imperfect channel estimation, we jointly optimize the amplitudes of the data and energy signals at the transmitter and the power splitting ratio at the receiver to maximize the harvested energy, while satisfying the data rate requirement. By adopting inequality relaxation, we then transform the original non-convex optimization problem into a geometric programming (GP) problem.
- Simulation results demonstrate the performance gain of our proposed OTFS-IDET over the conventional OFDM-IDET in high-mobility scenarios.

3.3.2.2 System Model

We study an OTFS-IDET system with a single pair of transmitter and receiver, as illustrated in Fig. 3.23. $x_{\rm I}$ is the $K \times L$ data signal matrix with the elements $x_{\rm I}[k,l] = z_{\rm I}[k,l]x_{\rm D}[k,l]$, where the $z_{\rm I} \sim \mathcal{CN}(0,1)$ is the random information matrix with the elements $z_{\rm I}[k,l]$ and the $x_{\rm D}$ is the matrix of power control with the element $x_{\rm D}[k,l]$. $x_{\rm E}$ is a deterministic $K \times L$ energy signal matrix with the elements $x_{\rm E}[k,l]$. At the transmitter, both are generated in the DD-domain, where $k = 0, 1, \cdots, N-1$ and $l = 0, 1, \cdots, M-1$ are the Doppler offset index and delay index, respectively. Here, M and N denote the number of the subcarriers and the number of the time slots, respectively. The superimposed DD-domain signal can be then expressed as

$$x[k,l] = x_{\rm I}[k,l] + x_{\rm E}[k,l]. \tag{3.66}$$

Next, the ISFFT converts the DD-domain signal to the TF-domain, which can be expressed as

$$\begin{aligned} X[n,m] &= \frac{1}{\sqrt{NM}} \sum_{k=0}^{N-1} \sum_{l=0}^{M-1} x[k,l] e^{j2\pi\left(\frac{nk}{N} - \frac{ml}{M}\right)} \\ &= \frac{1}{\sqrt{NM}} \sum_{k=0}^{N-1} \sum_{l=0}^{M-1} (x_{\rm I}[k,l] + x_{\rm E}[k,l]) e^{j2\pi\left(\frac{nk}{N} - \frac{ml}{M}\right)} \\ &= X_{\rm I}[n,m] + X_{\rm E}[n,m], \end{aligned} \tag{3.67}$$

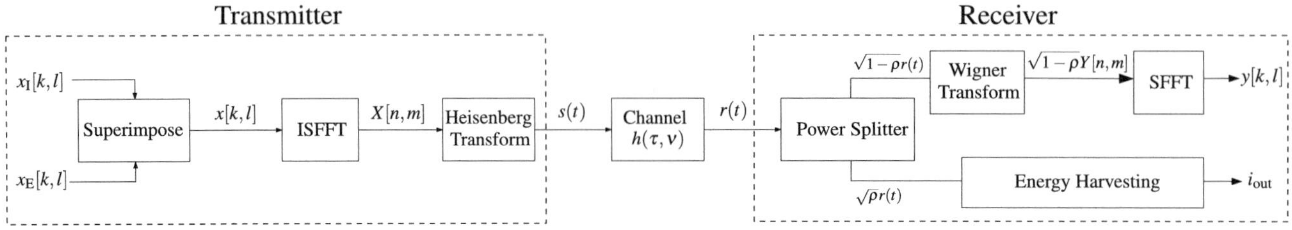

Fig. 3.23 System transmitter and receiver

for $n = 0, 1, \cdots, N-1$ and $m = 0, 1, \cdots, M-1$. X_{I} is the data signal matrix with the elements $X_{\mathrm{I}}[n, m]$, and X_{E} is the energy signal matrix with the elements $X_{\mathrm{E}}[n, m]$ in the TF-domain. The transmit signal $s(t)$ can be obtained through the Heisenberg transform.

The channel[3] fading coefficient $h(\tau, \nu)$ with the imperfect channel estimation in DD-domain is modeled as [74]

$$h(\tau, \nu) = \sum_{i=0}^{P-1} \left(\hat{h}_i + e_i\right) \delta(\tau - \tau_i) \delta(\nu - \nu_i), \tag{3.68}$$

where P denotes the number of propagation paths, and $\hat{h}_i$ represents the estimated channel coefficient on the i-th path with an estimation error of $e_i \sim \mathcal{CN}(0, MN\sigma_e^2/P)$. Moreover, τ_i and ν_i represent delay and Doppler offset of the i-th path, respectively. The parameters $\tau = \frac{1}{M\Delta f}$ and $\nu = \frac{1}{NT}$ represent the resolutions of the DD-domain. Note that $\{M, \Delta f, N, T\}$ are set in the TF-domain, which are the same as the OFDM counterpart. Specifically, T and Δf represent the symbol duration and the subcarrier frequency interval, respectively. Accordingly, the signal bandwidth is $M\Delta f$, while the total time duration is NT.

The corresponding channel coefficient in the TF-domain can be then expressed as

$$H[n, m] = \int_\tau \int_\nu h(\tau, \nu) e^{j2\pi(\nu nT - m\Delta f \tau)} d\tau d\nu. \tag{3.69}$$

By considering the channel estimation error in the DD-domain, $H[n, m]$ can be further expressed as

$$H[n, m] = \hat{H}[n, m] + e[n, m], \tag{3.70}$$

where estimated channel coefficient in the TF-domain is $\hat{H}[n, m]$ with the channel estimation error $e[n, m] \sim \mathcal{CN}(0, \sigma_e^2)$.

The ideal pulse shape[4] is considered in this chapter. By invoking the Wigner transform on the received signal $r(t)$, the channel input–output relation in the TF-domain can be expressed as

$$\begin{aligned} Y[n, m] &= X[n, m]H[n, m] + N_0[n, m], \\ &= (X_{\mathrm{I}}[n, m] + X_{\mathrm{E}}[n, m]) \left(\hat{H}[n, m] + e[n, m]\right) + N_0[n, m], \\ &= Y_{\mathrm{I}}[n, m] + Y_{\mathrm{E}}[n, m] + N_0[n, m], \end{aligned} \tag{3.71}$$

where $N_0[n, m]$ is the additive white Gaussian noise (AWGN), and we can get

$$Y_{\mathrm{I}}[n, m] = X_{\mathrm{I}}[n, m]\left(\hat{H}[n, m] + e[n, m]\right) = \hat{Y}_{\mathrm{I}}[n, m] + X_{\mathrm{I}}[n, m]e[n, m], \tag{3.72}$$

$$Y_{\mathrm{E}}[n, m] = X_{\mathrm{E}}[n, m]\left(\hat{H}[n, m] + e[n, m]\right) = \hat{Y}_{\mathrm{E}}[n, m] + X_{\mathrm{E}}[n, m]e[n, m]. \tag{3.73}$$

By denoting the power splitting ratio as $\rho \in [0, 1]$ at the receiver, we have a portion of received signal $\sqrt{\rho}Y[n, m]$ for EH, while the remaining $\sqrt{1-\rho}Y[n, m]$ for data recovery. After performing SFFT, the received signal $y[k, l]$ in the DD-domain is expressed as

$$y[k, l] = \frac{1}{\sqrt{NM}} \sum_{n=0}^{N-1} \sum_{m=0}^{M-1} \sqrt{1-\rho}Y[n, m]e^{-j2\pi\left(\frac{nk}{N} - \frac{ml}{M}\right)}. \tag{3.74}$$

[3] By adopting some beamformers and combiners, we can readily convert a multiple-input multiple-output (MIMO) channel into an equivalent single input single output (SISO) channel. The OTFS signal can be then designed in the DD-domain by using our method.

[4] The practical pulse shape of OTFS may introduce extra inter-carrier and inter-symbol interference in our model [75], which would be considered in our future work.

3.3.2.3 OTFS-IDET Signal Design

In this chapter, we adopt a diode based non-linear[5] EH model energy harvester [51], which leverages the output DC to evaluate the WET performance. We ignore the contribution of the noise to the WET performance, since it is far lower than the signal power. For clarity, Y, Y_{I}, Y_{E}, H, and $\hat{H}$ are defined as the matrices with the elements $Y[n,m]$, $Y_{\mathrm{I}}[n,m]$, $Y_{\mathrm{E}}[n,m]$, $H[n,m]$, and $\hat{H}[n,m]$, respectively.

First, the random information matrix Z_{I} in the TF-domain is obtained through z_{I} by Eq. (3.67). Second, the corresponding power control matrix for Z_{I} in the TF-domain is defined as X_{D}, which satisfies: $Z_{\mathrm{I}} \otimes X_{\mathrm{D}} = \mathrm{ISFFT}\,(z_{\mathrm{I}} \otimes x_{\mathrm{D}}) = X_{\mathrm{I}}$. Third, the received signal for WDT is expressed as $Y_{\mathrm{I}} = X_{\mathrm{I}} \otimes H = Z_{\mathrm{I}} \otimes X_{\mathrm{D}} \otimes H$. Finally, the receive power control matrix for WDT is obtained as $Y_{\mathrm{D}} = X_{\mathrm{D}} \otimes H$. By considering the channel estimation error, we have $\hat{Y}_{\mathrm{D}} = X_{\mathrm{D}} \otimes \hat{H}$.

The relationship between the output DC i_{out} and the input RF signal Y is then expressed as

$$i_{\mathrm{out}} = k_2 R \psi\left(\left(\sqrt{\rho}Y\right)^2\right) + k_4 R^2 \psi\left(\left(\sqrt{\rho}Y\right)^4\right), \tag{3.75}$$

where k_2, k_4 and R are the circuit parameters, and $\psi(\cdot)$ is the DC component of the received signal. Since random and deterministic signals behave differently in non-linear EH model, Eq. (3.75) can be further reformulated as

$$\begin{aligned} i_{\mathrm{out}} &= k_2 \rho R \left(\psi\left(Y_{\mathrm{E}}^2\right) + \psi\left(Y_{\mathrm{D}}^2\right)\right) \\ &+ k_4 \rho^2 R^2 \left(\psi\left(Y_{\mathrm{E}}^4\right) + \psi\left(Y_{\mathrm{D}}^4\right) + 6\psi\left(Y_{\mathrm{E}}^2\right) \times \psi\left(Y_{\mathrm{D}}^2\right)\right), \end{aligned} \tag{3.76}$$

where we have

$$\psi\left(Y_{\mathrm{D}}^2\right) = \frac{1}{2}\sum_{n=0}^{N-1}\sum_{m=0}^{M-1}\left(\left|\hat{Y}_{\mathrm{D}}[n,m]\right|^2 + |X_{\mathrm{D}}[n,m]|^2\sigma_e^2\right), \tag{3.77a}$$

$$\psi\left(Y_{\mathrm{E}}^2\right) = \frac{1}{2}\sum_{n=0}^{N-1}\sum_{m=0}^{M-1}\left(\left|\hat{Y}_{\mathrm{E}}[n,m]\right|^2 + |X_{\mathrm{E}}[n,m]|^2\sigma_e^2\right), \tag{3.77b}$$

$$\psi\left(Y_{\mathrm{E}}^4\right) = \frac{3}{8}\sum_{n=0}^{N-1}\sum_{\substack{m_0,m_1,m_2,m_3\\ m_0+m_1=m_2+m_3}}^{M-1}\left(\prod_{j=0}^{3}\left|\hat{Y}_{\mathrm{E}}[n,m_j]\right| + \prod_{j=0}^{3}|X_{\mathrm{D}}[n,m_j]|\sigma_e\right). \tag{3.77c}$$

$$\psi\left(Y_{\mathrm{D}}^4\right) = \frac{3}{4}\sum_{n=0}^{N-1}\left(\sum_{m=0}^{M-1}\left|\hat{Y}_{\mathrm{D}}[n,m]\right|^2 + |X_{\mathrm{D}}[n,m]|^2\sigma_e^2\right)^2. \tag{3.77d}$$

The WDT performance of the OTFS-IDET system is characterized in terms of the data rate. At the transmitter, randomly modulated signals in the DD-domain are converted into the TF-domain via the ISFFT, which still obey Gaussian distribution.

The received signal is split into the power-domain at the receiver, and the remaining signal power $P_{\mathrm{D}}[n,m]$ for data recovery is expressed as

$$P_{\mathrm{D}}[n,m] = (1-\rho)\left(|\hat{Y}_{\mathrm{D}}[n,m]|^2 + |X_{\mathrm{D}}[n,m]|^2\sigma_e^2\right), \tag{3.78}$$

where $\hat{P}_{\mathrm{D}}[n,m] = (1-\rho)|\hat{Y}_{\mathrm{D}}[n,m]|^2$ is the useful signal power, and $(1-\rho)X_{\mathrm{D}}[n,m]\sigma_e^2$ is the interference caused by the channel estimation error. Similarly, the power $P_{\mathrm{E}}[n,m]$ of the received energy signal that may interfere the data recovery is expressed as

[5] The linear EH model cannot accurately capture the relationship between the output DC i_{out} and the input signal Y, and we decide to design our OTFS signal by considering the nonlinear EH model of Eq. (3.76).

$$P_{\mathrm{E}}[n,m] = (1-\rho)\left(|\hat{Y}_{\mathrm{E}}[n,m]|^2 + |X_{\mathrm{E}}[n,m]|^2\sigma_e^2\right). \tag{3.79}$$

The parameter λ is defined as the remaining interference caused by the energy single on the data recovery at the receiver. Hence, the interference at the n-th time slot and the m-th subcarrier can be expressed as $\lambda P_{\mathrm{E}}[n,m]$. As a result, the data rate C_n at the n-th time slot can be characterized by the classic Shannon–Hartley capacity, which is expressed as

$$C_n = \sum_{m=0}^{M-1} \log_2\left(1 + \left(\frac{\lambda(1-\rho)\left(|\hat{Y}_{\mathrm{E}}[n,m]|^2 + |X_{\mathrm{E}}[n,m]|^2\sigma_e^2\right)}{(1-\rho)|\hat{Y}_{\mathrm{D}}[n,m]|^2} + \frac{(1-\rho)|X_{\mathrm{D}}[n,m]|^2\sigma_e^2 + P_{\mathrm{noise}}}{(1-\rho)|\hat{Y}_{\mathrm{D}}[n,m]|^2}\right)^{(-1)}\right), \tag{3.80}$$

where P_{noise} represents the Gaussian noise power. Hence, the average data rate C per time slot can be expressed as

$$C = \frac{\sum_{n=0}^{N-1} C_n}{N}. \tag{3.81}$$

Given a budget of total transmit power P_{tx}, the amplitudes of transmit data signal x_{D} and the energy signal x_{E} in the DD-domain at the transmitter and the power splitting ratio ρ at the receiver should be jointly designed for maximizing the output DC i_{out}, while satisfying the minimum data rate requirement $R_{\min}$. The fixed power of the pilot used for the channel estimation is assumed as P_{p}. Hence, the corresponding optimization problem is formulated as

$$(\text{P3.6.7}): \arg\max_{x_{\mathrm{D}}, x_{\mathrm{E}}, \rho} \quad i_{\mathrm{out}}(x_{\mathrm{D}}, x_{\mathrm{E}}, \rho) \tag{3.82}$$

$$\text{s.t} \quad C(x_{\mathrm{D}}, x_{\mathrm{E}}, \rho) \geq R_{\min}, \tag{3.82a}$$

$$\frac{1}{2}\left[||x_{\mathrm{D}}||_{m_2}^2 + ||x_{\mathrm{E}}||_{m_2}^2\right] \leq P_{\mathrm{o}}, \tag{3.82b}$$

$$\sum_{m=0}^{M-1} f_1[n,m] \leq P_{\mathrm{peak}}, \forall n, \tag{3.82c}$$

$$\rho \in [0,1], \tag{3.82d}$$

where $P_{\mathrm{o}} = P_{\mathrm{tx}} - P_{\mathrm{p}}$ is the power limitation for the data and energy transfer, and P_{peak} is the maximum peak power in each time slot that can be achieved by the power amplifier at the transmitter. The $f_1[n,m] = |X_{\mathrm{D}}[n,m]|^2 + |X_{\mathrm{E}}[n,m]|^2$, and m_2 represents the Frobenius norm.

Unfortunately, this is a non-convex optimization problem. Then, we transform (P3.6.7) to a GP [76] to obtain a solution with a reasonable computational complexity. Note that both the WDT and WET performance can be expressed in the TF-domain. And, the optimization variables $x_{\mathrm{D}}[k,l]$ and $x_{\mathrm{E}}[k,l]$ can be derived by the $X_{\mathrm{D}}[n,m]$ and $X_{\mathrm{E}}[n,m]$, respectively.

Hence, the objective of (P3.6.7) becomes $i_{\mathrm{out}}(X_{\mathrm{D}}, X_{\mathrm{E}}, \rho)$. Let $g_{k_1}(X_{\mathrm{D}}, X_{\mathrm{E}}, \rho)$ represent the monomial terms of it, where $i_{\mathrm{out}}(X_{\mathrm{D}}, X_{\mathrm{E}}, \rho) = \sum_{k_1=0}^{K_1-1} g_{k_1}(X_{\mathrm{D}}, X_{\mathrm{E}}, \rho)$. In the same way, the $g_{k_2}(X_{\mathrm{D}}, X_{\mathrm{E}}, \rho)$ is defined as the momomial of $\kappa[n,m]$ satisfying $\kappa[n,m] = \sum_{k_2=0}^{K_2-1} g_{k_2}(X_{\mathrm{D}}, X_{\mathrm{E}}, \rho) = \lambda P_{\mathrm{E}}[n,m] + P_{\mathrm{D}}[n,m] + P_{\mathrm{noise}}$. The K_1 and K_2 respectively represent that the number of monomials in $i_{\mathrm{out}}(X_{\mathrm{D}}, X_{\mathrm{E}}, \rho)$ and $\kappa[n,m]$.

Then the data rate constraint (3.82a) can be expressed as

$$2^{R_{\min}} \prod_{m=0}^{M-1} \left(\lambda P_{\mathrm{E}}[n,m] + (1-\rho)X_{\mathrm{D}}[n,m]\sigma_e^2 + P_{\mathrm{noise}}\right)/\kappa[n,m] \leq 1, \forall n. \tag{3.83}$$

Furthermore, by introducing the parameter $\gamma[m]$ and

$$f_2[n,m] = \left(\lambda P_{\mathrm{E}}[n,m] + (1-\rho)X_{\mathrm{D}}[n,m]\sigma_e^2 + P_{\mathrm{noise}}\right)/\gamma[m], \tag{3.84}$$

Algorithm 3.6 Signal design for OTFS-IDET

Input:

The DD-domain channel estimation $\hat{h}(\tau,\nu)$; Minimum data rate $R_{\min}$;Total transmission power P_{tx}; Pilot power P_{p}; Maximum peak power P_{peak}; Required accuracy ϵ;

Output:

The OTFS data signal and energy signal amplitude $x_{\text{D}}[k,l]$ and $x_{\text{E}}[k,l]$; The power splitting ratio ρ;

1: Calculate the $\hat{H}[n,m]$ through $\hat{h}(\tau,\nu)$; Calculate the power for optimization $P_{\text{o}} = P_{\text{tx}} - P_{\text{p}}$; the average fading of channel: $h_{\text{L}} = \sum_{n=1}^{N}\sum_{m=1}^{M}(\hat{H}[n,m]^2+\sigma^2)/MN$; the estimation of maximum transmission: $C_{\max} = \log(1+P_{\text{o}}h_{\text{L}}/P_{\text{noise}})$; the initialise factor: $\zeta = R_{\min}/C_{\max}$.

2: Initialise $X_{\text{D}}[n,m]=\sqrt{P_{\text{o}}\zeta/MN}$, $X_{\text{E}}[n,m]=\sqrt{P_{\text{o}}(1-\zeta)/MN}$, $\rho=1-\zeta$, $\bar{\rho}=\zeta$, $i_{\text{out}}^{(0)}=0$, $i=0$, $t_{k_1}=1/K_1$, $w_{k_2}=1/K_2$;

3: **repeat**

4: Calculate $\hat{i}_{\text{out}}$ and $\hat{\kappa}[n,m]$ by the formula Eq. (3.85) and Eq. (3.86);

5: Update X_{D}, X_{E}, ρ, $\bar{\rho}$ by solving (P3.6.8);

6: Calculate $i_{\text{out}}^{(i+1)}$ by the formula Eq. (3.75);

7: Update t_{k_1},w_{k_2} by the formula Eq. (3.88);

8: Update $i \leftarrow i+1$

9: **until** $|i_{\text{out}}^{(i)} - i_{\text{out}}^{(i-1)}| < \epsilon$

10: $x_{\text{D}}[k,l] = \text{SFFT}(X_{\text{D}}[n,m])$ and $x_{\text{E}}[k,l] = \text{SFFT}(X_{\text{E}}[n,m])$

the constraint can be divided into two parts: $2^{R_{\min}}\prod_{m=0}^{M-1} f_2[n,m] \le 1, \forall n$ and $\gamma[m]/\kappa[n,m] \le 1, \forall n, \forall m$.

With the variable η, the original maximization problem can be transformed into a minimization of $1/\eta$ with the constraint $\eta/\text{i}_{\text{out}}(\text{X}_{\text{D}},\text{X}_{\text{E}},\boldsymbol{\rho}) \le 1$. In order to satisfy the form of GP, the summation in the denominator is converted into the multiplication via the arithmetic mean-geometric mean (AM–GM) inequality as

$$\hat{i}_{\text{out}}(X_{\text{D}},X_{\text{E}},\rho) = \prod_{k_1=0}^{K_1-1}\left(\frac{g_{k_1}(X_{\text{D}},X_{\text{E}},\rho)}{t_{k_1}}\right)^{-t_{k_1}} \le i_{\text{out}}(X_{\text{D}},X_{\text{E}},\rho), \tag{3.85}$$

$$\hat{\kappa}[n,m] = \prod_{k_2=0}^{K_2-1}\left(\frac{g_{k_2}(X_{\text{D}},X_{\text{E}},\rho)}{w_{k_2}}\right)^{-w_{k_2}} \le \kappa[n,m], \tag{3.86}$$

where t_{k_1}, w_{k_2} satisfying $\sum t_{k_1}=1$, $\sum w_{k_2}=1$ are the weights of the above two inequalities. Now, the standard GP can be formulated as

$$(\text{P3.6.8}):\ \underset{X_{\text{D}},X_{\text{E}},\rho,\bar{\rho}}{\arg\min}\ \frac{1}{\eta} \tag{3.87}$$

$$\text{s.t}\ \ \eta/\hat{\text{i}}_{\text{out}}(X_{\text{D}},X_{\text{E}},\boldsymbol{\rho}) \le 1, \tag{3.87a}$$

$$2^{R_{min}}\prod_{m=0}^{M-1} f_2[n,m] \le 1, \forall n, \tag{3.87b}$$

$$\gamma[m]/\hat{\kappa}[n,m] \le 1, \forall n, \forall m, \tag{3.87c}$$

$$\frac{1}{2}\left[||X_{\text{D}}||_{m_2}^2 + ||X_{\text{E}}||_{m_2}^2\right] \le P_{\text{o}}, \tag{3.87d}$$

$$\sum_{m=0}^{M-1} f_1[n,m] \le P_{\text{peak}}, \forall n, \tag{3.87e}$$

$$\rho + \bar{\rho} \le 1, \tag{3.87f}$$

where $\bar{\rho}$ represents the splitting ratio for data recovery.

Table 3.2 Simulation parameters

Parameter	Value
Carrier frequency f_c (GHz)	27
Subcarrier frequency interval Δf (KHz)	15
OTFS frame size $[N, M]$	[12,12]
Number of paths P	3
UE speed (Kmph)	30,150,300
Channel	Ideal and imperfect

The optimization algorithm is summarized in Algorithm 3.1, in which the gap between the intermediate variables $\left\{\hat{i}_{\text{out}}, \hat{\kappa}[n,m]\right\}$ and the upper bound $\{i_{\text{out}}, \kappa[n,m]\}$ will be narrow by constantly modifying the weights t_{k_1} and w_{k_2} to approach the optimal value. The updated formulas of t_{k_1} and w_{k_2} are given as

$$t_{k_1} = \frac{g_{k_1}(X_D, X_E, \rho)}{i_{\text{out}}(X_D, X_E, \rho)}, \qquad w_{k_2} = \frac{g_{k_2}(X_D, X_E, \rho)}{\kappa[n,m]}. \tag{3.88}$$

By invoking the Ellipsoid method [77], (P3.6.8) can be solved with the complexity $\mathcal{O}\left((2MN+1)^2\right)$. Suppose that δ iterations are required to let Algorithm 3.1 converge, the total complexity becomes $\mathcal{O}\left(\delta\,(2MN+1)^2\right)$.

3.3.2.4 Numerical Results

The OFDM-IDET system is considered as the benchmark of our OTFS-IDET system. The parameters of the non-linear energy harvester are set as $k_2 = 0.0034$, $k_4 = 0.3859$, and $R = 50\ \Omega$ [51]. Additionally, the transmit power P_{tx} and P_p are set as 36.1 dBm and 20 dBm; the receive antenna gain is 2 dBi; the large-scale path-loss is −50 dB. The system parameters employed in our simulations are summarized in Table 3.2 as [78]. Especially, the maximum delay index is set as $l_{\tau_m} = \tau_m M \Delta f = 6$ with the maximum delay $\tau_m = 0.3$ ms. The maximum Doppler index is set as $k_{\nu_m} = (f_c V_m N)/(V_c \Delta f) = 6$, where $V_m = 300$ Kmph is the maximum speed in the scenarios and V_c is the speed of the light. Additionally, the Doppler tap of the i-th path is generated with equal probabilities from the set $\{-k_{\nu_m}, -k_{\nu_m}+1, \ldots, 0, 1, \ldots, k_{\nu_m}\}$, while the delay tap set is $\{1, \ldots, l_{\tau_m}\}$ and $\tau_0 = 0$.

First, the impact of the mobility on the OTFS-IDET and OFDM-IDET system is shown in Fig. 3.24a. Since the Doppler offset cuasey speed makes the TF channel $H[n,m]$ fluctuate, the appropriate speed is beneficial to the OTFS-IDET system. Doppler offset leads to serious inter-carrier interference (ICI) in the OFDM-IDET system. The OTFS signal can be designed by considering the channel characteristics in the DD-domain, while the OFDM signal can only be designed in the frequency domain. Therefore, the OTFS signal design with additional degree of freedom is expected to have a better IDET performance than the OFDM counterpart. Therefore, the OTFS-IDET system outperforms the OFDM-IDET counterpart in different speed scenarios. Observe from Fig. 3.24b that the output DC i_{out} gradually increases as we have more iterations. Normally, Algorithm 3.1 converges after 7 iterations.

Second, we investigate the impact of the residual energy signal interference on the OTFS-IDET system in Fig. 3.25. Since a larger value of λ indicates more residual interference imposed by the energy signals on the data recovery at the receiver, the IDET performance is degraded, as shown in Fig. 3.25. When the maximum data transmission rate is attained, the transmitter only sends the data signals without energy signal; in order to obtain the maximum DC, only the energy signals are transmitted, which results in varying λ makes no difference in the output DC performance with no requirement of data rate. Hence, all the curves share the same start and end points in the coordinates.

Finally, the OTFS-IDET performance with the imperfect channel estimation is portrayed in Fig. 3.26. Observe from Fig. 3.26 the performance is reduced, as the variance of the channel estimation error increases from $\sigma^2 = 0$ to $\sigma^2 = 0.05$. This is because the OTFS signal designed with the inaccurate channel estimation sacrifices the optimality to some extent. Due to the estimation error, the performance of the OTFS-IDET system is reduced in terms of EH and data transmission. When the gap between the estimation channel and the perfect channel becomes larger, the optimized parameters become more unreasonable. Therefore, the performance of OTFS-IDET with the estimation error $\sigma^2 = 0.05$ is worse than the counterpart with $\sigma^2 = 0.01$.

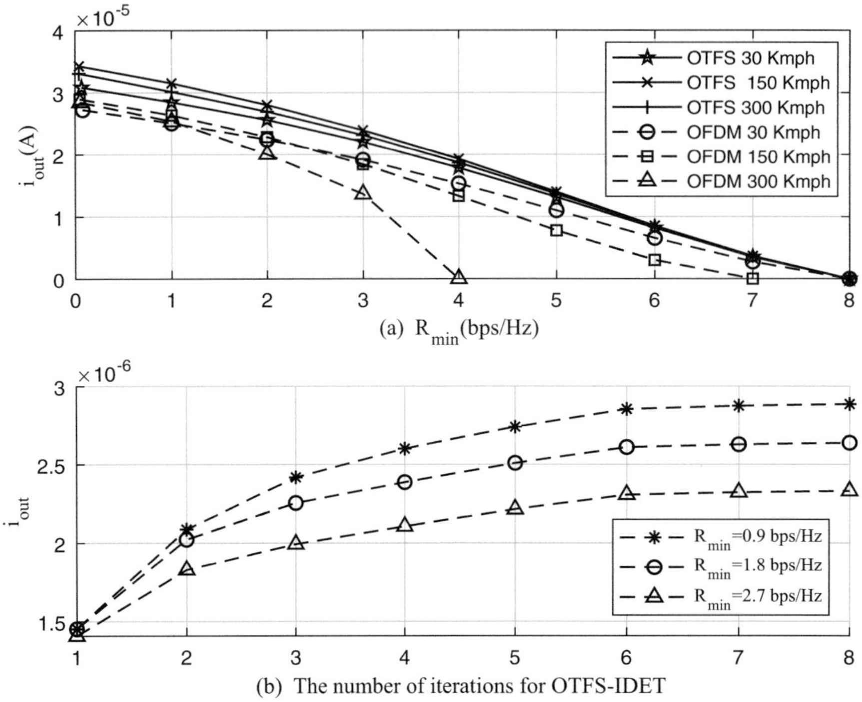

Fig. 3.24 (**a**) The performance comparison between OTFS-IDET and OFDM-IDET with different speeds, where $\sigma^2 = 0$; (**b**) The EH performance of the OTFS-IDET system under the imperfect channel estimation

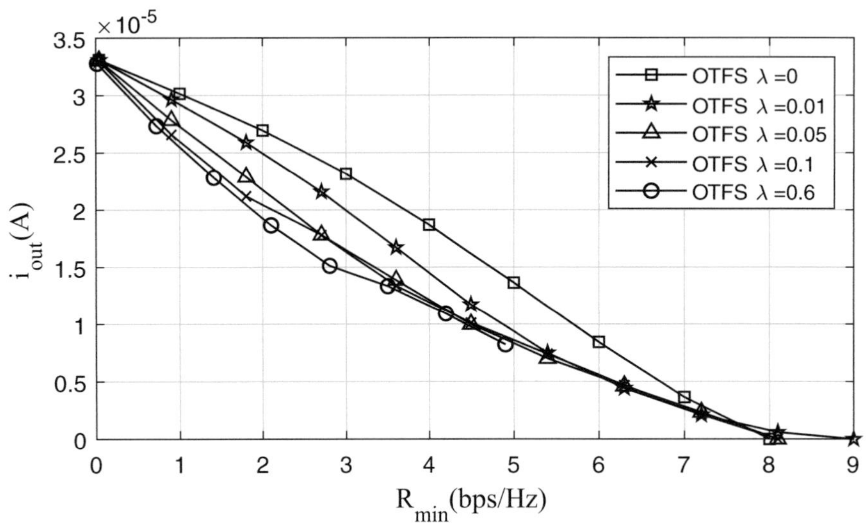

Fig. 3.25 The performance comparison between OTFS-IDET and OFDM-IDET with different λ

3.4 Exercises

1. Find an alternate set of orthonormal basis functions for the space spanned by $cos(2\pi t/T)$ and $sin(2\pi t/T)$.
2. Consider a 4-PSK constellation with $d_{min} = \sqrt{2}$. What is the additional energy required to send one extra bit (8-PSK) while keeping the same minimum distance (and thus with the same bit error probability)?
3. Show that the average power of a square signal constellation with l bits per dimension, P_l, is proportional to $4^l/3$ and that the average power for one more bit per dimension, keeping the same minimum distance, is $P_{l+1} \approx 4P_l$. Find P_l for $l = 2$ and compute the average energy of MPSK and MPAM constellations with the same number of bits per symbol.

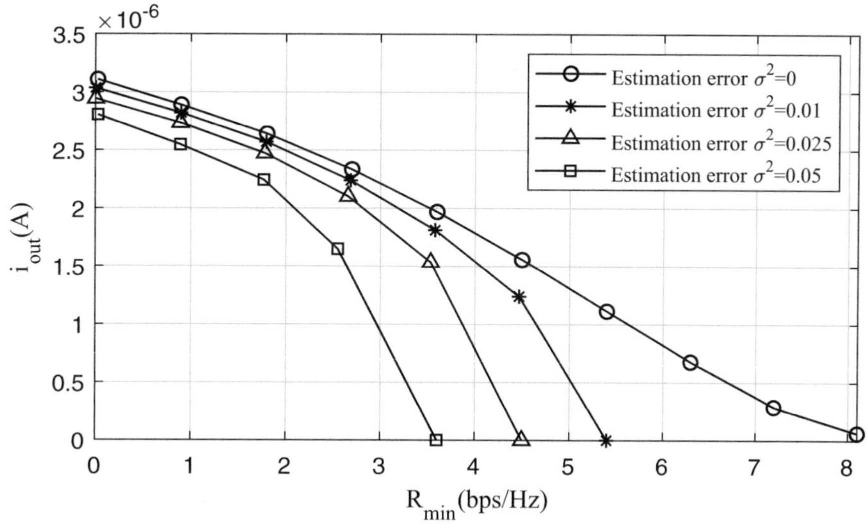

Fig. 3.26 The EH performance of the OTFS-IDET system under the channel estimation error

4. For MPSK with differential modulation, let $\Delta\phi$ denote the phase drift of the channel over a symbol time T_s. In the absence of noise, how large must $\Delta\phi$ be in order for a detection error to occur?
5. Suppose the four subchannels in 802.11a used for pilot estimation could be used for data transmission by taking advantage of blind estimation techniques. Assuming the same modulation and coding formats are available, what maximum and minimum data rates could be achieved by including these extra subchannels?
6. Consider a data signal with a bandwidth of 0.5 MHz and a data rate of 0.5 Mbps. The signal is transmitted over a wireless channel with a delay spread of 10 μs. If multi-carrier modulation with non-overlapping subchannels is used to mitigate the effects of ISI, approximately how many subcarriers are needed? What is the data rate and symbol time on each subcarrier? (We do not need to eliminate the ISI completely, so $T_s = T_m$ is sufficient for ISI mitigation.)

References

1. Zhang, G., Yang, K., Liu, P., & Wei, J. (2015). Power allocation for full-duplex relaying-based D2D communication underlaying cellular networks. *IEEE Transactions on Vehicular Technology, 64*(10), 4911–4916. ISSN: 0018–9545.
2. Wang, K., Yang, K., & Magurawalage, C. S. (2018). Joint energy minimization and resource allocation in C-RAN with mobile cloud. *IEEE Transactions on Cloud Computing, 6*(3), 760–770. ISSN: 2168–7161.
3. Xu, G., Li, H., Dai, Y., Yang, K., & Lin, X. (2019). Enabling efficient and geometric range query with access control over encrypted spatial data. *IEEE Transactions on Information Forensics and Security, 14*(4), 870–885.
4. Ding, Z., Lei, X., Karagiannidis, G. K., Schober, R., Yuan, J., & Bhargava, V. K. (2017). A survey on non-orthogonal multiple access for 5G networks: Research challenges and future trends. *IEEE Journal on Selected Areas in Communications, 35*(10), 2181–2195. ISSN: 0733–8716.
5. Ding, Z., Liu, Y., Choi, J., Sun, Q., Elkashlan, M., Chih-Lin, I., & Poor, H. V. (2017). Application of non-orthogonal multiple access in LTE and 5G networks. *IEEE Communications Magazine, 55*(2), 185–191.
6. Huang, Y., Zhang, C., Wang, J., Jing, Y., Yang, L., & You, X. (2018). Signal processing for MIMO-NOMA: Present and future challenges. *IEEE Wireless Communications, 25*(2), 32–38. ISSN: 1536–1284.
7. Zeng, M., Yadav, A., Dobre, O. A., & Poor, H. V. (2018). Energy-efficient power allocation for MIMO-NOMA with multiple users in a cluster. *IEEE Access, 6*, 5170–5181.
8. Zhang, L., Liu, J., Xiao, M., Wu, G., Liang, Y. C., & Li, S. (2017). Performance analysis and optimization in downlink NOMA systems with cooperative full-duplex relaying. *IEEE Journal on Selected Areas in Communications, 35*(10), 2398–2412. ISSN: 0733–8716.
9. Zhou, X., Ho, C. K., & Zhang, R. (2016). Wireless power meets energy harvesting: A joint energy allocation approach in OFDM-based system. *IEEE Transactions on Wireless Communications, 15*(5), 3481–3491. ISSN: 1536–1276.
10. Hu, J., Kun, Y., Guangjun, W., & Lajos, H. (2018). Integrated data and energy communication network: A comprehensive survey. *IEEE Communications Surveys Tutorials, 20*(4), 3169–3219.
11. Lee, H., Lee, K., Kim, H., & Lee, I. (2018). Joint transceiver optimization for MISO SWIPT systems with time switching. *IEEE Transactions on Wireless Communications, 17*(5), 3298–3312. ISSN: 1558–2248.
12. Hu, Z., Yuan, C., & Gao, F. (2017). Maximizing harvested energy for full-duplex SWIPT system with power splitting. *IEEE Access, 5*, 24 975–24 987.

13. Guo, S., Zhang, H., Wang, Y., & Yuan, D. (2016). Spatial modulated simultaneous wireless information and power transfer. In *2016 IEEE Global Communications Conference (GLOBECOM)* (pp. 1–6).
14. Yang, Z., Ding, Z., Fan, P., & Al-Dhahir, N. (2017). The impact of power allocation on cooperative non-orthogonal multiple access networks with SWIPT. *IEEE Transactions on Wireless Communications, 16*(7), 4332–4343. ISSN: 1536–1276.
15. Liu, Y., Ding, Z., Elkashlan, M., & Poor, H. V. (2016). Cooperative non-orthogonal multiple access with simultaneous wireless information and power transfer. *IEEE Journal on Selected Areas in Communications, 34*(4), 938–953. ISSN: 0733–8716.
16. Xu, Y., Shen, C., Ding, Z., Sun, X., Yan, S., Zhu, G., & Zhong, Z. (2017). Joint beamforming and power-splitting control in downlink cooperative SWIPT NOMA systems. *IEEE Transactions on Signal Processing, 65*(18), 4874–4886. ISSN: 1053–587X.
17. Diamantoulakis, P. D., Pappi, K. N., Ding, Z., & Karagiannidis, G. K. (2016). Wireless-powered communications with non-orthogonal multiple access. *IEEE Transactions on Wireless Communications, 15*(12), 8422–8436. ISSN: 1536–1276.
18. Alsaba, Y., Leow, C. Y., & Rahim, S. K. A. (2018). Full-duplex cooperative non-orthogonal multiple access with beamforming and energy harvesting. *IEEE Access, 6*, 19 726–19 738. ISSN: 2169–3536.
19. Hedayati, M., & Kim, I. (2018). On the performance of NOMA in the two-user SWIPT system. *IEEE Transactions on Vehicular Technology, 67*(11), 11 258–11 263. ISSN: 0018–9545.
20. Dai, L., Wang, B., Peng, M., & Chen, S. (2019). Hybrid precoding-based millimeter-wave massive MIMO-NOMA with simultaneous wireless information and power transfer. *IEEE Journal on Selected Areas in Communications, 37*(1), 131–141. ISSN: 0733–8716.
21. Wu, W., Yin, X., Deng, P., Guo, T., & Wang, B. (2019). Transceiver design for downlink SWIPT NOMA systems with cooperative full-duplex relaying. *IEEE Access, 7*, 33 464–33 472. ISSN: 2169–3536.
22. Hu, J., Zhao, Y., & Yang, K. (2019). Modulation and coding design for simultaneous wireless information and power transfer. *IEEE Communications Magazine, 57*(5), 124–130. ISSN: 1558–1896.
23. Zhang, M., & Cheng, X. (2017). Spatial-modulation-based wireless-powered communication for achievable rate enhancement. *IEEE Communications Letters, 21*(6), 1365–1368. ISSN: 1089–7798.
24. Mohjazi, L., Muhaidat, S., & Dianati, M. (2016). Performance analysis of differential modulation in SWIPT cooperative networks. *IEEE Signal Processing Letters, 23*(5), 620–624. ISSN: 1070–9908.
25. Jiang, C., & Wu, Z. (2017). A novel uplink noma scheme based on low density superposition modulation. In *2017 IEEE 86th Vehicular Technology Conference (VTC-Fall)* (pp. 1–5).
26. Niharika, G. (2018). Performance comparison of modulation schemes for downlink NOMA. In *2018 2nd International Conference on Inventive Systems and Control (ICISC)* (pp. 1431–1433).
27. Liu, Y., Yang, L., & Hanzo, L. (2018). Spatial modulation aided sparse code-division multiple access. *IEEE Transactions on Wireless Communications, 17*(3), 1474–1487. ISSN: 1536–1276.
28. Li, Y., & Moon, J. (2007). Error probability bounds for bit-interleaved space–time trellis coding over block-fading channels. *IEEE Transactions on Information Theory, 53*(11), 4285–4292. ISSN: 0018–9448.
29. Ning, J., & Fu, M. (2009). Design of serially concatenated continuous phase modulation with symbol-wise interleaving. *IEEE Communications Letters, 13*(10), 785–787. ISSN: 1089–7798.
30. Ye, N., Wang, A., Li, X., Liu, W., Hou, X., & Yu, H. (2018). On constellation rotation of NOMA with SIC receiver. *IEEE Communications Letters, 22*(3), 514–517. ISSN: 1089–7798.
31. Lin, C., Shieh, S., Chi, T., & Chen, P. (2019). Optimal inter-constellation rotation based on minimum distance criterion for uplink NOMA. *IEEE Transactions on Vehicular Technology, 68*(1), 525–539. ISSN: 0018–9545.
32. Narayanan, S., Shikh-Bahaei, M. Hou, J. & Flanagan, M. F. (2018). Wireless-powered distributed spatial modulation with energy recycling and finite-energy storage. *IEEE Transactions on Wireless Communications, 17*(10), 6645–6662. ISSN: 1558–2248.
33. Simon, M. K., & Alouini, M. S. (2000). *Digital communication over fading channels: A unified approach to performance analysis*. Wiley.
34. Belebi, M., & Arslan, H. (2015). Theoretical analysis of the co-existence of LTE-A signals and design of an ML-SIC receiver. *IEEE Transactions on Wireless Communications, 14*(8), 4626–4639. ISSN: 1536–1276.
35. Yuan, J.-S., & Kritchanchai, E. (2016). RF energy harvesting using emerging TFET technology. In *2016 13th IEEE International Conference on Solid-State and Integrated Circuit Technology (ICSICT)* (pp. 49–52).
36. Bezdek, J. C., & Hathaway, R. J. (2003). Convergence of alternating optimization. *Neural Parallel and Scientific Computations, 11*(4), 351–368.
37. Timotheou, S., & Krikidis, I. (2015). Fairness for non-orthogonal multiple access in 5G systems. *IEEE Signal Processing Letters, 22*(10), 1647–1651. ISSN: 1070–9908.
38. Hu, J., Wang, Q., & Yang, K. (2021). Energy self-sustainability in full-spectrum 6G. *Accepted by IEEE Wireless Communications, 28*(1), 104–111.
39. Yue, Q., Hu, J., Yang, K., & Huang, C. (2020). Transceiver design for simultaneous wireless information and power multicast in multi-user mmwave mimo system. *IEEE Transactions on Vehicular Technology, 69*(10), 11 394–11 407.
40. Zhu, F., Gao, F., Eldar, Y. C., & Qian, G. (2019). Robust simultaneous wireless information and power transfer in beamspace massive MIMO. *IEEE Transactions on Wireless Communications, 18*(9), 4199–4212.
41. Zhao, Y., Hu, J., Xie, A., Yang, K., & Wong, K. (2020). Receive spatial modulation aided simultaneous wireless information and power transfer with finite alphabet. *IEEE Transactions on Wireless Communications, 19*(12), 8039–8053.
42. Pan, C., Ren, H., Wang, K., Elkashlan, M., Nallanathan, A., Wang, J., & Hanzo, L. (2020). Intelligent reflecting surface aided MIMO broadcasting for simultaneous wireless information and power transfer. *IEEE Journal on Selected Areas in Communications, 38*(8), 1719–1734.
43. Zhao, Y., Hu, J., Yang, K., & Wong, K.-K. (2021). Unary coding design for simultaneous wireless information and power transfer with practical M-QAM. *IEEE Transactions on Wireless Communications, 20*(5), 2850–2862. ISSN: 1558–2248.
44. Hu, J., Li, M., Yang, K., Ng, S. X., & Wong, K. (2020). Unary coding controlled simultaneous wireless information and power transfer. *IEEE Transactions on Wireless Communications, 19*(1), 637–649.
45. Hu, J., Liang, G., Yu, Q., Yang, K., & Lu, X. (2022). Simultaneous wireless information and power transfer with fixed and adaptive modulation. *Digital Communications and Networks, 8*(3), 303–313. ISSN: 2352–8648.

46. Zhao, Y., Hu, J., Ding, Z., & Yang, K. (2019). Joint interleaver and modulation design for Multi-User SWIPT-NOMA. *IEEE Transactions on Communications, 67*(10), 7288–7301.
47. Clerckx, B., & Bayguzina, E. (2016). Waveform design for wireless power transfer. *IEEE Transactions on Signal Processing, 64*(23), 6313–6328. ISSN: 1053–587X.
48. Kim, D. I., Moon, J. H., & Park, J. J. (2016). New SWIPT using PAPR: How it works. *IEEE Wireless Communications Letters, 5*(6), 672–675.
49. Krikidis, I., & Psomas, C. (2019). Tone-index multisine modulation for SWIPT. *IEEE Signal Processing Letters, 26*(8), 1252–1256.
50. Zhao, X., Zhang, Y., Geng, S., Du, F., Zhou, Z., & Yang, L. (2020). Hybrid precoding for an adaptive interference decoding SWIPT system with full-duplex IoT devices. *IEEE Internet of Things Journal, 7*(2), 1164–1177.
51. Clerckx, B. (2018). Wireless information and power transfer: Nonlinearity, waveform design, and rate-energy tradeoff. *IEEE Transactions on Signal Processing, 66*(4), 847–862.
52. Kim, J., Clerckx, B., & Mitcheson, P. D. (2019). Experimental analysis of harvested energy and throughput trade-off in a realistic SWIPT system. In *2019 IEEE Wireless Power Transfer Conference (WPTC)*. IEEE (pp. 1–5).
53. Buckley, R. F., & Heath, R. W. (2021). System and design for selective OFDM SWIPT transmission. *IEEE Transactions on Green Communications and Networking, 5*(1), 335–347.
54. Khormuji, M. N., Popović, B. M., & Perotti, A. G. (2019). Enabling SWIPT via OFDM-DC. In *2019 IEEE Wireless Communications and Networking Conference (WCNC)* (pp. 1–6).
55. Xu, D., & Zhu, H. (2019). Outage minimized resource allocation for multiuser OFDM systems with swipt. *IEEE Access, 7*, 79 714–79 725.
56. Tusha, A., Doğan, S., & Arslan, H. (2019). Performance analysis of frequency domain IM schemes under CFO and IQ imbalance. In *2019 IEEE 30th Annual International Symposium on Personal, Indoor and Mobile Radio Communications (PIMRC)* (pp. 1–5).
57. Savazzi, S., & Spagnolini, U. (2009). Optimizing training lengths and training intervals in time-varying fading channels. *IEEE Transactions on Signal Processing, 57*(3), 1098–1112.
58. Chun, C., Kang, J., & Kim, I. (2018). Adaptive rate and energy harvesting interval control based on reinforcement learning for SWIPT. *IEEE Communications Letters, 22*(12), 2571–2574. ISSN: 1089–7798.
59. Zheng, J., Zhang, J., Björnson, E., & Ai, B. (2021). Impact of channel aging on cell-free massive mimo over spatially correlated channels. *IEEE Transactions on Wireless Communications, 20*(10), 6451–6466.
60. Clerckx, B., Zhang, R., Schober, R., Ng, D. W. K., Kim, D. I., & Poor, H. V. (2019). Fundamentals of wireless information and power transfer: From RF energy harvester models to signal and system designs. *IEEE Journal on Selected Areas in Communications, 37*(1), 4–33. ISSN: 1558–0008.
61. Medard, M. (2000). The effect upon channel capacity in wireless communications of perfect and imperfect knowledge of the channel. *IEEE Transactions on Information Theory, 46*(3), 933–946.
62. Boyd, S., Kim, S. J., Vandenberghe, L., & Hassibi, A. (2007). A tutorial on geometric programming. *Optimization & Engineering, 8*(1), 67.
63. Grant, M. (2015). CVX: Matlab software for disciplined convex programming. http://cvxr.com/cvx/
64. Bland, R. G., Goldfarb, D., & Todd, M. J. (1981). Ellipsoid method: A survey. *Operations Research, 29*(6), 1039–1091.
65. Zhang, L. (2018). A gravitational artificial bee colony optimization algorithm and application. In *2018 Eighth International Conference on Instrumentation Measurement, Computer, Communication and Control (IMCCC)* (pp. 1839–1842).
66. Cheng, L., Yu, M., Yang, J., & Wang, Y. (2019). An improved artificial bee colony algorithm based on beetle antennae search. In *2019 Chinese Control Conference (CCC)* (pp. 2312–2316).
67. Kim, J., Clerckx, B., & Mitcheson, P. D. (2020). Signal and system design for wireless power transfer: Prototype, experiment and validation. *IEEE Transactions on Wireless Communications, 19*(11), 7453–7469.
68. Meredith, J. (2016). Study on channel model for frequency spectrum above 6 GHz. 3GPP TR 38.900, Jun, Technical Report.
69. Lu, W., Gong, Y., Wu, J., Peng, H., & Hua, J. (2017). Simultaneous wireless information and power transfer based on joint subcarrier and power allocation in OFDM systems. *IEEE Access, 5*, 2763–2770.
70. Zhao, Y., Hu, J., Ding, Z., & Yang, K. (2018). Constellation rotation aided modulation design for the multi-user swipt-noma. In *2018 IEEE International Conference on Communications (ICC)* (pp. 1–6).
71. Xia, C., Jia, R., Shi, Y., Hu, A. P., & Zhou, Y. (2021). Simultaneous wireless power and information transfer based on phase-shift modulation in ICPT system. *IEEE Transactions on Energy Conversion, 36*(2), 629–639. ISSN: 1558–0059.
72. Hou, Y., Hu, J., & Yang, K. (2021). Waveform shaping for wireless information and energy provision with carrier frequency offset. In *2021 IEEE International Conference on Communications Workshops (ICC Workshops)* (pp. 1–6).
73. Hadani, R., Rakib, S., Tsatsanis, M., Monk, A., & Calderbank, R. (2017). Orthogonal time frequency space modulation. In *IEEE Wireless Communications & Networking Conference*.
74. Naikoti, A., & Chockalingam, A. (2022). Performance analysis of otfs with imperfect delay-doppler channel state information. In *2022 IEEE 95th Vehicular Technology Conference: (VTC2022-Spring)* (pp. 1–6).
75. Raviteja, P., Phan, K. T., Hong, Y., & Viterbo, E. (2018). Interference cancellation and iterative detection for orthogonal time frequency space modulation. *IEEE Transactions on Wireless Communications, 17*(10), 6501–6515.
76. Ogura, M., Kishida, M., & Lam, J. (2020). Geometric programming for optimal positive linear systems. *IEEE Transactions on Automatic Control, 65*(11), 4648–4663.
77. Bland, R. G., Goldfarb, D., & Todd, M. J. (1981). The ellipsoid method: A survey. *Operations Research, 29*(6), 1039–1091.
78. Xiang, L., et al. (2021). Gaussian approximate message passing detection of orthogonal time frequency space modulation. *IEEE Transactions on Vehicular Technology, 70*(10), 10 999–11 004.

Chapter 4
Data and Energy Beamforming with Multiple Antennas

4.1 Overview of Classic Beamforming

Low-power Internet of Everything (IoE) devices for sensing, small-scale computation, and communication can be wireless powered by ratio frequency (RF) signals in wearable medical applications and environments monitoring. The number of IoE devices (e.g., electronic tablets, sensors, wearables, and so on) is anticipated to grow from 7 billion in 2018 to 22 billion by 2025. A tremendous number of wireless devices thus require a scalable solution for perpetual energy supply, which can be realized by RF-based wireless power transfer (WPT) in the future energy self-sustainable 6G [1]. With the rapid growth of the low-power IoE devices, energy supply of these are usually powered by embedded batteries. However, quickly drained batteries largely limit performance of IoE devices. Therefore, RF signals are reliable to transfer wireless power to these devices. However, due to the severe channel attenuation during the wireless signal propagation, the transmitter has to suffer from very high energy consumption in order to maintain satisfactory WPT in a reasonable coverage [2, 3]. For the sake of counteracting the channel attenuation, multiple-input multiple-output (MIMO) system has been exploited for providing substantial spatial gains in WPT.

Specifically, Zhong et al. [4] studied WPT in the beamspace of a large-scale MIMO system by conceiving a linear energy harvester. However, Boshkovska et al. [5] proposed a practical model for characterizing the saturation and nonlinearity of diode based energy harvesters. Xiong et al. [6] explored the rate-energy region of SWIPT in a MIMO aided broadcasting system with these nonlinear energy harvesters. By considering the saturation phenomenon of a practical nonlinear energy harvester, Lu et al. [7] proposed a novel power splitting-based energy receiver in a MISO-SWIPT system. The received RF power was split into several portions to avoid the saturation region of nonlinear energy harvesters. As for a MIMO-SWIPT system, how to effectively utilize the spatial gain in improving the WPT with a practical energy harvester is still an open problem.

Most of the existing works focus on either the transmit beamformer or the precoder design for dedicated WPT and the SWIPT [8]. However, signal processing for the performance improvement of the energy harvesters has been largely ignored. The key question is how do we deal with the multiple energy flows gleaned by multiple antennas of energy harvester. No matter what kinds of energy harvesters we conceive, linear or nonlinear one, the rectified output DC power is mainly determined by the input RF power. Most of MIMO-SWIPT or MIMO-WPT related works [9, 10] presumed that the total RF power is the sum of the RF powers gleaned by every antenna. However, this is impractical in terms of hardware implementation. In order to reduce the hardware complexity, Linnartz et al. [11] proposed a more practical energy harvester with multiple antennas. By adjusting the phases of the RF signals received by the multiple antennas in the analog domain, they were constructively combined and rectified by a single energy harvester. Shanpu et al. [12] considered the RF combining by adopting the nonlinear energy harvester. However, this energy harvester is modeled by Taylor expansion, which is suitable for only a small region. Therefore, the optimal design of the energy combiner by considering a more accurate energy harvesting model is still worth further discussion.

Millimeter wave (mmWave) system with large-scale antenna arrays has a range of merits for both WIT and WPT. For example, wide bandwidth provided by mmWave-MIMO system is sufficient to satisfy the ultra-high throughput of IUs [13]. A huge number of uncorrelated antennas can be compacted in small areas at both transmitters and receivers for efficient signal transmission and reception [13]. Directional beams formed by large-scale antennas are capable of overcoming energy loss during signal propagation [14]. Therefore, mmWave-MIMO is regarded as a feasible solution for simultaneous wireless information and power transfer (SWIPT) [14].

J. Hu, K. Yang, *Integrated Data and Energy Transfer in Wireless Networks*, Textbooks in Telecommunication Engineering,
https://doi.org/10.1007/978-3-031-60514-7_4

Transceiver design of mmWave-MIMO system for pure WIT has attracted a lot of attention in the past few years. Due to the limited number of RF chains, instead of full-digital solution, hybrid beamforming schemes consisting of both digital and analog components are widely adopted in mmWave system [14]. Unfortunately, the hybrid beamforming scheme is difficult to be achieved due to the non-convex constraint imposed by the phase-shifter (PS)-based analog beamformer. When considering continuous PSs, the hybrid beamformer for maximizing the spectral efficiency can be found by minimizing the Euclidean distance between the hybrid solution and its full-digital counterpart having the same objective [15–17]. The most popular approach for finding the hybrid beamformer and for maximizing the spectral efficiency is the codebook-based hybrid beamformer design, where the columns of an analog precoder were selected from a predesigned codebook [18, 19].

Furthermore, apart from a limited number of RF chains, limited resolution of PSs also degrades the WIT performance of a practical mmWave system. A common approach is to ignore the resolution constraints of PSs by considering continuous alternatives and obtain the continuous solution. Then based on the continuous solution, the low-resolution-based solution can be obtained by quantizing the continuous counterpart [19–21].

Apart from point-to-point mmWave-MIMO system, hybrid beamforming has also been studied for wireless information multicast (WIM) in multiuser mmWave-MIMO systems [22, 23]. The beamformer design in WIM is usually formulated as a max-min fairness (MMF) problem, which aims for maximizing the minimum SNR or the corresponding minimum throughput of all IUs. Specifically, Gui et al. [22] studied a mmWave based multigroup WIM system, and they designed the hybrid beamformer to maximize the spectral efficiency. Wang et al. [23] studied the energy efficient analog beamformer design for mmWave based WIM system. They proposed a low-complexity algorithm to find the low-resolution analog beamformer and digital beamformer, while analyzing the performance of the asymptotically full-digital beamformer and the analog beamformer. However, the receivers in [22] and [23] only had a single antenna. How to design the combiners for receivers with multiple antennas is still an open problem.

In order to simultaneously satisfy low-power devices' appetite for both information and energy, RF signal aided SWIPT has attracted much attention from academia for years [24–26]. Yang et al. [27] studied WPT in non-orthogonal multiple access (NOMA). They showed the theoretical power regions of NOMA with fixed data requirement. Gao et al. [28] proposed a contract design for resource allocation in the time-domain, and they also studied a pricing scheme in a WPT aided network. In order to exploit both spatial and diversity gains, multiple antennas aided transceiver has been introduced in SWIPT systems. Ding et al. [29] found that having multiple antennas at both transmitters and receivers substantially improves both WIT and WPT performance. The transmit covariance was designed by maximizing the spectral efficiency, while satisfying the QoS of EUs. Zhang et al. [30] studied a MIMO-SWIPT system having a pair of IU and EU. A linear energy harvester is conceived by the EU. They found a closed-form solution for the transmit covariance design, which aims for maximizing the spectral efficiency of the IU, while satisfying the energy requirement of the EU. Furthermore, practical nonlinear energy harvesters have been widely considered in the design of both WPT and SWIPT for [25, 31]. In our system, we will fully exploit the broadcast nature of wireless channels. The transmitter simultaneously serves multiple EUs during a single transmission frame. Transceivers will be optimally designed for simultaneously satisfying the energy requirements of multiple EUs, which yields wireless power multicast (WPM).

When massive transmit antennas are implemented at a SWIPT transmitter, we still have to carefully design the corresponding hybrid beamformer due to the hardware constraints. For instance, Dai et al. [32] studied a massive MIMO aided SWIPT system operating in the mmWave band, when the non-orthogonal multiple access (NOMA) technique was adopted. The hybrid beamformer was carefully designed for maximizing the sum rate of all the IUs, while satisfying the energy requirements of all the EUs. Moreover, Wang et al. [23] investigated the MMF problem in mmWave-based multicast system for SWIPT. They assumed that every information and energy user is equipped with a single antenna. However, the joint design of the MIMO-based transceiver has been largely ignored for simultaneous wireless information and power multicast (SWIPM).

There are works on WPT in THz band, and Sei et al. [33] developed a 303 GHz rectenna and tested its antenna gain and RF–DC conversion efficiency. Hanh et al. [34] proposed an interface for sub-THz wireless power transfer under wafer-level testing. Benjamin et al. [35] exploited a novel form of beam diversity for near-field WPT and analyzed the WPT power budget under the channel model they formulated. Fortunately, the wavelength reduces as frequency increases. Ultra-massive antennas can be compactly implemented at transceivers, which may result in abundant spatial gains [36–38]. However, full-digital transceivers cannot be implemented in THz bands due to the unaffordable hardware cost and the energy consumption [39]. Alternatively, hybrid beamforming with a limited number of radio frequency (RF) chains is conceived [40–42]. There are many works focusing on the hybrid beamforming design in THz bands. Specifically, Yuan et al. [43] considered a multi-carrier system in THz band. To improve the average SINR, they designed the hybrid beamforming that overcomes frequency selective fading and a robust version against imperfect channel state information (CSI). Moreover, Lin et al. [44] designed a distance-aware-based hybrid beamforming for multiuser THz system. To minimize the usage of active RF components under constraint of power bugdet and data transmission rates, the distance-aware scheme takes full advantage of transmission windows at different distances. Busari et al. [45] investigated the vehicular connectivity in THz bands, in which they presented a novel generalized structure for hybrid beamforming to improve system performance. Yan et al. [46] proposed a widely spaced multi-subarray hybrid beamforming architecture to improve the spatial gain. They maximized the spectral efficiency constraint to transmit power. Elbir et al. [47] designed a hybrid beamforming for joint radar communications in THz band, and they maximized the spectral efficiency of the communications link and provided maximized SNR toward the radar targets under transmit power constraint. Zhai and Dai et al. [48, 49] insert several time delay devices into hybrid beamforming structure. The devices can provide frequency-dependent phase shifts to combat the beam squint in THz band.

Intelligent reflecting surface (IRS)-assisted wireless communication emerges as a key enabling technique for next generation of wireless communications [50]. Pan et al. [51] introduced IRS from different aspects, such as applications, challenges, and future research directions. An IRS is composed of a large number of low-cost reflecting elements, each of which is capable of passively reflecting RF signals by adjusting their phases in the analog domain [52]. Therefore, the resultant passive beam aims at receivers, which may substantially improve wireless communication performance by exploiting additional spatial gains. Compared to a traditional MIMO-based relaying system, the IRS-aided MIMO system only passively reflects RF signals, which are emitted by a transmitter. It does not invoke any active power-consuming RF chains, which substantially reduces its hardware complexity and energy consumption. Furthermore, due to the sophisticate signal processing at relays, a conventional relaying system mostly spent additional transmission frames for forwarding information to the destination, which may thus reduce the end-to-end throughput. However, the RF signals can be instantaneously reflected by the IRS, which do not need another transmission frame. Moreover, due to the severe path-loss of wireless channels, line of sight (LoS) is preferred by both the WPT and the SWIPT.

Specifically, Bai et al. [53] studied optimal resource allocation in an IRS-aided mobile edge computing system. Pan et al. [54] studied an IRS-aided MIMO-SWIPT system. They maximized the weighted sum rate of multiple information receivers by using the transmit precoder at the transmitter and the passive reflecting beamformer at the IRS, while satisfying the energy harvesting requirements of all the energy receivers. Lyu et al. [55] investigated an IRS-assisted wireless powered communication network to improve both the downlink WPT and the uplink wireless information transfer (WIT) performance. An uplink sum-rate maximization problem was formulated by jointly optimizing the time scheduling and the passive reflecting beamformer of the IRS. Wu et al. [56] studied an IRS optimized MISO-SWIPT system. They aimed to maximize the weighted sum power harvested by all the energy users (EUs) by jointly optimizing the transmit precoder and the passive reflecting beamformer of the IRS, while satisfying the individual signal to interference plus noise ratio constraints of all the information users.

Some effective algorithms were proposed for designing the discrete phase shifters of the IRS-aided system [57–60]. Specifically, Wu et al. [57] converted the discrete phase optimization on passive reflecting beamformer of the IRS to a linear integer programming (LIP). Gong et al. [58] reformulated the discrete phase optimization to an approximate "quadratic assignment programming" (QAP) problem. However, it did not satisfy the real positive quadratic coefficient requirement of the original QAP. Then, they converted the QAP into the LIP, which could be solved by a classic branch-and-bound method. However, the complexity of the branch-and-bound method exponentially grew as the number of the reflectors in the IRS increased, which had almost the same complexity with the exhaustive search. Therefore, the standard branch-and-bound-based algorithm was only suitable when the number of reflectors in the IRS was fewer than 30 [58]. Therefore, some sub-optimal solutions were investigated in [59, 60]. They developed an iterative algorithm for alternatively optimizing every reflector, when the other peers are fixed.

Moreover, some recent works investigated the IRS-assisted design in THz bands. Specifically, Chapala et al. [61] considered an IRS-assisted THz system. They proposed the $\alpha - \mu$ fading and proved that IRS can improve the performance over the combined effect of pointing errors and $\alpha - \mu$ fading. Pan et al. [62] studied the IRS-aided UAVs THz communication. They maximize the minimum average achievable rate of all users by designing the UAV's trajectory, power control, the phase shifter of the IRS, and the allocation of THz sub-bands. Qiao et al. [63] studied the secure transmission for an IRS-assisted THz system. Under the constraint of transmit power and discrete phase shift, they jointly optimized the transmit and IRS beamformer to maximize the secrecy rate. Ning et al. [64] considered massive MIMO and IRS in THz multiuser communication system where every user is served by multiple IRS and the transmitter. They develop a cooperation beam training scheme to estimate the channel with IRS. Then the hybrid beamforming is designed based on the training results. Wang et al. [65] proposed a downlink beam alignment method in a THz system utilizing the sparsity of station-RIS-user cascaded channel. Lu et al. [66] deployed an IRS at the transmitter, and they investigated a sum-rate maximization problem and proposed a deep-learning-based hybrid precoding scheme. Liu et al. [67] considered an IRS-aided integrated sensing and communications system. They proposed a primal–dual proximal policy optimization and its multi-threading version to jointly optimize the transmitter and the IRS. Qiao et al. [68] investigated the securing communication in IRS-assisted THz system. A joint transmitter and IRS beamforming design based on zero-forcing and iterative algorithm is proposed to improve the secrecy performance.

4.2 Hybrid Beamforming for Integrated Data and Energy Transfer in mmWave

4.2.1 System Model and Problem Formulation

4.2.1.1 System Model

The transceiver architecture of a mmWave-based MIMO-SWIPM system is portrayed in Fig. 4.1, which consists of a single transmitter, K IUs and M EUs in a single cell. All these users are equipped with N_r antennas, while the transmitter adopts N_t antennas both for WIM to the IUs and for WPM to the EUs. The transmitter is equipped with N_t^{RF} RF chains, which satisfies $1 \leq N_t^{RF} \leq N_t$. As shown in Fig. 4.1, the baseband digital beamfomer $\mathbf{F}_{BB} \in \mathbb{C}^{N_t^{RF} \times 1}$ converts the information flow into the N_t^{RF} RF chains. The corresponding passband analog beamfomer $\mathbf{F}_{RF} \in \mathbb{C}^{N_t \times N_t^{RF}}$ converts the information flows output by the RF chains into the N_t antennas. The fully connected topology is adopted, i.e., a single RF chain is connected to all the N_t antennas via N_t analog phase shifters. Note that the analog beamformer $\mathbf{F}_{RF}$ is realized by conceiving a number of analog phase shifters whose resolution is B bits. Therefore, the elements in the matrix $\mathbf{F}_{RF}$ are chosen from a set $\mathcal{F} \triangleq \{\frac{1}{\sqrt{N_t}} e^{j\frac{2\pi b}{2^B}} | b = 1, 2, \cdots, 2^B\}$. The transmit signal is expressed as $\mathbf{x} = \mathbf{F_{RF}} \mathbf{F_{BB}} s$, where s is an $N_s \times 1$ complex symbol of the information flow requested by the IUs satisfying $s \sim \mathcal{CN}(0, 1)$. The signals received by the k-th IUs and the m-th EUs can be expressed as

$$\mathbf{y}_k = \sqrt{P}\mathbf{H}_k \mathbf{F_{RF}} \mathbf{F_{BB}} s + \mathbf{n}_k, \quad k = 1, 2, \cdots K, \tag{4.1}$$

$$\mathbf{y}_m = \sqrt{P}\mathbf{G}_m \mathbf{F_{RF}} \mathbf{F_{BB}} s + \mathbf{n}_m, \quad m = K + 1, \cdots K + M, \tag{4.2}$$

respectively, where $\mathbf{n}_k, \mathbf{n}_m \in \mathbb{C}^{N_r \times 1}$ are Gaussian noise and all their elements are independently and identically Gaussian distributed complex random variables having a zero mean and a unity variance, while $\mathbf{H}_k$ ($\mathbf{G}_m$) is the channel matrix of IU_k (EU_m). We assume that the channel state information (CSI) is perfectly known by both the transmitter and the receivers.

On the receiver side, Roth et al. [69] proposed a mmWave massive MIMO equalization with low-resolution analog-to-digital converters (ADCs) method. This low-complexity method has good performance. It can be readily implemented in our system for addressing the equalization issue. Furthermore, it is robust to channel estimation errors. Zhu et al. [70] studied a novel beamforming strategy for directional frame timing synchronization in mmWave systems operating with low-resolution ADCs, which can be relied upon for tackling the synchronization challenge. At the k-th IUs, the passband analog RF combiner $\mathbf{W}_{k,RF} \in \mathbb{C}^{N_r \times N_r^{RF}}$ and the baseband digital combiner $\mathbf{W}_{k,BB} \in \mathbb{C}^{N_r^{RF} \times 1}$ are adopted to process the received signal y_k, which yields

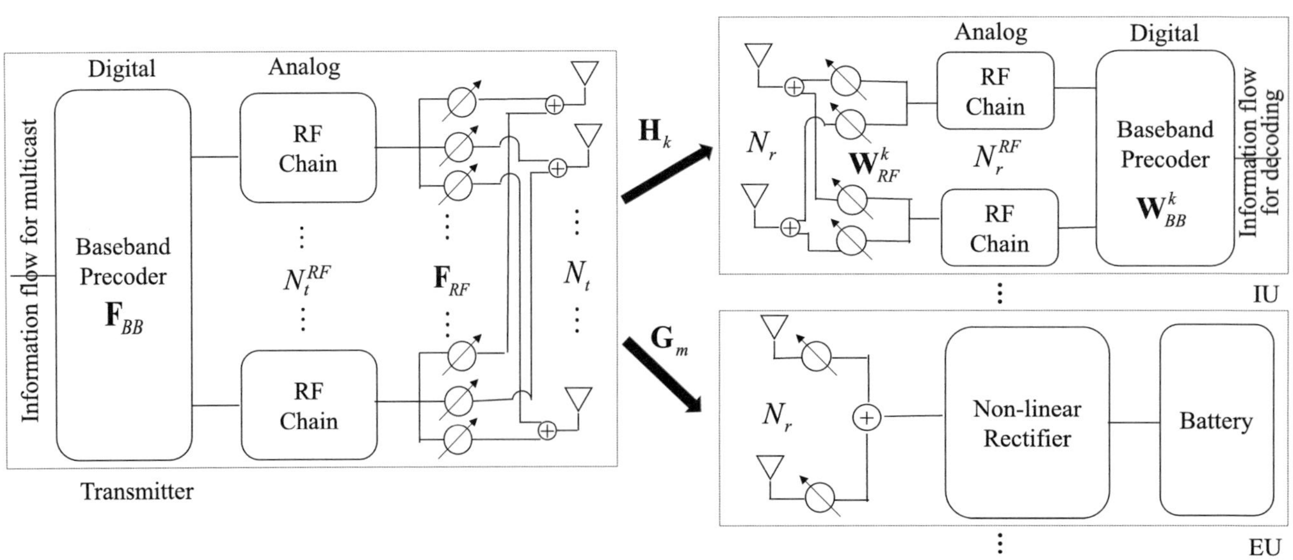

Fig. 4.1 mmWave MIMO system with hybrid beamforming structure at transmitter and receiver

$$\hat{y}_k = \sqrt{P}\mathbf{W}_{k,BB}^H\mathbf{W}_{k,RF}^H\mathbf{H}_k\mathbf{F}_{RF}\mathbf{F}_{BB}s+ \\ \mathbf{W}_{k,BB}^H\mathbf{W}_{k,RF}^H\mathbf{n}_k, \quad k = 1, 2. \cdots K. \tag{4.3}$$

The fully connected topology between the N_r antennas and N_r^{RF} RF chains is also conceived at all the IUs. All the passband analog combiners $\{\mathbf{W}_{k,RF}|\ k = 1, 2. \cdots K\}$ also rely upon a number of analog PSs having a resolution of B bits. Therefore, the elements in the matrix $\mathbf{W}_{k,RF}$ are chosen from a set $\mathcal{W}_{IU} \triangleq \{\frac{1}{\sqrt{N_r}}e^{j\frac{2\pi b}{2^B}}|b = 1, 2, \cdots, 2^B\}$.

The spectral efficiency of the k-th IU is then formulated as

$$R_{IU_k} = \log\Big|1 + P\mathbf{C}_k^{-1}\mathbf{W}_{k,BB}^H\mathbf{W}_{k,RF}^H\mathbf{H}_k\mathbf{F}_{RF}\mathbf{F}_{BB} \\ \times\mathbf{F}_{BB}^H\mathbf{F}_{RF}^H\mathbf{H}_k^H\mathbf{W}_{k,RF}\mathbf{W}_{k,BB}\Big|, \quad k = 1, 2, \cdots, K, \tag{4.4}$$

where $\mathbf{C}_k = \sigma_n^2\mathbf{W}_{k,BB}^H\mathbf{W}_{k,RF}^H\mathbf{W}_{k,RF}\mathbf{W}_{k,BB}$ is the covariance matrix of noise after the signal combining of IU_K and P represents the average transmit power. The "WIM Spectral Efficiency" R is defined as the minimum spectral efficiency of IUs, i.e., $R = \min_{1\le k\le K} R_{IU_k}$.

The geometric channel model is adopted for characterizing multipath fading in the mmWave band. The channel matrix $\mathbf{H}_k(\mathbf{G}_m)$ between the transmitter and IU_k (EU_m) is assumed to be the sum of the signals scattered by N_{cl} clusters. Each cluster has N_p independent signal propagation paths. Therefore, the discrete-time channel matrix $\mathbf{H}_k(\mathbf{G}_m)$ can be characterized by

$$\mathbf{H}_k = \sqrt{\frac{N_t N_r}{N_{cl}N_p}}\sum_{i=1}^{N_{cl}}\sum_{l=1}^{N_p}\beta_{il}\mathbf{a}_r(\theta_{il}^r)\mathbf{a}_t(\theta_{il}^t), \tag{4.5}$$

where $\beta_{il} \sim \mathcal{CN}(0, \sigma_i^2)$ is the complex channel gain of the l-th path in the i-th cluster, while σ_i^2 is the average power of the i-th cluster. Moreover, θ_{il}^r and θ_{il}^t are the azimuth arrival angle and departure angle of the l-th path in the i-th cluster, while $\mathbf{a}_r(\theta_{il}^r) \in \mathbb{C}^{N_r\times 1}$ and $\mathbf{a}_t(\theta_{il}^t) \in \mathbb{C}^{N_t\times 1}$ are the corresponding normalized transmit and receive response vectors, respectively, which are determined by the structure of the transmit and receive antenna arrays as well as the azimuth angles θ_{il}^r and θ_{il}^t. When the uniform liner antenna arrays are conceived, the channel response vectors $\mathbf{a}_r(\theta_{il}^r)$ and $\mathbf{a}_t(\theta_{il}^t)$ are expressed as $\mathbf{a}_r(\theta_{il}^r) = \frac{1}{\sqrt{N_r}}\left[1, e^{j\frac{2\pi}{\lambda}dsin(\theta_{il}^r)}, \cdots, e^{j(N_r-1)\frac{2\pi}{\lambda}dsin(\theta_{il}^r)}\right]$, and $\mathbf{a}_t(\theta_{il}^t) = \frac{1}{\sqrt{N_t}}\left[1, e^{j\frac{2\pi}{\lambda}dsin(\theta_{il}^t)}, \cdots, e^{j(N_t-1)\frac{2\pi}{\lambda}dsin(\theta_{il}^t)}\right]$, respectively, where λ is the wavelength and d is the distance between two adjacent antennas.

4.2.1.2 Energy Harvesting

Clerckx et al. [25] found that the rectifying efficiency of a diode based energy harvester is nonlinear. The saturation nonlinear model of energy harvesters is characterized as

$$\Psi(P) = \frac{M}{X(1 + \exp(-a(P - b)))} - Y, \qquad [Watt], \tag{4.6}$$

where we have $X = \frac{\exp(ab)}{1+\exp(ab)}$ and $Y = \frac{M}{\exp(ab)}$. Moreover, in Eq. (4.6), M denotes the maximum limitation on the output direct current (DC) power, while constants a and b represent the joint effects of resistances, capacitances, and circuit sensitivities.

Furthermore, Wang et al. [71] studied a multi-antenna aided energy receiver, where the RF signal collected by every antenna is aggregated first and then rectified by the energy harvester, which is denoted as Structure 1.

The architecture of our energy receivers is illustrated in Fig. 4.1, which is denoted as Structure 2. Multiple RF signals collected by the antennas are optimally combined at first, which is then rectified by a RF-DC rectifier. As portrayed in Fig. 4.1, the analog combiner of EU is also constituted by a number of analog phase shifters. Specifically, for EU_m, the combined RF signal $\hat{y}_m$ is expressed as

$$\hat{y}_m = \mathbf{w}_m^H \mathbf{y}_m, \ m = K+1, \cdots, K+M, \tag{4.7}$$

where the analog combiner $\mathbf{w}_m \in \mathbb{C}^{1\times N_r}$ is a row vector. The elements in $\mathbf{w}_m$ are chosen from $\mathcal{W}_{EU} \triangleq \{e^{j\frac{2\pi b}{2^B}} | b = 1, 2, \cdots, 2^B\}$. Given the combined signal $\hat{y}_m$, the DC energy finally harvested by EU_m can be formulated as $E_{out_m} = \Psi\left(\left[\hat{y}_m^H \hat{y}_m\right]\right)$.

Proposition 4.1 *The proposed Structure 2 is more efficient.*

Proof The energy harvested by Structure 1 is formulated as $E'_{out_m} = \Psi(\left[\mathbf{y}_m^H \mathbf{y}_m\right])$. Since energy harvesting function is the monotonously increasing function, we just need to compare $\hat{y}_m^H \hat{y}_m$ and $\mathbf{y}_m^H \mathbf{y}_m$, whose difference is formulated as $\hat{y}_m^H \hat{y}_m - \mathbf{y}_m^H \mathbf{y}_m = 2\sum \mathbf{w}_m(i)^H \mathbf{y}_m(i)^H \mathbf{w}_m(j) \mathbf{y}_m(j)$. A suitable combiner $\mathbf{w}_m$ can be carefully chosen to guarantee $2\sum \mathbf{w}_m(i)^H \mathbf{y}_m(i)^H \mathbf{w}_m(j) \mathbf{y}_m(j) > 0$. Therefore, we demonstrate that our energy receiver architecture outperforms the counterpart in Structure 1.

Since the energy harvesting function $\Psi(\cdot)$ is a monotonically increasing, we only need to consider the input energy $E_m = \mathbf{w}_m^H \mathbf{y}_m \mathbf{y}_m^H \mathbf{w}_m$ in our transceiver design.

4.2.1.3 Problem Formulation

The spectral efficiency of wireless information multicast (WIM) is defined as the minimum among all the IUs. Our objective function in (P4.2.1) is to maximize the WIM spectral efficiency by designing the hybrid transmit beamformer $\{\mathbf{F}_{BB}, \mathbf{F}_{RF}\}$ and receive combiners $\{\mathbf{W}_{1,BB}, \mathbf{W}_{1,RF}, \cdots \mathbf{W}_{K,BB}, \mathbf{W}_{K,RF}\}$ of the IUs and $\{\mathbf{w}_1, \cdots, \mathbf{w}_M\}$ of the EUs. By considering low-resolution analog phase shifters, all the elements in the analog beamformer $\mathbf{F}_{RF}$ satisfy $\{\mathbf{F}_{RF}(i, j) \in \mathcal{F}, \ \forall i, \ \forall j\}$, where $\mathcal{F} \triangleq \{\frac{1}{\sqrt{N_t}} e^{j\frac{2\pi b}{2^B}} | b = 1, 2, \cdots, 2^B\}$. The transmitter's total power constraint is enforced by normalizing $\mathbf{F}_{BB}$ such that $||\mathbf{F}_{BB}\mathbf{F}_{RF}||_F^2 = 1$. At the IU_k, similarly, $\mathbf{W}_{k,RF}$ is constrained by low-resolution phase shifters. Therefore, all its elements have to satisfy $\{\mathbf{W}_{k,RF}(i, j) \in \mathcal{W}_{IU}, \ \forall i, \ \forall j\}$, where $\mathcal{W}_{IU} \triangleq \{\frac{1}{\sqrt{N_r}} e^{j\frac{2\pi b}{2^B}} | b = 1, 2, \cdots, 2^B\}$. The receiver's total power constraint is enforced by normalizing $\mathbf{W}_{k,BB}$ such that $||\mathbf{W}_{k,RF}\mathbf{W}_{k,BB}||_F^2 = 1$. Since the hybrid combiners $\mathbf{W}_{k,RF}$ and $\mathbf{W}_{k,BB}$ are operated on both signals and noise, the SNR of IU_k stays constant when we change the norm of hybrid combiners. Therefore, the norm of IUs' hybrid combiners does not affect the WIM efficiency. This constraint is conceived only for obtaining exact values of IUs' digital combiners in practical engineering. The EU_m has its minimum energy harvesting requirement and uses analog phase shifters to process the received signal, which is expressed as $\hat{y}_m = \mathbf{w}_m^H \mathbf{y}_m$. The energy combiner $\mathbf{w}_m$ is also constrained by low-resolution phase shifters. Therefore, all its elements have to satisfy $\{\mathbf{w}_m(i) \in \mathcal{W}_{EU}, \ \forall i\}$, where $\mathcal{W}_{EU} \triangleq \{e^{j\frac{2\pi b}{2^B}} | b = 1, 2, \cdots, 2^B\}$.

Our joint transceiver design can be formulated as

$$(\text{P4.2.1}): \max_{\substack{\mathbf{F}_{BB}, \mathbf{F}_{RF}, \mathbf{W}_{1,BB}, \mathbf{W}_{1,RF}, \cdots \\ \mathbf{W}_{K,BB}, \mathbf{W}_{K,RF}, \mathbf{w}_{K+1}, \cdots, \mathbf{w}_{K+M}}} \min_{1\le k\le K} R_k, \tag{4.8}$$

$$\text{s. t. } ||\mathbf{F}_{BB}\mathbf{F}_{RF}||_F^2 = 1, \tag{4.8a}$$

$$E_m \ge E_0, \ m = K+1, \cdots, K+M, \tag{4.8b}$$

$$\mathbf{F}_{RF}(i, j) \in \mathcal{F}, \ \forall i, \ \forall j, \tag{4.8c}$$

$$\mathbf{W}_{k,RF}(i, j) \in \mathcal{W}_{IU}, \ \forall i, \ \forall j, \ k = 1, \cdots, K \tag{4.8d}$$

$$\mathbf{w}_m(i) \in \mathcal{W}_{EU}, \ \forall i, \ m = K+1, \cdots, K+M, \tag{4.8e}$$

$$||\mathbf{W}_{k,RF}\mathbf{W}_{k,BB}||_F^2 = 1, \ k = 1, \cdots, K. \tag{4.8f}$$

The objective of (P4.2.1) aims for maximizing the WIM spectral efficiency of all the IUs by optimizing the hybrid transmit beamformer $\{\mathbf{F}_{BB}, \mathbf{F}_{RF}\}$ and the receive combiners $\{\mathbf{W}_{1,BB}, \mathbf{W}_{1,RF}, \cdots \mathbf{W}_{K,BB}, \mathbf{W}_{K,RF}\}$ of the IUs and $\{\mathbf{w}_1, \cdots, \mathbf{w}_M\}$ of the EUs. The norm constraint (4.8a) on the transmit hybrid beamformer represents the limitation on the transmit power, while (4.8f) represents the constraint on the IUs' received power. Note that the constraint on the IUs' combiners actually does not have any impact on the WIM efficiency. This constraint is conceived only for obtaining exact values of IUs' digital combiners in practical engineering. The constraint (4.8b) represents the minimum energy harvesting requirement E_0 of all the EUs. The energy harvested by EU_m is $E_m = \mathbf{w}_m^H \mathbf{y}_m \mathbf{y}_m^H \mathbf{w}_m$. The elements in $\mathbf{F}_{RF}$, $\mathbf{W}_{k,RF}$, and $\mathbf{w}_m$ are chosen from $\mathcal{F}$, $\mathcal{W}_{IU}$, and $\mathcal{W}_{EU}$, respectively, due to the implementation of low-resolution analog phase shifters as shown in the constraints (4.8c)–(4.8e).

4.2.2 Asymptotically Optimal Transceiver Design

Before solving (P4.2.1), we first consider a joint design of the asymptotically optimal full-digital beamformer $\mathbf{F}$ and combiners $\{\mathbf{w}_1, \cdots, \mathbf{w}_K\}$ of all the IUs as well as asymptotically optimal analog combiners $\{\mathbf{w}_K + 1, \cdots, \mathbf{w}_{K+M}\}$ of all the EUs, which is formulated as

$$\text{(P4.2.2):} \quad \max_{\substack{\mathbf{F}, \mathbf{w}_1, \cdots \mathbf{w}_K, \\ \mathbf{w}_{K+1}, \cdots, \mathbf{w}_{K+M}}} \min_{1 \leq k \leq K} \log \left(1 + \frac{P}{\sigma^2} \mathbf{w}_k^H \mathbf{H}_k \mathbf{F} \mathbf{F}^H \mathbf{H}_k^H \mathbf{w}_k \right), \tag{4.9}$$

$$\text{s. t. } ||\mathbf{F}||_F^2 = 1, \tag{4.9a}$$

$$E_m \geq E_0, \ m = K+1, \cdots, K+M, \tag{4.9b}$$

$$\mathbf{w}_m(i) \in \mathcal{W}_{EU}, \ \forall i, \ m = K+1, \cdots, K+M \tag{4.9c}$$

$$||\mathbf{w}_k||_F^2 = 1, \ k = 1, \cdots, K. \tag{4.9d}$$

(P4.2.2) also aims for maximizing the WIM spectral efficiency, while satisfying the energy requirements of EUs. In Sect. 4.2.2, we assume that the perfect CSI is known at the transmitter and channel hardening effect is exploited.

In Sect. 4.2.2, we mainly solve (P4.2.2) by following the steps below:

- Step 1: We find the structure of the asymptotically optimal full-digital transmit beamformer by assuming arbitrary combiners of IUs and EUs.
- Step 2: Based on the structure of the asymptotically optimal full-digital transmit beamformer, we obtain the asymptotically optimal analog combiners of EUs by considering phase shifters of low resolution.
- Step 3: Based on this structure, we obtain the asymptotically optimal full-digital receive combiners of IUs.
- Step 4: Given the combiners derived in Steps 2 and 3 as well as its structure obtained in Step 1, the asymptotically optimal full-digital transmit beamformer is obtained.

4.2.2.1 Structure of Asymptotically Optimal Full-Digital Beamformer

In conventional MIMO beamforming for WIM, the optimal digital beamformer design is a NP-hard problem. Normally, the sub-optimal solution is obtained by exploiting SDR with a complexity as high as $O(N_t^7)$. Sadeghi et al. [72, 73] and Xiang et al. [74] have studied the asymptotical property of the digital beamformer for massive MIMO aided WIM in Gaussian channels. In our massive MIMO aided SWIPM system operating in the mmWave band, transmitter only has a single information flow to multicast to all the IUs. Given arbitrary combiners $\mathbf{w}_k$ of IU_k, the receive signal of IU_k is expressed as

$$\begin{aligned} \hat{y}_k =& \sqrt{P} \mathbf{w}_k^H \mathbf{H}_k \mathbf{F} s + \mathbf{w}_k^H \mathbf{n}_k \\ =& \sqrt{P} \hat{\mathbf{h}}_k^H \mathbf{F} s + \hat{n}_k \quad k = 1, 2. \cdots K, \end{aligned} \tag{4.10}$$

where $\mathbf{F}$ is the full-digital transmit beamformer and $\hat{\mathbf{h}}_k = \mathbf{w}_k^H \mathbf{H}_k$. Similarly, we define the equivalent channel $\hat{\mathbf{g}}_m$ as $\hat{\mathbf{g}}_m = \mathbf{w}_m^H \mathbf{G}_m$.

Remark 4.1 According to Eq. (4.10), the mmWave channel is generated as the sum of a number of random variables. Therefore, the elements in the vector $\hat{\mathbf{h}}_k$ are correlated with each other.

Remark 4.2 Since a mmWave channel is generated by different random variables, the k_1-th IU and k_2-th IU equivalent channel vectors $\hat{\mathbf{h}}_{k_1}, \hat{\mathbf{h}}_{k_2}$ are independent. We then have the following relationships:

$$\lim_{N_t \to \infty} \frac{\hat{\mathbf{h}}_{k_1}\hat{\mathbf{h}}_{k_1}}{N_t} = \eta_{k_1} > 1, \qquad \lim_{N_t \to \infty} \frac{\hat{\mathbf{h}}_{k_1}\hat{\mathbf{h}}_{k_2}}{N_t} = 0, \tag{4.11}$$

where η_{k_1} is a function of the receiver's combiner.

Proposition 4.2 *The asymptotically optimal beamformer* $\mathbf{F}$ *has a form of* $\mathbf{F} = \sum \omega_k \hat{\mathbf{h}}_k + \sum \omega_m \hat{\mathbf{g}}_m$, *where the component* $\sum \omega_k \hat{\mathbf{h}}_k$ *contributes to WIM to the IUs, while* $\sum \omega_m \hat{\mathbf{g}}_m$ *contributes to WPM to the EUs.*

Proof For IU_k, the received signal is formulated as

$$\begin{aligned} \hat{y}_k &= \sqrt{P}\hat{\mathbf{h}}_k^H \mathbf{F} s + \hat{n}_k \\ &= \sqrt{P}\omega_k \hat{\mathbf{h}}_k^H \hat{\mathbf{h}}_k s + \sum_{k' \neq k} \omega_{k'} \hat{\mathbf{h}}_k^H \hat{\mathbf{h}}_{k'} s + \sum \omega_{m'} \hat{\mathbf{h}}_k^H \hat{\mathbf{g}}_{m'} s + \hat{n}_k, \end{aligned} \tag{4.12}$$

for $\forall k = 1, \cdots, K$. Observe from Eq. (4.12) that only the component $\hat{\mathbf{h}}_k$ in the full-digital beamformer $\mathbf{F}$ contributes to the communication performance of IU_k. Similarly, by analyzing the received signal $\widehat{y}_m$ of EU_m, we find that only $\hat{\mathbf{g}}_m$ contributes to the energy harvesting performance of EU_m. The power of the received signal $\hat{y}_m$ can be expressed as $E'_m = (\hat{y}_m)^2 = \omega_m^2 P \mathrm{tr}(\hat{\mathbf{g}}_m^H \hat{\mathbf{g}}_m \hat{\mathbf{g}}_m^H \hat{\mathbf{g}}_m)$.

Any vector can be represented by a linear combination of several orthogonal bases. We assume that the asymptotically optimal full-digital beamformer has a form of $\mathbf{F} = \sum \omega_k \hat{\mathbf{h}}_k + \sum \omega_m \hat{\mathbf{g}}_m + \sum \omega_j \mathbf{u}_j$, where $\{\mathbf{u}_j\}$ is another group of the orthogonal basis of the complementary space to $\{\hat{\mathbf{g}}_m\}$ and $\{\hat{\mathbf{h}}_k\}$. The received signal $\hat{y}_k$ of IU_k can be reformulated as

$$\begin{aligned} \hat{y}_k &= \sqrt{P}\hat{\mathbf{h}}_k^H \mathbf{F} s + \hat{n}_k \\ &= \sqrt{P}\omega_k \hat{\mathbf{h}}_k^H \hat{\mathbf{h}}_k s + \sqrt{P}(\sum_{k' \neq k} \omega_{k'} \hat{\mathbf{h}}_k^H \hat{\mathbf{h}}_{k'} + \sum \omega_{m'} \hat{\mathbf{h}}_k^H \hat{\mathbf{g}}_{m'} \\ &\quad + \sum \omega_j \hat{\mathbf{h}}_k^H \mathbf{u}_j) s + \hat{n}_k, \text{ for } \forall k = 1, 2. \cdots K. \end{aligned} \tag{4.13}$$

Since $\mathbf{u}_j$ is orthogonal with $\hat{\mathbf{h}}_k$ and $\hat{\mathbf{g}}_m$, the component $\sum \omega_j \mathbf{u}_j$ neither contributes to IU_k's communication nor contributes to EU_k's energy harvesting. Therefore, we may always find a better solution expressed as $\mathbf{F} = \sum \gamma \omega_k \hat{\mathbf{h}}_k + \sum \omega_m \hat{\mathbf{g}}_m$, where $\gamma = \sqrt{(\sum \omega_k^2 + \sum \omega_j^2)/\sum \omega_k^2} > 1$. Since the EU-related components in beamformer $\mathbf{F}$ do not change, the energy requirements of all the EUs can be satisfied by both the beamformers $\mathbf{F}'$ and $\mathbf{F}$. By contrast, the beamformer $\mathbf{F}$ may result in a higher information transmission performance of IU_k than $\mathbf{F}'$, since we have $\gamma \omega_k \geq \omega_k$. Because $\mathbf{F}$ achieves a better information transmission performance than $\mathbf{F}'$, $\mathbf{F}'$ is no longer an optimal solution, which contradicts to our assumption. Therefore, the assumption does not hold. The proof of Proposition 4.2 is completed.

The "asymptotically optimal full-digital beamformer" is obtained by exploiting the law of large numbers when the number of transmit antennas tends to infinity. In practice, we cannot implement an infinite number of transmit antennas. The obtained asymptotically optimal full-digital beamformer is a sub-optimal solution, when we have a finite number of transmit antennas. However, according to the law of large numbers, when the number of antennas is sufficiently large, the performance of the asymptotically optimal solution gets arbitrarily close to that of the optimal one.

Let us now discuss the parameters of the full-digital beamformer $\mathbf{F}$, whose normalized version is expressed as

$$\bar{\mathbf{F}} = \frac{\sum \omega_k \hat{\mathbf{h}}_k + \sum \omega_m \hat{\mathbf{g}}_m}{\beta \sqrt{N_t}}, \tag{4.14}$$

where β is the normalized factor formulated as

$$\beta = \frac{|| \sum \omega_k \hat{\mathbf{h}}_k + \sum \omega_m \hat{\mathbf{g}}_m ||_F}{\sqrt{N_t}}. \tag{4.15}$$

Then, when we mention full-digital beamformer, it means this normalized version beamformer $\bar{\mathbf{F}}$. The asymptotic analysis on β^2 is obtained as

$$\begin{aligned}\lim_{N_t\to\infty}\beta^2 &= \lim_{N_t\to\infty}\frac{\left(\sum_{k=1}^{K}\omega_k\hat{\mathbf{h}}_k+\sum_{m=K+1}^{K+M}\omega_m\hat{\mathbf{g}}_m\right)^H}{N_t}\\ &\quad\times\left(\sum_{k=1}^{K}\omega_k\hat{\mathbf{h}}_k+\sum_{m=K+1}^{K+M}\omega_m\hat{\mathbf{g}}_m\right)\\ &= \lim_{N_t\to\infty}\frac{\sum_{m=K+1}^{K+M}\sum_{m'=K+1}^{K+M}\omega_m\omega_{m'}\hat{\mathbf{g}}_m^H\hat{\mathbf{g}}_{m'}}{N_t}\\ &\quad+\frac{2\sum_{k=1}^{K}\sum_{m=K+1}^{K+M}\omega_k\omega_m\hat{\mathbf{h}}_k^H\hat{\mathbf{g}}_m+\sum_{k=1}^{K}\sum_{k'=1}^{K}\omega_k\omega_{k'}\hat{\mathbf{h}}_k^H\hat{\mathbf{h}}_{k'}}{N_t}\\ &=\sum_{k=1}^{K}\omega_k^2\eta_k+\sum_{m=K+1}^{K+M}\omega_m^2\eta_m. \end{aligned}\tag{4.16}$$

Moreover, the beamformer $\bar{\mathbf{F}}$ has to satisfy the WPM energy constrains of Eq. (4.8b). For EU_m, the power of the received signal obeys the following inequality:

$$\begin{aligned}&\lim_{N_t\to\infty}P\hat{\mathbf{g}}_m^H\bar{\mathbf{F}}\bar{\mathbf{F}}^H\hat{\mathbf{g}}_m\geq E_m\\ &\Rightarrow\frac{P\omega_m^2N_t\eta_m^2}{\sum_{k=1}^{K}\omega_k^2\eta_k+\sum_{m'=K+1}^{K+M}\omega_{m'}^2\eta_{m'}}\geq E_m\\ &\Rightarrow\zeta_m\geq E_m/(PN_t\eta_m),\end{aligned}\tag{4.17}$$

where ζ_m is defined as

$$\zeta_m=\frac{\omega_m^2\eta_m}{\sum_{k=1}^{K}\omega_k^2\eta_k+\sum_{m'=K+1}^{K+M}\omega_{m'}^2\eta_{m'}},\quad m=K+1,\cdots,K+M.\tag{4.18}$$

According to Eq. (4.18), whether the energy constraints of EU_m can be satisfied or not is only determined by the weight ω_m in the full-digital beamformer $\bar{\mathbf{F}}$.

Similarly, we define ζ_k as

$$\zeta_k=\frac{\omega_k^2\eta_k}{\sum_{k'=1}^{K}\omega_{k'}^2\eta_{k'}+\sum_{m=K+1}^{K+M}\omega_m^2\eta_m},\quad k=1,\cdots,K.\tag{4.19}$$

Observe from Eqs. (4.18) and (4.19) that we have

$$\sum_{k=1}^{K}\zeta_k+\sum_{m=K+1}^{K+M}\zeta_m=1.\tag{4.20}$$

According to Eqs. (4.18), (4.19), and (4.20), ζ_k and ζ_m are equivalent to ω_k and ω_m, which can be then substituted into Eq. (4.14) of the asymptotically optimal normalized beamformer $\bar{\mathbf{F}}$.

Since the transmitter only multicast a single information flow to all the IUs, maximizing the spectrum efficiency of IU_k is equivalent to maximizing its SNR. The asymptotic SNR of the k-th IU is expressed as

$$\begin{aligned} \lim_{N_t\to\infty} SNR_k &= \lim_{N_t\to\infty} \frac{P\hat{\mathbf{h}}_k^H\bar{\mathbf{F}}\bar{\mathbf{F}}^H\hat{\mathbf{h}}_k}{\sigma^2} \\ &= \frac{P\omega_m^2 N_t\eta_k^2}{\left(\sum\limits_{k'=1}^{K}\omega_{k'}^2\eta_{k'} + \sum\limits_{m=K+1}^{K+M}\omega_m^2\eta_m\right)\sigma^2} \\ &= \frac{\zeta_k P N_t \eta_k}{\sigma^2}. \end{aligned} \tag{4.21}$$

Observe from Eq. (4.21) that the SNR and the corresponding spectrum efficiency of IU_k only depend on ζ_k.

The following two lemmas further discuss how we obtain the asymptotically optimal beamformer $\bar{\mathbf{F}}$ for the SWIPM.

Lemma 4.1 *The asymptotically optimal beamformer* $\bar{\mathbf{F}}$ *is obtained when energy harvested by* EU_m *is equal to its minimum requirement, which indicates that* $\zeta_m = E_0/(PN_t\eta_m) \quad \forall m = K+1,\cdots,K+M$.

Proof Suppose that $\{\zeta_k\}$ and $\{\zeta_m\}$ are the weights of asymptotically optimal beamformer, which are expressed as

$$\begin{aligned} \zeta_k &= \frac{\omega_k^2\eta_k}{\left(\sum\limits_{k=1}^{K}\omega_k^2\eta_k + \sum\limits_{m=K+1}^{K+M}\omega_m^2\eta_m\right)}, \quad \forall k = 1,\cdots,K \\ \zeta_m &= \frac{\omega_m^2\eta_m}{\left(\sum\limits_{k=1}^{K}\omega_k^2\eta_k + \sum\limits_{m=K+1}^{K+M}\omega_m^2\eta_m\right)}, \forall m = K+1,\cdots,K+M. \end{aligned} \tag{4.22}$$

There is at least a single $EU_{m'}$ harvesting more power than its minimum requirement, which can be expressed as $\zeta_{m'} = E_0/(PN_t\eta_{m'}) + \Delta$. We can always find another group of $\{\zeta_k'\}$ and $\{\zeta_m'\}$ satisfying the energy harvesting constraints while achieving higher WIM spectral efficiency, which are expressed as

$$\begin{aligned} \zeta_k' &= \left(1 + \frac{\Delta(\sum\limits_{k=1}^{K}\omega_k^2\eta_k + \sum\limits_{m=K+1}^{K+M}\omega_m^2\eta_m)^2}{\sum\limits_{k=1}^{K}\omega_k^2\eta_k}\right)\zeta_k, \forall k = 1,\cdots,K, \\ \zeta_{m'}' &= E_0/(PN_t\eta_{m'}), \text{ for } m = m', \\ \zeta_m' &= \zeta_m, \text{ for } \forall m \neq m'. \end{aligned}$$

The new beamformer reduces the power allocation for EUs while satisfying all the energy constraints. Therefore, more power will be allocated to the WIM toward IUs, which results in a higher WIM spectral efficiency.

Lemma 4.2 *The asymptotically optimal beamformer* $\bar{\mathbf{F}}$ *is obtained when all IUs have the same SNR,* $SNR_1 = SNR_2 = \cdots = SNR_K$.

Proof The beamformer leads to that the minimum SNR of IU_{k1} is not equal to the SNR of IU_{k2}, i.e., $SNR_{k1} \neq SNR_{k2}$. Therefore, the WIM spectral efficiency is $\log(1 + SNR_{k1})$. Denote the $SNR^* = \sum_{k=1}^{K} SNR_k/K$. By letting $SNR_1' = ,\cdots, = SNR_K' = SNR^*$, the WIM spectral efficiency is $\log(1 + SNR^*)$, which is bigger than precious.

Based on Lemmas 4.1 and 4.2, the asymptotically optimal beamformer $\bar{\mathbf{F}}$ can be obtained by solving the following equations:

$$\begin{cases} \frac{\omega_m^2 \eta_m}{\left(\sum\limits_{k=1}^{K} \omega_k^2 \eta_k + \sum\limits_{m'=K+1}^{K+M} \omega_{m'}^2 \eta'_m\right)} = E_m/(PN_t\eta_m), \\ \forall m = K+1, \cdots, K+M, \\ \frac{\omega_{k_1}^2 \eta_{k_1}^2}{\left(\sum\limits_{k'=1}^{K} \omega_{k'}^2 \eta_{k'} + \sum\limits_{m=K+1}^{K+M} \omega_m^2 \eta_m\right)} = \frac{\omega_{k_2}^2 \eta_{k_2}^2}{\left(\sum\limits_{k'=1}^{K} \omega_{k'}^2 \eta_{k'} + \sum\limits_{m=K+1}^{K+M} \omega_m^2 \eta_m\right)} \\ \forall k_1, k_2 \in \{1, 2, \cdots, K\} \\ \bar{\mathbf{F}}\bar{\mathbf{F}}^H = 1. \end{cases} \tag{4.23}$$

In order to obtain the solution to Eq. (4.23), we have to find the analog combiners $\mathbf{w}_K + 1 \cdots \mathbf{w}_{K+M}$ for EUs and digital combiners $\mathbf{w}_1 \cdots \mathbf{w}_K$ for IUs, which will be discussed in previous section.

4.2.2.2 Asymptotically Optimal Analog Combiners of EUs

The energy harvesting function Eq. (4.6) is an increasing function with respect to the power of the received RF signals. Therefore, we only need to maximize the received RF power. In Sect. 4.2.2.1, we analyze the structure of asymptotically optimal beamformer. We then consider a fixed transmit beamformer having this structure. The combiner of EU_m for maximizing its own energy harvesting performance is then formulated as

$$(\text{P4.2.3}): \max_{\mathbf{w}_m} ||\mathbf{w}_m^H \mathbf{y}_m||_F^2, \tag{4.24}$$

$$\text{s. t. } \mathbf{w}_m(i) \in \mathcal{W}_{EU}, \ \forall i, \tag{4.24a}$$

where the constraint (4.24a) is imposed by limited resolution of the analog phase shifters. Since the multipath fading channel causes phase offset of the RF signal, directly aggregating RF signals received by multiple antennas may result in destructive signal combination, which severely degrades the energy harvesting performance. By contrast, careful design of the transmit beamformer and the receive combiner may lead to constructive signal combination at the EUs, which is capable of increasing the energy harvesting performance.

Proposition 4.3 *The transmit beamformer $\bar{\mathbf{F}}$ and the energy combiner $\mathbf{w}_m$ of EU_m are capable of constructively combining the RF signals received by multiple antennas.*

Proof By substituting Eqs. (4.14) and (4.18) into Eq. (4.10), the received signal after the energy combiner of EU_m is expressed as

$$\begin{aligned} \hat{y}_m &= \sqrt{P}\mathbf{w}_m^H \mathbf{G}_m \bar{\mathbf{F}} s + \mathbf{w}_m \mathbf{n}_m \\ &= \sqrt{P}\zeta_m \mathbf{w}_m^H \mathbf{G}_m \mathbf{G}_m^H \mathbf{w}_m + \sum_{m' \neq m} \mathbf{w}_m^H \mathbf{G}_m \mathbf{G}_{m'}^H \mathbf{w}_{m'} \\ &+ \mathbf{w}_m \mathbf{n}_m. \end{aligned} \tag{4.25}$$

The term $\sum_{m' \neq m} \mathbf{w}_m^H \mathbf{G}_m \mathbf{G}_{m'}^H \mathbf{w}_{m'}$ does not contribute to the received signal due to asymptotically orthogonality of Eq. (4.11). Since the first term $\sqrt{P}\zeta_m \mathbf{w}_m^H \mathbf{G}_m \mathbf{G}_m^H \mathbf{w}_m$ is positive, we are able to cancel the adverse effect of the phase offset.

Furthermore, by relaxing the non-convex constraints (4.24a), (P4.2.3) can be reformulated as

$$(\text{P4.2.4}): \max_{\mathbf{w}'_m} \mathbf{w}'^H_m \mathbf{G}_m \mathbf{G}_m^H \mathbf{w}'_m, \tag{4.26}$$

$$\text{s. t. } ||\mathbf{w}'_m||_F^2 = N_r. \tag{4.26a}$$

Proposition 4.4 *Letting* $\mathbf{u}$ *denote the eigenvector corresponding to the maximum eigenvalue of the matrix* $\mathbf{G}_m\mathbf{G}_m^H$*, the optimal solution of P(4.2.4) is obtained as* $\mathbf{w}_m' = N_r\mathbf{u}$.

Proof The eigenvalue decomposition of $\mathbf{G}_m\mathbf{G}_m^H$ is expressed as $\mathbf{G}_m\mathbf{G}_m^H = \mathbf{U}_m^H\mathbf{\Lambda}_m\mathbf{U}_m$, where $\mathbf{u}_{m1}, \mathbf{u}_{m2}, \cdots, \mathbf{u}_{mN_r}$ are the column vector of $\mathbf{U}_m$, and $\lambda_{m1} \geq \lambda_{m2} \geq \cdots \geq \lambda_{mN_r}$ are the corresponding eigenvalues. The optimal solution to (P4.2.4) is then formulated as $\mathbf{w}_m' = N_r \sum_{i=1}^{N_r} l_i \mathbf{u}_{mi}$, where we have $\sum_{i=1}^{N_r} l_i^2 = 1$. Hence, the norm of the combiner $\mathbf{w}_m$ is equal to N_r.

The objective function (4.26) of (P4.2.4) is then transformed as $\mathbf{w}_m'^H\mathbf{G}_m\mathbf{G}_m^H\mathbf{w}_m' = N_r\sum_{i=1}^{N_r} l_i^2\lambda_i$. The maximum value of objective function (4.26) is λ_{m1} when $l_1 = 1$. The optimal solution $\mathbf{w}_m'$ is obtained when the eigenvector corresponding to the largest eigenvalue of the $\mathbf{G}_m\mathbf{G}_m^H$ is chosen.

The asymptotically optimal solution to (P4.2.3) is obtained by minimizing Euclidean distance between $\mathbf{w}_m$ and $\mathbf{w}_m'$, which is formulated as

$$\text{(P4.2.5): } \min_{\mathbf{w}_m} ||\mathbf{w}_m - \mathbf{w}_m'||_F^2, \tag{4.27}$$

$$\text{s. t. } \mathbf{w}_m(i) \in \mathcal{W}_{EU}, \ \forall i. \tag{4.27a}$$

The objective function (4.27) of (P4.2.5) can be further reformulated as

$$||\mathbf{w}_m - \mathbf{w}_m'||_F^2 = \sum_{i=1}^{N_r} ||\mathbf{w}_m(i) - \mathbf{w}_m(i)'||_F^2. \tag{4.28}$$

Since every constraint of Eq. (4.27a) only includes a single variable $\mathbf{w}_m(i)$, (P4.2.5) is then decoupled into a number of subproblems as

$$\text{(P4.2.6): } \min_{\mathbf{w}_m(i)} ||\mathbf{w}_m(i) - \mathbf{w}_m(i)'||, \tag{4.29}$$

$$\text{s. t. } \mathbf{w}_m(i) \in \mathcal{W}_{EU}. \tag{4.29a}$$

Observe from (P4.2.6) that since $\mathbf{w}_m(i)$ and $\mathbf{w}_m(i)'$ are complex numbers, the objective function (4.29) represents the Euclidean distance between this pair of numbers. The optimal EU combiners with B-bits phase shifter are then obtained by $\mathbf{w}_m(i) = \arg\min ||\mathbf{w}_m(i) - \mathbf{w}_m(i)'||$.

4.2.2.3 Asymptotically Optimal Full-Digital Combiners of IUs

The design of IUs' full-digital combiners aims for maximizing their own attainable spectrum efficiency, which is equivalent to maximize the received SNR of the IUs. We consider a fixed transmit beamformer having a structure of Eq. (4.14). Therefore, for IU_k, the digital combiner $\mathbf{w}_k$ can be obtained by solving the following problem:

$$\text{(P4.2.7): } \max_{\mathbf{w}_k} \frac{||\mathbf{w}_k^H\mathbf{y}_k||_F^2}{\sigma^2}, \tag{4.30}$$

$$\text{s. t. } ||\mathbf{w}_k||_F^2 = 1, \tag{4.30a}$$

which is similar to (P4.2.4). Therefore, we have the following proposition.

Proposition 4.5 *The asymptotically optimal digital combiner* $\mathbf{w}_k$ *in (P4.2.7) is the eigenvector of the* $\mathbf{H}_k\mathbf{H}_k^H$*'s largest eigenvalue.*

Algorithm 4.1 Design of asymptotically optimal beamformer and combiners

Require: The IUs channel matrices $\mathbf{H}_1, \cdots, \mathbf{H}_k$ and the EUs channel matrices $\mathbf{G}_1, \cdots, \mathbf{G}_m$.
Ensure: Asymptotically optimal transmit beamformer $\bar{\mathbf{F}}$, asymptotically optimal IUs combiners $\mathbf{w}_1, \cdots, \mathbf{w}_k$ and asymptotically optimal EUs combiners $\mathbf{w}_{K+1}, \cdots, \mathbf{w}_{K+M}$
1: Do SVD for the IUs channel matrices $\mathbf{H}_1, \cdots, \mathbf{H}_k$ and the EUs channel matrices $\mathbf{G}_1, \cdots, \mathbf{G}_m$;
2: Obtain the asymptotically optimal EUs combiners by sloving (P4.2.3);
3: Obtain the asymptotically optimal IUs combiners by Proposition 4.5;
4: Obtain the asymptotically optimal beamformer by Eq.(4.31);
5: **return** $\mathbf{F}, \mathbf{w}_1, \cdots, \mathbf{w}_K$ and $\mathbf{w}_{K+1}, \cdots, \mathbf{w}_{K+M}$.

4.2.2.4 Asymptotically Optimal Full-Digital Beamformer

Given the asymptotically optimal full-digital combiners $\{\mathbf{w}_1, \cdots, \mathbf{w}_K\}$ of IUs in Sect. 4.2.2.2, the asymptotically optimal analog combiners $\{\mathbf{w}_K + 1, \cdots, \mathbf{w}_{K+M}\}$ of EUs and its structure, the asymptotically optimal full-digital beamformer $\mathbf{F}$ is then designed.

We define the singular vectors as $\mathbf{h}_k$ and $\mathbf{g}_m$ corresponding to the largest singular value λ_k of the channel $\mathbf{H}_k$ and λ_m of the channels $\mathbf{G}_m$, respectively. By solving Eq. (4.23), the asymptotically optimal full-digital beamformer is expressed as

$$\bar{\mathbf{F}} = \sum_{k=1}^{K} \underbrace{\sqrt{(1 - \sum_{m=K+1}^{K+M} w_m^2)\frac{1}{\sum_{k=1}^{K}\frac{1}{\lambda_k^2}}}}_{\omega_k} \mathbf{h}_k + \sum_{m=K+1}^{K+M} \omega_m \mathbf{g}_m, \tag{4.31}$$

where $\omega_m = \sqrt{E_0/(PN_t\lambda_m^2)}$). For EU_m, the weight ω_m reduces as we increase λ_m. Since an increasing λ_m represents an improved channel quality, the asymptotically optimal full-digital transmit beamformer $\bar{\mathbf{F}}$ may focus less energy toward EU_m to satisfy its energy requirements, according to Eq. (4.31). For IU_k, its corresponding weight ω_k reduces as we increase λ_k. Similarly, since an increasing λ_k represents an improved quality of the channel to IU_k, $\bar{\mathbf{F}}$ may allocate less power for the downlink information transmission toward IU_k, while allocating more power for IUs having poor channel qualities. Therefore, the WIM spectrum efficiency can be optimized.

The asymptotically optimal beamformer and combiner can be obtained via Algorithm 4.1. The complexity of the proposed Algorithm 4.1 is $O(KN_t^2N_r + MN_t^2N_r)$, which is far lower than the conventional SDR having a complexity of $O((N_t^2 + M + K)^3.5N_{ior})$, where N_{ior} is the number of iterations.

4.2.3 *Hybrid Beamformer and Combiner Design*

We have so far discussed the asymptotically optimal solution by conceiving a full-digital beamformer of the transmitter, full-digital combiners of IUs and analog combiners with limited resolution of EUs. The hybrid beamformers have the minimum Euclidean distance to the asymptotically optimal solution. The difference between the optimal hybrid beamformers and its asymptotically optimal counterpart results in the performance loss. If the asymptotically optimal solution satisfies the energy harvesting requirements of EUs, e.g., Eq. (4.8b), the hybrid one may not. For obtaining the hybrid beamformer, the following two problems need to be carefully solved:

- How can we design a hybrid beamformer having the closest Euclidean distance to the asymptotically optimal counterpart in full-digital domain?
- How can we design a hybrid beamformer satisfying the energy harvesting requirements of EUs?

Moreover, the design of hybrid combiners of IUs has to address similar problems.

4.2.3.1 Hybrid Beamformer Design

When a full-digital beamformer $\bar{\mathbf{F}}$ is given, the hybrid counterpart can be designed as

Algorithm 4.2 The low-resolution hybrid beamformer design

Require: Full-digital beamformer $\bar{\mathbf{F}}$; matrix of array response vectors $\mathbf{A}$
Ensure: Baseband beamformer $\mathbf{F}_{BB}$ and analog beamformer $\mathbf{F}_{RF}$

Initialise $\mathbf{F}_0 \leftarrow \bar{\mathbf{F}}$
for $i = 1 : N_{RF}$ **do**
  Obtain $\boldsymbol{\varphi} \leftarrow \mathbf{A}\mathbf{f}_0$;
  Obtain $l \leftarrow \arg\max diag(\boldsymbol{\varphi} * \boldsymbol{\varphi}^H)$
  Update $\mathbf{F}_{RFC} \leftarrow [\mathbf{F}_{RFC}|F(:, l)]$;
  Update $\mathbf{F}_{BB} \leftarrow (\mathbf{F}_{RFC}^H \mathbf{F}_{RFC})^{-1} \mathbf{F}_{RFC}^H \bar{\mathbf{F}}$;
  Update $\mathbf{F}_0 \leftarrow \frac{\bar{\mathbf{F}} - \mathbf{F}_{RFC}\mathbf{F}_{BB}}{||\bar{\mathbf{F}} - \mathbf{F}_{RFC}\mathbf{F}_{BB}||_F^2}$;
end for
Obtain $\mathbf{F}_{BB} \leftarrow \frac{\mathbf{F}_{BB}}{||\mathbf{F}_{BB}||_F^2}$
for $j = 1 : N_{RF}$ **do**
  for $l = 1 : N_t$ **do**
    $\mathbf{F}_{RF}(j, l) \leftarrow \arg\min ||\mathbf{F}_{RF}(j, l) - \mathbf{F}_{RFC}(j, l)'||$;
  end for
end for
return $\mathbf{F}_{BB}, \mathbf{F}_{RF}$.

$$\text{(P4.2.7)}: \min_{\mathbf{F}_{BB}, \mathbf{F}_{RF}} ||\bar{\mathbf{F}} - \mathbf{F}_{RF}\mathbf{F}_{BB}||_F^2, \quad \text{s.t. (4.8a),(4.8c).} \tag{4.32}$$

Due to the constraint of (8c), (P7) is a discrete optimization, which cannot be solved in polynomial time. By relaxing the discrete constraint of (8c), i.e., the limited resolution PSs become continuous ones, (P7) can be then converted into a optimization problem with continuous variables as

$$\text{(P4.2.8)}: \min_{\mathbf{F}_{BB}, \mathbf{F}_{RFC}} ||\bar{\mathbf{F}} - \mathbf{F}_{RFC}\mathbf{F}_{BB}||_F^2, \quad \text{s.t.} |\mathbf{F}_{RFC}(i, j)| = 1/N_t, \text{ (4.8a),} \tag{4.33}$$

where the elements in $\mathbf{F}_{RFC}$ are continuous. This problem can be solved by orthogonal matching pursuit (OMP) [18]. Furthermore, the elements in the analog beamformer $\mathbf{F}_{RF}$ having limited resolution are given as

$$\mathbf{F}_{RF}(i, j) = \arg\min ||\mathbf{F}_{RF}(i, j) - \mathbf{F}_{RFC}(i, j)||. \tag{4.34}$$

Algorithm 4.2 solves (P4.2.8) by obtaining the baseband beamformer $\mathbf{F}_{BB}$ and analog beamformer $\mathbf{F}_{RF}$ with limited resolution.

4.2.3.2 Hybrid Combiner Design of IUs

The hybrid combiners of IU_k can be designed with a similar approach, which is formulated as

$$\text{(P4.2.9)}: \min_{W_{k,BB}, \mathbf{W}_{k,RF}} ||\mathbf{w}_k - \mathbf{W}_{RF} W_{BB}||_F^2, \quad \text{s. t. (4.8d), (4.8f).} \tag{4.35}$$

Since this problem is similar to (P4.2.7), Algorithm 4.2 can also be relied upon for solving (P4.2.9).

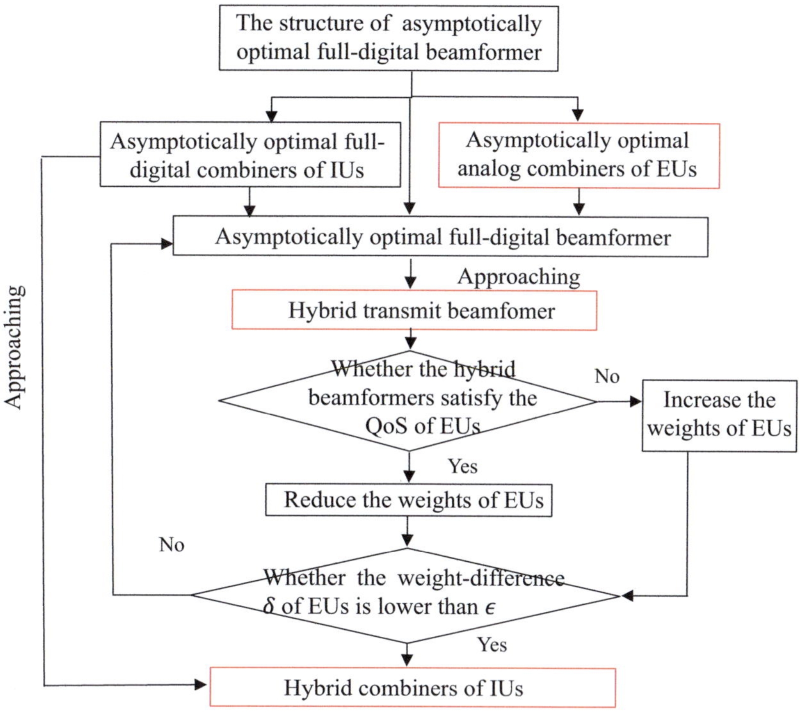

Fig. 4.2 The flow chart of Algorithm 4.3

Algorithm 4.3 Joint hybrid beamformer and combiners design

Require: The IUs channel matrices $\mathbf{H}_1, \cdots, \mathbf{H}_K$ and the EUs channel matrices $\mathbf{G}_{K+1}, \cdots, \mathbf{G}_{K+M}$, The QoS E_0, the error ϵ.

Ensure: Baseband beamformer $\mathbf{F}_{BB}$, analog beamformer $\mathbf{F}_{RF}$, baseband combiners of IUs $\mathbf{W}_{k,BB}$ $k = 1, \cdots, K$, analog combiners of IUs $\mathbf{W}_{k,RF}$ $k = 1, \cdots, K$ and analog combiners of EUs $\mathbf{w}_m$ $m = K+1, \cdots, K+M$

1: Initialise $\omega_m^{(1)} \leftarrow 1/\sqrt{M}$ $(m = K+1, \cdots, K+M)$, $\omega_m^{(2)} \leftarrow 0$ $(m = K+1, \cdots, K+M)$.
2: Obtain the EU combiners $\mathbf{w}_{K+1}, \cdots, \mathbf{w}_{K+M}$ and IU full-digital combiner $\mathbf{w}_1, \cdots, \mathbf{w}_K$ by Algorithm 4.1.
3: **while** $\delta > \epsilon$ **do**
4: $\quad \omega_m = (\omega_m^{(1)} + \omega_m^{(2)})/2$ $(m = K+1, \cdots, K+M)$
5: $\quad$ Obtain the full-digital beamformer $\bar{\mathbf{F}}$ by substituting ω_m $(m = K+1, \cdots, K+M)$ into Eq. (4.31) ;
6: $\quad$ Obtain the low resolution hybrid beamformer $\mathbf{F}_{BB}, \mathbf{F}_{RF}$ by Algorithm 4.2 with the full-digital beamformer $\bar{\mathbf{F}}$;
7: $\quad$ **if** all $E_m > E_0$ **then**
8: $\quad\quad$ Update $\omega_m^{(1)} = \omega_m$;
9: $\quad$ **else**
10: $\quad\quad$ Update $\omega_m^{(2)} = \omega_m$;
11: $\quad$ **end if**
12: $\quad$ Calculate $\delta = \sum_{i=K+1}^{K+M} (\omega_m^{(1)} - \omega_m^{(2)})^2$;
13: **end while**
14: Obtain the baseband combiners of IUs $\mathbf{W}_{k,BB}$ $k = 1, \cdots, K$, analog combiners of IUs $\mathbf{W}_{k,RF}$ $k = 1, \cdots, K$ by Algorithm 4.2.

4.2.3.3 Joint Hybrid Beamformer and Combiners Design

According to Proposition 4.2, the full-digital beamformer has a structure of $\bar{\mathbf{F}} = (\sum \omega_k \hat{\mathbf{h}}_k + \sum \omega_m \hat{\mathbf{g}}_m)/(\sqrt{N_t}\beta)$. For the m-th EU, when weight ω_m is higher, the energy harvested becomes higher. If the hybrid beamformer is obtained by the full-digital counterpart having a higher weight ω_m, m-th EU harvested more energy.

Based on this positive correlation between the energy harvesting performance of EU_m and the weight ω_m, we may use a one-dimensional search for finding feasible weights for all the EUs. Therefore, a full-digital beamformer can be found, whose corresponding hybrid counterpart satisfies the energy harvesting requirements of the EUs.

Figure 4.2 illustrates how we design the hybrid transceiver in a massive MIMO aided SWIPM system in Algorithm 4.3, which is summarized as below:

- Step 1: Initialize two groups of weights for EUs, namely the upper bound weights and the lower bound weights$\omega_{K+1}^{(2)}, \cdots, \omega_{K+M}^{(2)}$, as shown in Line 1 of Algorithm 4.3.

- Step 2: Obtain the EUs' asymptotically optimal analog combiners by EUs' channel matrices $\mathbf{G}_{K+1}, \cdots, \mathbf{G}_{K+M}$ and obtain the IUs' asymptotically optimal full-digital combiners by IUs' channel matrices $\mathbf{H}_1, \cdots, \mathbf{H}_K$, as detailed in Line 2 of Algorithm 4.3.
- Step 3: Obtain the middle of upper bound weights and lower bound weights, namely middle weights. Calculate the full-digital beamformer $\bar{\mathbf{F}}$ by the IUs' channel matrices $\mathbf{H}_1, \cdots, \mathbf{H}_K$, the EUs' channel matrices $\mathbf{G}_{K+1}, \cdots, \mathbf{G}_{K+M}$ and the middle weights. Then, obtain its corresponding hybrid beamformer by Algorithm 4.2, as detailed in Lines 4–6 of Algorithm 4.3.
- Step 4: Check whether the hybrid beamformer satisfies the energy harvesting requirements of the EUs. For EU_m, if it satisfies, update the upper bound weight; if it does not, update the lower bound, as detailed in Lines 7–11 of Algorithm 4.3.
- Step 5: Obtain the IUs' hybrid combiners $\{\mathbf{W}_{k,BB}, \mathbf{W}_{k,RF} | \, k = 1, \cdots, K\}$, as detailed in Line 14 of Algorithm 4.3.

The complexity of Algorithm 4.3 are attributed to two parts: (1) the hybrid beamformer and combiner design with continuous phase shifters having a complexity of $O(N_{RF}N_t^2)$ and $O(N_{RF}N_r^2)$; (2) the design with low-resolution phase shifters having a complexity of $O(N_{RF}N_t)$ and $O(N_{RF}N_r)$. Since the number N_{iter} of iterations for the binary search is normally small, the total complexity of the hybrid beamformer and combiner design is $O((N_{RF}N_t^2+(K+M)N_{RF}N_r^2)N_{iter})$.

4.2.4 Imperfect CSI Design

In the previous sections, the beamformer and combiners are designed by considering the perfect CSI at the transmitter. However, it is difficult to obtain the perfect CSI due to the channel estimation error, channel feedback delay, and channel quantification error, etc. As a result, we investigate the impact of the imperfect CSI (ICSI) on our SWIPM system in this section. Similarly, the full-digital beamformer will be derived first, and then the hybrid counterparts are obtained. The relationship between the estimated CSI $\widetilde{\mathbf{H}}_k$ and the actual CSI $\mathbf{H}_k$ of the channel to IU_k is expressed as

$$\widetilde{\mathbf{H}}_k = \mathbf{H}_k + \Delta_k, \tag{4.36}$$

where $\Delta_k \sim \mathcal{CN}(0, \sigma_k \mathbf{I})$ represents a random channel uncertainty independent of $\mathbf{H}_k$. Similarly, the ICSI $\widetilde{\mathbf{G}}_m$ of EU_m is expressed as

$$\widetilde{\mathbf{G}}_m = \mathbf{G}_m + \Delta_m, \tag{4.37}$$

where $\mathbf{G}_m$ is the actual CSI and $\Delta_m \sim \mathcal{CN}(0, \sigma_k \mathbf{I})$ is channel uncertainty independent of $\mathbf{G}_m$. When the IU_k and EU_m adopt an arbitrary combiner $\widetilde{\mathbf{w}}_k$ and $\widetilde{\mathbf{w}}_m$, respectively, the equivalent channels of IU_k and EU_m are $\widetilde{\mathbf{h}}_{k_{ICSI}} = \widetilde{\mathbf{w}}_k \widetilde{\mathbf{H}}_k$ and $\widetilde{\mathbf{g}}_m = \widetilde{\mathbf{w}}_m \widetilde{\mathbf{G}}_m$. According to the estimated CSI and Proposition 4.2, the full-digital beamformer with ICSI is expressed as

$$\widetilde{\mathbf{F}}_{ICSI} = \frac{\sum \omega_k \widetilde{\mathbf{h}}_{k_{ICSI}} + \sum \omega_m \widetilde{\mathbf{g}}_{m_{ICSI}}}{\beta_{ICSI}\sqrt{N_t}}, \tag{4.38}$$

where β_{ICSI} is the normalized factor to let $||\widetilde{\mathbf{F}}_{ICSI}||_F^2 = 1$. The asymptotic analysis on β_{ICSI} is obtained as

$$\begin{aligned}
\lim_{N_t\to\infty} \beta_{ICSI}^2 = \lim_{N_t\to\infty} & \frac{\left(\sum_{k=1}^{K} \omega_k \widetilde{\mathbf{h}}_{k_{ICSI}} + \sum_{m=K+1}^{K+M} \omega_m \widetilde{\mathbf{g}}_{m_{ICSI}}\right)^H}{N_t} \\
& \times \left(\sum_{k=1}^{K} \omega_k \widetilde{\mathbf{h}}_{k_{ICSI}} + \sum_{m=K+1}^{K+M} \omega_m \widetilde{\mathbf{g}}_{mICSI}\right) \\
& = \sum_{k=1}^{K} \omega_k^2(\eta_k + \phi_k) + \sum_{m=K+1}^{K+M} \omega_m^2(\eta_m + \phi_m),
\end{aligned} \tag{4.39}$$

where $\phi_k = (\widetilde{\mathbf{w}}_k \Delta_k \Delta_k^H \widetilde{\mathbf{w}}_k^H)/N_t$ and $\phi_m = (\widetilde{\mathbf{w}}_m \Delta_m \Delta_m^H \widetilde{\mathbf{w}}_m^H)/N_t$ representing the channel estimation uncertainties. Then the SNR of IU_k with ICSI can be further expressed as

$$\begin{aligned}
&\lim_{N_t\to\infty} SNR_{k_{ICSI}} \\
&= \lim_{N_t\to\infty} \frac{P\hat{\mathbf{h}}_k^H \widetilde{\mathbf{F}}_{ICSI}\widetilde{\mathbf{F}}_{ICSI}^H \hat{\mathbf{h}}_k}{\sigma^2} \\
&= \frac{P\omega_m^2 N_t \eta_k^2}{\left(\sum\limits_{k'=1}^{K} \omega_{k'}^2(\eta_{k'}+\phi_{k'}) + \sum\limits_{m=K+1}^{K+M} \omega_m^2(\eta_m+\phi_m)\right)\sigma^2}.
\end{aligned} \tag{4.40}$$

Observe from Eq. (4.40) that the resultant SNR has a similar form as Eq. (4.21) with perfect CSI. The only difference in the denominator is incurred by the terms $\sum\limits_{k=1}^{K}\omega_k\phi_k + \sum\limits_{m=K+1}^{K+M}\omega_m\phi_m$ depending on the variances of channel estimation uncertainties. The asymptotic analysis on the received RF power of EU_m is derived as

$$\begin{aligned}
&\lim_{N_t\to\infty} P\hat{\mathbf{g}}_m^H \widetilde{\mathbf{F}}_{ICSI}\widetilde{\mathbf{F}}_{ICSI}^H \hat{\mathbf{g}}_m \ge E_m \\
&\Rightarrow \frac{P\omega_m^2 N_t \eta_m^2}{\sum\limits_{k=1}^{K} \omega_k^2(\eta_k+\phi_k) + \sum\limits_{m'=K+1}^{K+M} \omega_{m'}^2(\eta_{m'}+\phi_{m'})} \ge E_m.
\end{aligned} \tag{4.41}$$

Similarly to Eq. (4.19), the asymptotically optimal full-digital beamformer with ICSI can be obtained by solving the following equations:

$$\begin{cases}
\frac{\omega_m^2\eta_m}{\sum\limits_{k=1}^{K}\omega_k^2(\eta_k+\phi_k)+\sum\limits_{m'=K+1}^{K+M}\omega_{m'}^2(\eta_{m'}+\phi_{m'})} = E_m/(PN_t\eta_m), \\
\forall m = K+1,\cdots,K+M, \\
\frac{\omega_{k_1}^2\eta_{k_1}^2}{(\sum\limits_{k'=1}^{K}\omega_{k'}^2(\eta_{k'}+\phi_{k'})+\sum\limits_{m=K+1}^{K+M}\omega_m^2(\eta_m+\phi_m))} \\
= \frac{\omega_{k_2}^2\eta_{k_2}^2}{(\sum\limits_{k'=1}^{K}\omega_{k'}^2(\eta_{k'}+\phi_{k'})+\sum\limits_{m=K+1}^{K+M}\omega_m^2(\eta_m+\phi_m)} \\
\forall k_1,k_2 \in \{1,2,\cdots,K\} \\
\widetilde{\mathbf{F}}_{ICSI}\widetilde{\mathbf{F}}_{ICSI}^H = 1.
\end{cases} \tag{4.42}$$

Similarly, the combiner $\widetilde{w}_k$ of IU_k is the eigenvector of the $\widetilde{H}_k\widetilde{H}_k^H$'s largest eigenvalue. And the combiner $\widetilde{w}_m$ of EU_m is obtained by approaching to the eigenvector of the $\widetilde{G}_m\widetilde{G}_m^H$'s largest eigenvalue, which is similar to (P4.2.5). Therefore the asymptotically optimal full-digital beamformer and combiners of IUs as well as are obtained. The hybrid beamformer can be then obtained by Algorithms 4.2 and 4.3.

4.2.5 Numerical Result

Numerical results of our transceiver design in MIMO aided SWIPM system operating in the mmWave band are provided in this section. The system has 2 IUs and 2 EUs. The transmitter is equipped with $N_t = 128$ antennas. The phase shifters of transmitter and receivers have the same resolution. All the IUs and EUs are equipped with $N_r = 32$ antennas. The transmitter and all IUs are equipped with 4 RF chains. We consider $N_p = 6$ paths and $N_{cl} = 10$ clusters in our wireless fading channel of Eq. (4.5). The distance between the transmitter and all the IUs and EUs is 10 m, while the path-loss is 60 dB [75]. The power of noise is -40 dBm. The transmit power is set to be $P = 30$ dBm. In the EH model of Eq. (4.6), the circuit parameters are $a = 1500$, $b = 0.0022$, and $M = 3.9$ mW provided by [5, 6, 76]. We use "ASY Optimal," "CPS," "PS B=1," and "PS B=2" to denote the asymptotically optimal full-digital solution, which is given in Sect. 4.2.2, the hybrid solution with the continuous PSs, 1-bit PSs and 2-bit PSs, which are given in Sect. 4.2.2.3, respectively.

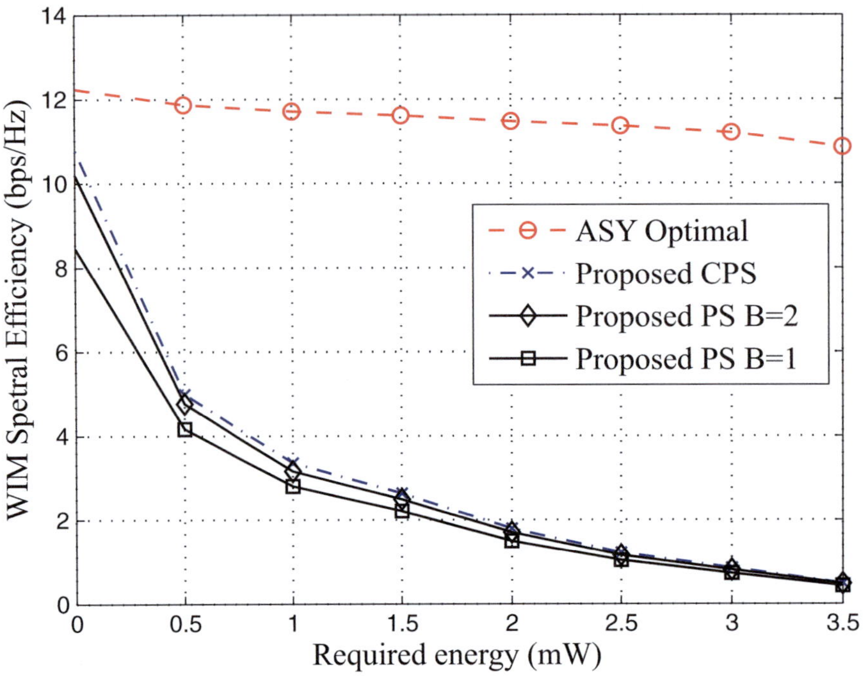

Fig. 4.3 The trade-off between received energy and WIM spectral efficiency

We characterize the trade-off between the WIM spectral efficiency and the energy harvested by EUs in Fig. 4.3. Observe from Fig. 4.3 that when EUs have a stringent energy harvesting requirement, the WIM spectral efficiency is reduced. When energy harvesting requirements of the EUs are all zeros, the WIM spectral efficiency of the asymptotically optimal solution, that of the hybrid solution with CPS, that of the hybrid solution with a phase-shifter resolution $B = 2$, and that with a phase-shifter resolution $B = 1$ are 12.23, 10.77, 10.18, and 8.45 bps/Hz, respectively. The WIM spectral efficiency of hybrid beamformers decreases more rapidly than their asymptotically optimal counterpart, as the energy harvesting requirement becomes stringent. Since the hybrid beamformer and combiners have performance degradation in both WIM and WPM compared to the optimal full-digital counterpart, in order to satisfy the energy requirement, the hybrid beamformer and combiners should allocate more transmit power to WPM, which results in the rapid reduction of their WIM spectral efficiency. Furthermore, our design by considering practical phase shifters having a resolution of B = 2 achieves almost the same performance as the continuous phase shifters.

Observe from Figs. 4.4 and 4.5 that the WIM spectral efficiency increases, as the transmit power increases, where we consider a single IU and a pair of EUs. The EUs' energy harvesting requirement is set to be 0.38 mW. An iterative phase match (IPM) algorithm for low-resolution mmWave MIMO hybrid beamformer and combiner design [21] is conceived as a benchmarker. As expected, WIM spectral efficiency of the asymptotically optimal full-digital design is higher than all the other hybrid designs. Observe from Fig. 4.4 that the IPM algorithm has a higher WIM spectral efficiency than our proposed algorithm in the region of high transmit power. For instance, WIM spectral efficiency of IPM algorithm is 3 bps/Hz higher than our algorithm, when the transmit power is 50 dBm. By contrast, our proposed algorithm is much better than IPM, when the transmit power is more than 20 dBm. By conceiving the CPS, PS B=2 and PS B=1, total energies harvested by EUs are 0.375, 0.388, and 0.433 mW, respectively, when the transmit power is 30 dBm as shown in Fig. 4.5. When our algorithm is implemented for different resolutions of the phase shifters, the energies harvested are the same at the transmit powers of 40 and 50 dBm, since the nonlinear energy harvesters operate in the saturation region.

We investigate the impact of phase shifters' resolution on the WIM spectral efficiency in Fig. 4.6. We simultaneously change the resolution of all phase shifters in the transmitter as well as all IUs and EUs. The minimum energy harvesting requirement of the EUs is 0.38 mW. Observe from Fig. 4.6 that increasing the resolution of phase shifters improves the WIM spectral efficiency. Furthermore, the phase shifters having $B = 3$ achieve almost the same WIM spectral efficiency as the continuous phase shifters. If we further increase the resolution of the phase shifters, the WIM spectral efficiency increases marginally.

We also investigate the impact of the number of antennas on the WIM spectral efficiency in Fig. 4.7. The EUs' minimum energy harvesting requirement is 0.18 mW. Observe from Fig. 4.7 that having more transmit antennas increases the WIM

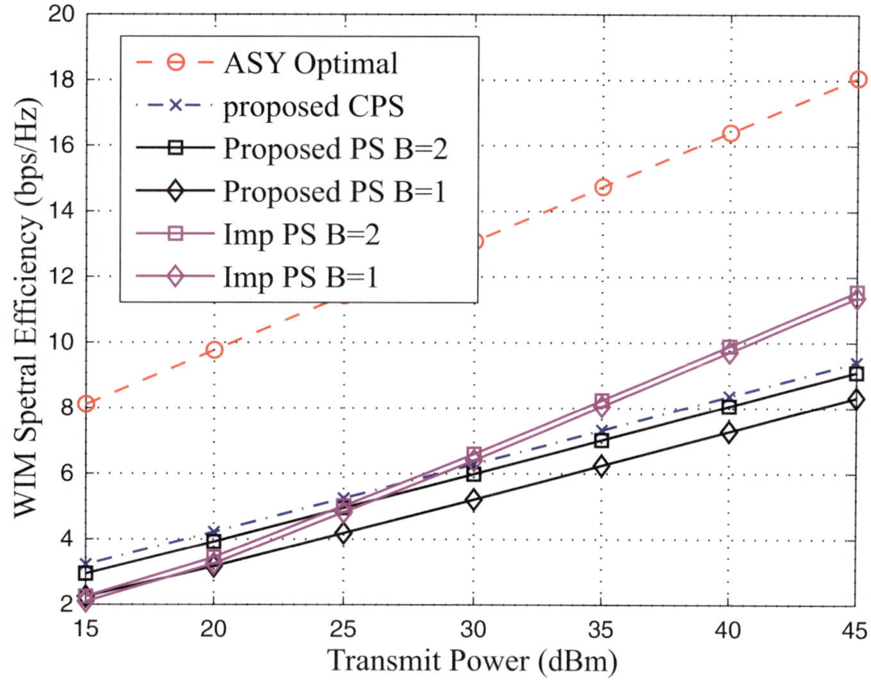

Fig. 4.4 WIM spectral efficiency versus transmit power

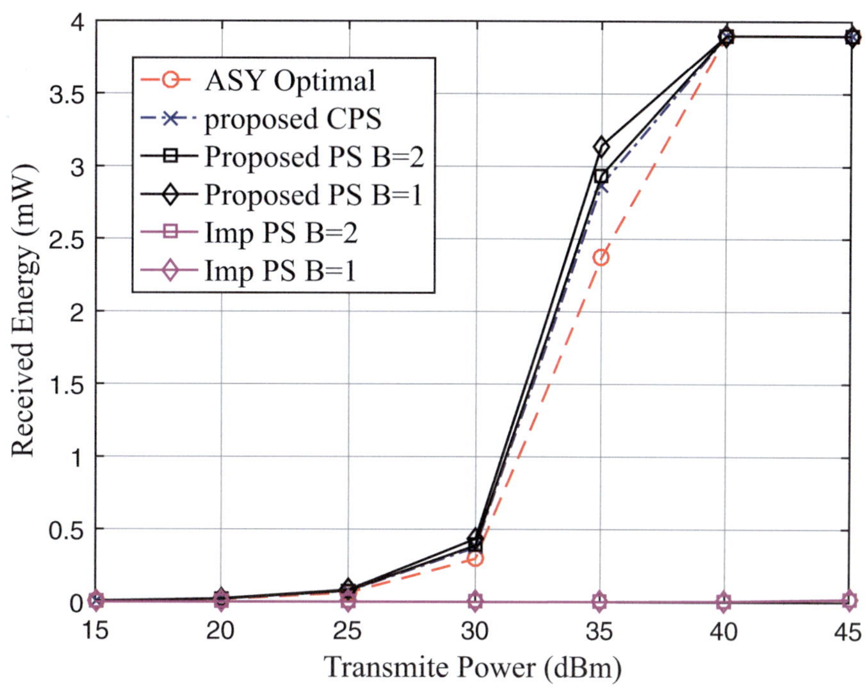

Fig. 4.5 Total energy harvested versus transmit power

spectral efficiency. The WIM spectral efficiency of the asymptotically optimal full-digital scheme grows smoothly. With more transmit antennas, the wireless channels become better. The minimum energy harvesting requirement of EUs can be readily satisfied. The growth of WIM spectral efficiency from 80 to 160 transmit antennas of hybrid beamformers with CPS, PS B = 2 and PS B = 1, are 0.52, 0.50, and 0.45 bit/s/Hz.

We then investigate the impact of number of RF chains and resolution of analog phase shifters on the WIM spectral efficiency in Fig. 4.8. The energy requirements of all the EUs are 0.18 mW. Observe from Fig. 4.8 that as we increase

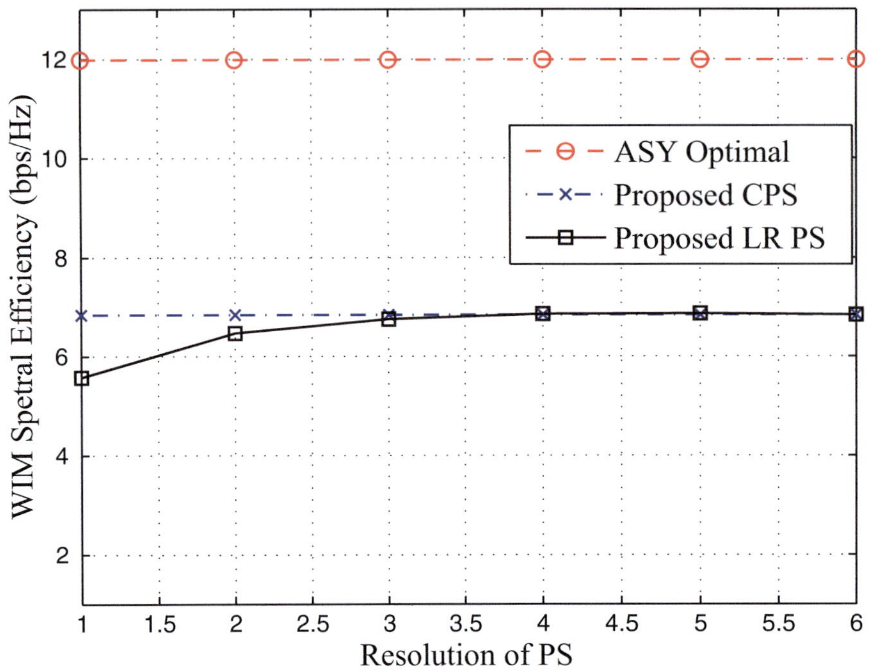

Fig. 4.6 WIM spectral efficiency versus the bit of phase shifters

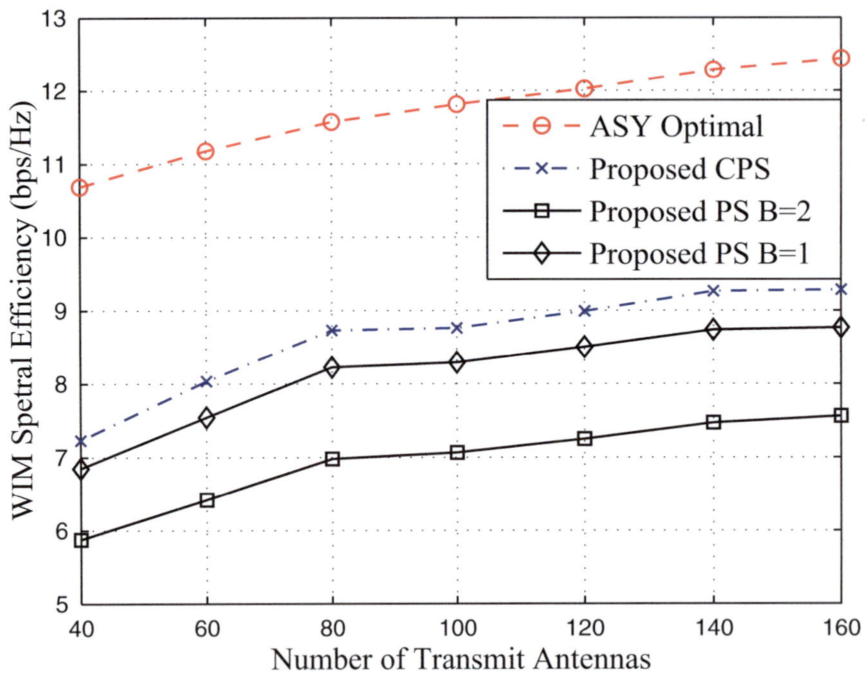

Fig. 4.7 WIM spectral efficiency versus the number of transmit antennas

the number of RF chains, the WIM spectral efficiency also increases in our hybrid transceiver design. However, the asymptotically optimal full-digital solution keeps unchanged. This is because, as expressed in (P4-1-1), our joint design only needs to satisfy the energy harvesting requirements of EUs. The rest of resources in the power- and spatial domain are then allocated to maximize the WIM spectral efficiency of IUs. However, when we have more than 4 RF chains, the WIM spectral efficiency increases more slowly. By jointly considering the hardware complexity and the attainable WIM spectral efficiency, implementing 4 RF chains with 128 antennas is a good choice. Similarly, increasing the resolution of phase shifters improves the WIM spectral efficiency. Furthermore, the phase shifters with 2-bit resolution achieve very similar performance

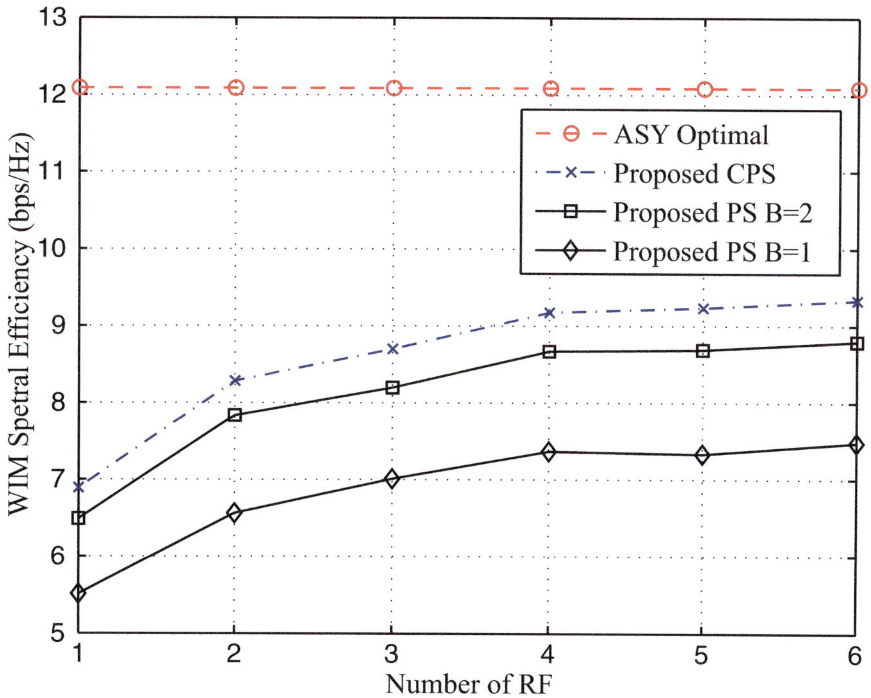

Fig. 4.8 WIM spectral efficiency versus the number of RF chains

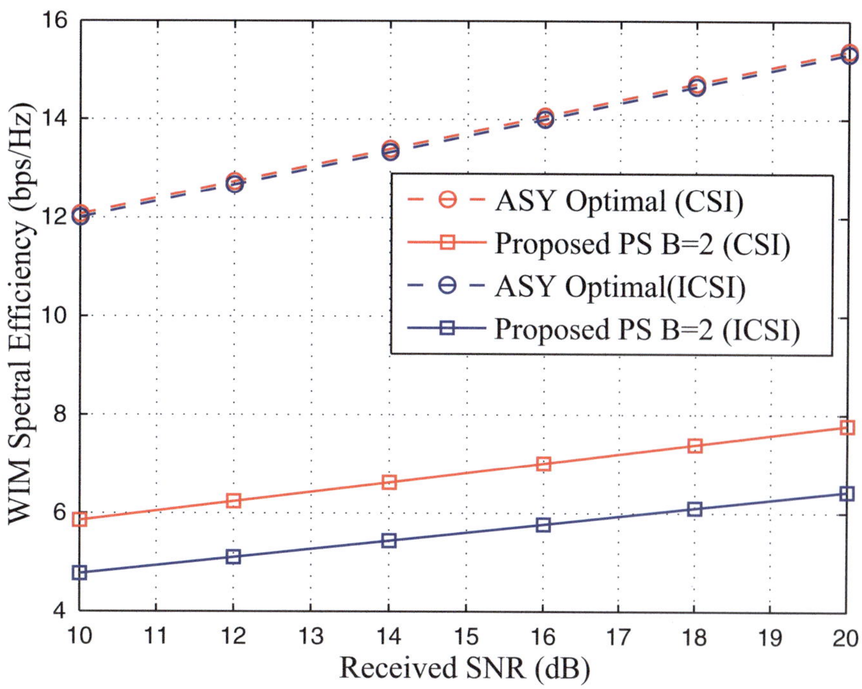

Fig. 4.9 WIM spectral efficiency versus the received SNR with ICSI

with the continuous shifter counterparts. By jointly considering the hardware complexity and the attainable WIM spectral efficiency, implementing phase shifters with 2 bits is a good choice.

We evaluate the impact of ICSI channel on the SWIPM performance in Fig. 4.9, where both perfect CSI and ICSI are considered. Moreover, we consider two IUs and two EUs in our system. The variance of the channel estimation uncertainty

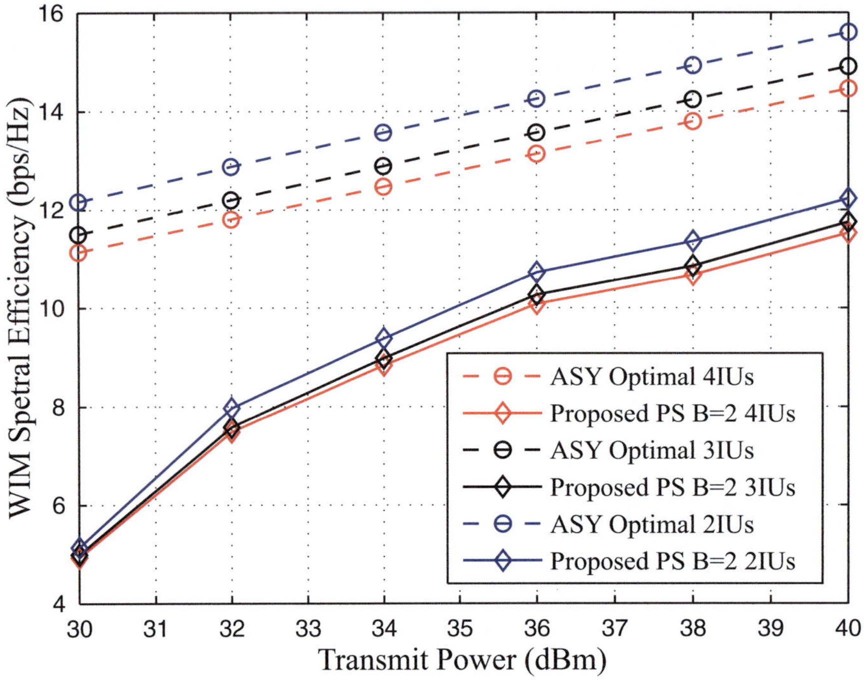

Fig. 4.10 WIM spectral efficiency versus the number of IUs

Δ is set to $\sigma = 0.1$. The requirement of EUs is 0.5 mW. Observe from Fig. 4.9 that having perfect CSI may achieve a better WIM performance than having ICSI.

We investigate the SWIPM performance for a different number of IUs in Fig. 4.10, when the number of EUs is set to 4. The energy requirement of every EU is 0.18 mW. Observe in Fig. 4.10 that when we have more IUs in the system, the WIM spectral efficiency of the asymptotically optimal design and that of hybrid design both decrease. In the asymptotically optimal scheme with a transmit power of 34 dBm, the WIM spectral efficiencies of 2 IUs are 0.7 and 1.1 bps/Hz higher than those of 3 IUs and 4 IUs, respectively. When we have a higher transmit power, the performance gap between the asymptotically optimal solution and the hybrid counterpart reduces. This is because the energy requirement of every EU can be readily satisfied by a high transmit power. The transmitter may allocate more power for improving the WIM spectral efficiency.

4.3 Exercises

1. What is beamforming and how does it improve SWIPT? Discuss the difference between analog beamforming, digital beamforming, hybrid beamforming, and IRS beamforming. Holographic beamforming is also popular in recent years, and you can look up information about it.
2. Maximum signal to noise ratio (MSNR), Maximum signal to interference and noise ratio (MSINR), minimum mean squared error ratio (MMSE), maximum likelihood (MLH), and linearly constrained minimum variance (LCMV) are general principles to design a beamforming. How these principles evolve when taking into account both hybrid beamforming and SWIPT?
3. In a SWIPT MIMO system, supposing that both the transmitter and receiver know the channel gain matrix $\mathbf{H}$, B=100 Hz, $P/\sigma^2 = 10$ dB, where P is the transmit power, σ is the noise variance at the receiver, channel gain matrix $\mathbf{H}$ and its SVD (singular value decomposition) defined as

$$\mathbf{H} = \begin{bmatrix} 0.1 & 0.3 & 0.7 \\ 0.5 & 0.4 & 0.1 \\ 0.2 & 0.3 & 0.2 \end{bmatrix}$$

$$= \begin{bmatrix} -0.72 & 0.65 & -0.22 \\ -0.55 & -0.74 & -0.39 \\ -0.42 & -0.16 & 0.89 \end{bmatrix} \begin{bmatrix} 0.9579 & 0 & 0 \\ 0 & 0.5086 & 0 \\ 0 & 0 & 0.0616 \end{bmatrix} \begin{bmatrix} -0.45 & -0.66 & -0.60 \\ -0.59 & -0.29 & 0.76 \\ -0.67 & 0.69 & -0.26 \end{bmatrix}. \quad (4.43)$$

Suppose the receiver is an IU, please use the water-filling algorithm to calculate the percentage of the transmitting power of each transmit antenna in the total power P_i/P and the channel capacity C, where P_i is the transmit power of the i-th transmit antenna. What if the receiver is an EU with $\sigma = -75$ dB? How much energy can this EU harvest when the nonlinear energy harvesting model is given by (4.6) in Sect. 4.2.1.2?

References

1. Hu, J., Wang, Q., & Yang, K. (2020). Energy self-sustainability in full-spectrum 6g. *Accepted by IEEE Wireless Communications.*
2. Visser, H. J., & Vullers, R. J. M. (2013). RF energy harvesting and transport for wireless sensor network applications: Principles and requirements. *Proceedings of the IEEE, 101*(6), 1410–1423.
3. Hu, J., Kun, Y., Guangjun, W., & Lajos, H. (2018). Integrated data and energy communication network: A comprehensive survey. *IEEE Communications Surveys Tutorials, 20*(4), 3169–3219.
4. Zhong, S., & Wang, X. (2019). Wireless power transfer by beamspace large-scale mimo with lens antenna array. *IEEE Transactions on Wireless Communications, 18*(2), 1390–1403.
5. Boshkovska, E., Ng, D. W. K., Zlatanov, N., & Schober, R. (2015). Practical non-linear energy harvesting model and resource allocation for SWIPT systems. *IEEE Communications Letters, 19*(12), 2082–2085.
6. Xiong, K., Wang, B., & Liu, K. J. R. (2017). Rate-energy region of SWIPT for MIMO broadcasting under nonlinear energy harvesting model. *IEEE Transactions on Wireless Communications, 16*(8), 5147–5161. ISSN: 1536–1276.
7. Lu, Y., Xiong, K., Fan, P., Ding, Z., Zhong, Z., & Letaief, K. B. (2019). Global energy efficiency in secure miso swipt systems with non-linear power-splitting eh model. *IEEE Journal on Selected Areas in Communications, 37*(1), 216–232.
8. Park, J., & Clerckx, B. (2015). Joint wireless information and energy transfer with reduced feedback in MIMO interference channels. *IEEE Journal on Selected Areas in Communications, 33*(8), 1563–1577. ISSN: 0733–8716.
9. Song, C., & Jeon, Y. (2018). Weighted MMSE precoder designs for sum-utility maximization in multi-user SWIPT network-MIMO with per-BS power constraints. *IEEE Transactions on Vehicular Technology, 67*(3), 2809–2813.
10. Mishra, D., & Alexandropoulos, G. C. (2018). Transmit precoding and receive power splitting for harvested power maximization in MIMO SWIPT systems. *IEEE Transactions on Green Communications and Networking, 2*(3), 774–786.
11. Linnartz, J. P. M. G., Wu, Y., Mare, J. G. A., & Matters-Kammerer, M. K. (2015). Multiple antenna rectifiers for radio frequency energy scavenging in wireless sensors. In *2015 IEEE International Symposium on Circuits and Systems (ISCAS)* (pp. 702–705).
12. Shen, S., & Clerckx, B. (2021). Beamforming optimization for MIMO wireless power transfer with nonlinear energy harvesting: RF combining versus DC combining. *IEEE Transactions on Wireless Communications, 20*(1), 199–213.
13. Pi, Z., & Khan, F. (2011). An introduction to millimeter-wave mobile broadband systems. *IEEE Communications Magazine, 49*(6), 101–107.
14. Swindlehurst, A. L., Ayanoglu, E., Heydari, P., & Capolino, F. (2014). Millimeter-wave massive MIMO: The next wireless revolution? *IEEE Communications Magazine, 52*(9), 56–62.
15. Yu, X., Shen, J. C., Zhang, J., & Letaief, K. B. (2016). Alternating minimization algorithms for hybrid precoding in millimeter wave MIMO systems. *IEEE Journal of Selected Topics in Signal Processing, 10*(3), 485–500.
16. Rusu, C., Mendez-Rial, R., Gonzalez-Prelcic, N., & Heath, R. W. (2016). Low complexity hybrid precoding strategies for millimeter wave communication systems. *IEEE Transactions on Wireless Communications, 15*(12), 8380–8393.
17. Ni, W., Dong, X., & Lu, W. S. (2017). Near-optimal hybrid processing for massive MIMO systems via matrix decomposition. *IEEE Transactions on Signal Processing, 65*(15), 3922–3933.
18. Ayach, O. E., Rajagopal, S., Abu-Surra, S., Pi, Z., & Heath, R. W. Jr. (2014). Spatially sparse precoding in millimeter wave MIMO systems. *IEEE Transactions on Wireless Communications, 13*(3), 1499–1513.
19. Chen, J. C. (2017). Hybrid beamforming with discrete phase shifters for millimeter-wave massive MIMO systems. *IEEE Transactions on Vehicular Technology, 66*(8), 7604–7608.
20. Sohrabi, F., & Wei, Y. (2016). Hybrid digital and analog beamforming design for large-scale antenna arrays. *IEEE Journal of Selected Topics in Signal Processing, 10*(3), 501–513.
21. Wang, Z., Member, S., Ming, L., & Member, S. (2018). Hybrid precoder and combiner design with low-resolution phase shifters in mmWave MIMO systems. *IEEE Journal of Selected Topics in Signal Processing, 12*(2), 256–269.
22. Gui, P., Zhao, M., Shi, Q., & Wu, Q. (2017). Max-min hybrid precoder design for Mmwave multi-group multicasting systems. In *2017 IEEE/CIC International Conference on Communications in China (ICCC)* (pp. 1–5).
23. Wang, Z., Liu, Q., Li, M., & Kellerer, W. (2019). Energy efficient analog beamformer design for mmWave multicast transmission. *IEEE Transactions on Green Communications and Networking, 3*(2), 552–564.
24. Grover, P., & Sahai, A. (2010). Shannon meets tesla: Wireless information and power transfer. In *2010 IEEE International Symposium on Information Theory* (pp. 2363–2367).
25. Clerckx, B. (2018). Wireless information and power transfer: Nonlinearity, waveform design, and rate-energy tradeoff. *IEEE Transactions on Signal Processing, 66*(4), 847–862.
26. Huang, Y., & Clerckx, B. (2016). Large-scale multi-antenna multi-sine wireless power transfer. *IEEE Transactions on Signal Processing, PP*(99), 5812–5827.
27. Yang, K., Yang, N., Ye, N., Jia, M., & Fan, R. (2018). Non-orthogonal multiple access: Achieving sustainable future radio access. *IEEE Communications Magazine, PP*(99), 1–6.
28. Gao, X., Niyato, D., Wang, P., Yang, K., & An, J. (2020). Contract design for time resource assignment and pricing in backscatter-assisted RF-powered networks. *IEEE Wireless Communications Letters, 9*(1), 42–46.

29. Ding, Z., Zhong, C., Wing Kwan Ng, D., Peng, M., Suraweera, H. A., Schober, R., & Poor, H. V. (2015). Application of smart antenna technologies in simultaneous wireless information and power transfer. *IEEE Communications Magazine, 53*(4), 86–93.
30. Zhang, R., & Ho, C. K. (2013). MIMO broadcasting for simultaneous wireless information and power transfer. *IEEE Transactions on Wireless Communications, 12*(5), 1989–2001. ISSN: 1536–1276.
31. Boshkovska, E., Ng, D. W. K., Zlatanov, N., Koelpin, A., & Schober, R. (2017). Robust resource allocation for MIMO wireless powered communication networks based on a non-linear EH model. *IEEE Transactions on Communications, 65*(5), 1984–1999.
32. Dai, L., Wang, B., Peng, M., & Chen, S. (2019). Hybrid precoding-based millimeter-wave massive MIMO-NOMA with simultaneous wireless information and power transfer. *IEEE Journal on Selected Areas in Communications, 37*(1), 131–141. ISSN: 0733–8716.
33. Mizojiri, S., & Shimamura, K. (2019). Recent progress of wireless power transfer via sub-THz wave. In *2019 IEEE Asia-Pacific Microwave Conference (APMC)* (pp. 705–707).
34. Dang-Ba, H., & Byun, G.-S. (2020). A sub-THz wireless power transfer for non-contact wafer-level testing. *Electronics, 9*(8), 1210.
35. Deutschmann, B. J. B., Wilding, T., Larsson, E. G., & Witrisal, K. (2022). Location-based initial access for wireless power transfer with physically large arrays. In *2022 IEEE International Conference on Communications Workshops (ICC Workshops)* (pp. 127–132).
36. Bjornson, E., Van der Perre, L., Buzzi, S., & Larsson, E. G. (2019). Massive mimo in sub-6 ghz and mmwave: Physical, practical, and use-case differences. *IEEE Wireless Communications, 26*(2), 100–108.
37. Molisch, A. F., Ratnam, V. V., Han, S., Li, Z., Nguyen, S. L. H., Li, L., & Haneda, K. (2017). Hybrid beamforming for massive mimo: A survey. *IEEE Communications Magazine, 55*(9), 134–141.
38. Gao, F., Wang, B., Xing, C., An, J., & Li, G. Y. (2021). Wideband beamforming for hybrid massive mimo terahertz communications. *IEEE Journal on Selected Areas in Communications, 39*(6), 1725–1740.
39. Lin, C., & Li, G. Y. (2016). Energy-efficient design of indoor mmwave and sub-THz systems with antenna arrays. *IEEE Transactions on Wireless Communications, 15*(7), 4660–4672.
40. Shah, S. H. A., Aditya, S., & Rangan, S. (2021). Power-efficient beam tracking during connected mode drx in mmwave and sub-THz systems. *IEEE Journal on Selected Areas in Communications, 39*(6), 1711–1724.
41. Sarieddeen, H., Alouini, M.-S., & Al-Naffouri, T. Y. (2019). Terahertz-band ultra-massive spatial modulation mimo. *IEEE Journal on Selected Areas in Communications, 37*(9), 2040–2052.
42. Krishna Moorthy, S., & Guan, Z. (2022). Beam learning in mmwave/THz-band drone networks under in-flight mobility uncertainties. *IEEE Transactions on Mobile Computing, 21*(6), 1945–1957.
43. Yuan, H., Yang, N., Yang, K., Han, C., & An, J. (2020). Hybrid beamforming for terahertz multi-carrier systems over frequency selective fading. *IEEE Transactions on Communications, 68*(10), 6186–6199.
44. Lin, C., & Li, G. Y. (2015). Adaptive beamforming with resource allocation for distance-aware multi-user indoor terahertz communications. *IEEE Transactions on Communications, 63*(8), 2985–2995.
45. Busari, S. A., Huq, K. M. S., Mumtaz, S., Rodriguez, J., Fang, Y., Sicker, D. C., Al-Rubaye, S., & Tsourdos, A. (2019). Generalized hybrid beamforming for vehicular connectivity using THz massive mimo. *IEEE Transactions on Vehicular Technology, 68*(9), 8372–8383.
46. Yan, L., Chen, Y., Han, C., & Yuan, J. (2022). Joint inter-path and intra-path multiplexing for terahertz widely-spaced multi-subarray hybrid beamforming systems. *IEEE Transactions on Communications, 70*(2), 1391–1406.
47. Elbir, A. M., Mishra, K. V., & Chatzinotas, S. (2021). Terahertz-band joint ultra-massive mimo radar-communications: Model-based and model-free hybrid beamforming. *IEEE Journal of Selected Topics in Signal Processing, 15*(6), 1468–1483.
48. Zhai, B., Zhu, Y., Tang, A., & Wang, X. (2020). Thzprism: Frequency-based beam spreading for terahertz communication systems. *IEEE Wireless Communications Letters, 9*(6), 897–900.
49. Dai, L., Tan, J., Chen, Z., & Poor, H. V. (2022). Delay-phase precoding for wideband THz massive mimo. *IEEE Transactions on Wireless Communications, 21*(9), 7271–7286.
50. Wu, Q., & Zhang, R. (2020). Towards smart and reconfigurable environment: Intelligent reflecting surface aided wireless network. *IEEE Communications Magazine, 58*(1), 106–112.
51. Pan, C., Ren, H., Wang, K., Kolb, J. F., Elkashlan, M., Chen, M., Di Renzo, M., Hao, Y., Wang, J., Swindlehurst, A. L., You, X., & Hanzo, L. (2021). Reconfigurable intelligent surfaces for 6G systems: Principles, applications, and research directions. *IEEE Communications Magazine, 59*(6), 14–20.
52. Cui, T. J., Qi, M. Q., Wan, X., Zhao, J., & Cheng, Q. (2014). Coding metamaterials, digital metamaterials and programmable metamaterials. *Light Science & Applications, 3*(10), e218.
53. Bai, T., Pan, C., Ren, H., Deng, Y., Elkashlan, M., & Nallanathan, A. (2021). Resource allocation for intelligent reflecting surface aided wireless powered mobile edge computing in OFDM systems. *IEEE Transactions on Wireless Communications, 20*(8), 5389–5407.
54. Pan, C., Ren, H., Wang, K., Elkashlan, M., Nallanathan, A., Wang, J., & Hanzo, L. (2020). Intelligent reflecting surface aided MIMO broadcasting for simultaneous wireless information and power transfer. *IEEE Journal on Selected Areas in Communications, 38*(8), 1719–1734.
55. Lyu, B., Hoang, D. T., Gong, S., & Yang, Z. (2020). Intelligent reflecting surface assisted wireless powered communication networks. In *2020 IEEE Wireless Communications and Networking Conference Workshops (WCNCW)* (pp. 1–6).
56. Wu, Q., & Zhang, R. (2020). Weighted sum power maximization for intelligent reflecting surface aided SWIPT. *IEEE Wireless Communications Letters, 9*(5), 586–590.
57. Wu, Q., & Zhang, R. (2020). Beamforming optimization for wireless network aided by intelligent reflecting surface with discrete phase shifts. *IEEE Transactions on Communications, 68*(3), 1838–1851.
58. Gong, S., Yang, Z., Xing, C., An, J., & Hanzo, L. (2021). Beamforming optimization for intelligent reflecting surface-aided SWIPT IoT networks relying on discrete phase shifts. *IEEE Internet of Things Journal, 8*(10), 8585–8602.
59. Abdullah, Z., Chen, G., Lambotharan, S., & Chambers, J. (2022). Low-complexity antenna selection and discrete phase-shifts design in IRS-assisted multiuser massive MIMO networks. *IEEE Transactions on Vehicular Technology, 71*(4), 3980–3994.
60. Hu, X., Jin, L., Huang, K., Sun, X., Zhou, Y., & Qu, J. (2021). Intelligent reflecting surface-assisted secret key generation with discrete phase shifts in static environment. *IEEE Wireless Communications Letters, 10*(9), 1867–1870.
61. Chapala, V. K., & Zafaruddin, S. M. (2021). Exact analysis of RIS-aided THz wireless systems over $\alpha - \mu$ fading with pointing errors. *IEEE Communications Letters, 25*(11), 3508–3512.

62. Pan, Y., Wang, K., Pan, C., Zhu, H., & Wang, J. (2021). UAV-assisted and intelligent reflecting surfaces-supported terahertz communications. *IEEE Wireless Communications Letters, 10*(6), 1256–1260.
63. Qiao, J., & Alouini, M.-S. (2020). Secure transmission for intelligent reflecting surface-assisted mmwave and terahertz systems. *IEEE Wireless Communications Letters, 9*(10), 1743–1747.
64. Ning, B., Chen, Z., Chen, W., Du, Y., & Fang, J. (2021). Terahertz multi-user massive mimo with intelligent reflecting surface: Beam training and hybrid beamforming. *IEEE Transactions on Vehicular Technology, 70*(2), 1376–1393.
65. Wang, P., Fang, J., Zhang, W., & Li, H. (2022). Fast beam training and alignment for irs-assisted millimeter wave/terahertz systems. *IEEE Transactions on Wireless Communications, 21*(4), 2710–2724.
66. Lu, Y., Hao, M., & Mackenzie, R. (2022). Reconfigurable intelligent surface based hybrid precoding for THz communications. *Intelligent and Converged Networks, 3*(1), 103–118.
67. Liu, X., Zhang, H., Long, K., Zhou, M., Li, Y., & Poor, H. V. (2022). Proximal policy optimization-based transmit beamforming and phase-shift design in an irs-aided ISAC system for the THz band. *IEEE Journal on Selected Areas in Communications, 40*(7), 2056–2069.
68. Qiao, J., Zhang, C., Dong, A., Bian, J., & Alouini, M.-S. (2022). Securing intelligent reflecting surface assisted terahertz systems. *IEEE Transactions on Vehicular Technology, 71*(8), 8519–8533.
69. Roth, K., & Nossek, J. A. (2018). Robust massive MIMO equilization for Mmwave systems with low resolution Adcs. In *2018 IEEE Wireless Communications and Networking Conference Workshops (WCNCW)* (pp. 113–118).
70. Zhu, D., Bendlin, R., Akoum, S., Ghosh, A., & Heath, R. W. (2019). Directional frame timing synchronization in wideband millimeter-wave systems with low-resolution ADCs. *IEEE Transactions on Wireless Communications, 18*(11), 5350–5366.
71. Wang, Y., Wang, Y., Zhou, F., Wu, Y., & Zhou, H. (2017). Resource allocation in wireless powered cognitive radio networks based on a practical non-linear energy harvesting model. *IEEE Access, 5*, 17618–17626.
72. Sadeghi, M., Björnson, E., Larsson, E. G., Yuen, C., & Marzetta, T. (2018). Joint unicast and multi-group multicast transmission in massive MIMO systems. *IEEE Transactions on Wireless Communications, 17*(10), 6375–6388.
73. Sadeghi, M., Bjrnson, E., Larsson, E. G., Yuen, C., & Marzetta, T. L. (2018). Max-min fair transmit precoding for multi-group multicasting in massive MIMO. *IEEE Transactions on Wireless Communications, 17*(2), 1358–1373.
74. Xiang, Z., Tao, M., & Wang, X. (2014). Massive MIMO multicasting in noncooperative cellular networks. *IEEE Journal on Selected Areas in Communications, 32*(6), 1180–1193.
75. Sulyman, A. I., Alwarafy, A., Maccartney, G., Rappaport, T., & Al-Sanie, A. (2016). Directional radio propagation path loss models for millimeter-wave wireless networks in the 28, 60, and 73 GHz bands. *IEEE Transactions on Wireless Communications, 15*(10), 6939–6947.
76. Guo, J., & Zhu, X. (2012). An improved analytical model for RF-DC conversion efficiency in microwave rectifiers. In *IEEE MTT-S International Microwave Symposium digest. IEEE MTT-S International Microwave Symposium* (pp. 1–3).

Chapter 5
Multidimensional Resource Allocation for Integrated Data and Energy Transfer

5.1 Resource Management: Fundamental and Theory

With the rapid development of wireless network technique and the widespread application of wireless devices, the network scale is rapidly expanding. The access of massive users makes the limited resources in wireless communication networks even more scarce. In order to ensure the quality of service for each user, wireless network resource management plays a very important role. We can achieve the goal of resource utilization maximization through reasonable scheduling design.

Nowadays, the optimization methods for resource allocation in wireless networks can be generally divided into two categories: offline strategy and online strategy. Online strategies can optimize resource scheduling in real time based on different states in networks such as channel state information. However, it limits the complexity of the algorithm. Specifically, high algorithm complexity can lead to an excessively long duration of updating policies. Therefore, outdated strategies do not improve network performance and may even reduce user experience. On the other hand, offline strategies can provide long-term stable network performance, and their requirements for algorithm complexity are more relaxed compared to online strategies. However, due to its inability to adjust policies in real time with changes of network states, it is bound to lose some network performance.

5.1.1 Resource Management in Integrated Data and Energy Networks

Compared with traditional wireless network resource allocation, wireless network resource allocation with energy harvesting requires the consideration of the randomness of harvested energy, which makes some traditional resource allocation strategies no longer optimal. For example, the power injection method is a classic power allocation algorithm that can maximize throughput, but it may not hold in integrated data and energy networks. This is because the feasible domain of optimization problems is changed by invoking energy harvesting. Specifically, WPT aided devices are limited by their harvested energy. The energy they use to emit information must not exceed their stored energy, which is generally ignored in conventional wireless devices with sufficient energy. Therefore, there is an essential condition having to be considered in integrated data and energy networks: energy causality constraints. Moreover, many researchers also consider more practical scenario, namely limited battery capacity in wireless powered devices. Generally, they can obtain more sufficient energy over a longer duration of WPT. However, they are unable to harvest any energy, while the battery is fully recharged.

As aforementioned, we consider energy causality in an integrated data and energy network. That is, the resource management in it no longer focuses on wireless communication but has two requirements, wireless energy and wireless communication. For example, energy limited devices transmit their data packets by consuming harvested energy. It is obvious that a long WPT duration is beneficial for wireless communication since the device stores more energy for their WIT. However, it undoubtedly squeezes wireless communication period. It may also result in the lack of timeliness in the networks. As a result, it is essential to balance the energy and data requirements by reasonable resource scheduling in integrated data and energy networks

J. Hu, K. Yang, *Integrated Data and Energy Transfer in Wireless Networks*, Textbooks in Telecommunication Engineering,
https://doi.org/10.1007/978-3-031-60514-7_5

5.1.2 A Brief Tutorial for Convex Optimization

The convex optimization theory is crucial for the resource management in both the conventional wireless communication networks and the integrated data and energy networks. In this part, we briefly introduce some essential foundations of convex optimization. Please refer to [1] for more information.

5.1.2.1 Some Basic Concepts in Convex Optimization

Definition 5.1.1 A set $\mathcal{C}$ is called a convex set, if it exists $\theta x_1 + (1-\theta)x_2 \in \mathcal{C}$ for both $\forall x_1, x_2 \in \mathcal{C}$ and $\forall\theta \in [0, 1]$.

Definition 5.1.2 A function $f(x)$ is a convex function if its domain is convex set and it satisfies $f(\theta x_1 + (1-\theta)x_2) \leq \theta f(x_1) + (1-\theta)f(x_2)$ for any two points x_1, x_2 in the domain and for $\forall\theta \in [0, 1]$. Besides, if it holds that $f(\theta x_1 + (1-\theta)x_2) = \theta f(x_1) + (1-\theta)f(x_2)$, $f(x)$ is an affine function. Finally, if it satisfies $f(\theta x_1 + (1-\theta)x_2) \geq \theta f(x_1) + (1-\theta)f(x_2)$, $f(x)$ is concave.

Definition 5.1.3 The perspective function of the function $f(x)$ is $g(x, t) = tf(x/t)$.

Property 5.1.1 A necessary and sufficient condition for $f(x)$ to be a convex function is that its Hessian matrix is a positive semi-definite matrix. Correspondingly, that for a concave function $f(x)$ is that the Hessian matrix of the function is a semi-negative definite matrix.

Property 5.1.2 $f(x)$ has the same convexity with its perspective function $g(x, t)$. If $f(x)$ is a convex function, its perspective function is also a convex function.

5.1.2.2 Convex Optimization Problems

Convex optimization problems have a very complete set of solving algorithms. If it can be proven that an optimization problem is a convex optimization problem, it can be convenient and fast to find the optimal solution.

Generally, the convex optimization problems have the following form:

$$\text{(P5.1.1)}: \min_{x} f_0(x), \tag{5.1}$$

$$\text{s. t. } f_i(x) \leq 0, \forall i = 1, \cdots, m, \tag{5.1a}$$

$$h_i(x) = 0, i = 1, \cdots, p. \tag{5.1b}$$

When the objective function and inequality constraint function are convex functions, and the equality constraint function is affine, (P5.1.1) is a convex optimization problem.

If the problem has the following form:

$$\text{(P5.1.2)}: \max_{x} f_0(x), \tag{5.2}$$

$$\text{s. t. } f_i(x) \geq 0, \forall i = 1, \cdots, m, \tag{5.2a}$$

$$h_i(x) = 0, i = 1, \cdots, p, , \tag{5.2b}$$

where $f_0(x)$ and $f_i(x)$ are concave and $h_i(x)$ is affine, (P5.1.2) is also a convex optimization problem.

5.1.2.3 Lagrangian Duality Method

The basic idea of the Lagrangian dual method is to consider the constraints of the optimization problem in the objective function. The weighted sum of the constraints is considered to obtain an augmented objective function. The Lagrange function for (P5.1.1) is defined as

$$\mathcal{L}(x, \lambda, \omega) = f_0(x) + \sum_{i=1}^{m} \lambda_i f_i(x) + \sum_{i=1}^{m} \omega_i h_i(x), \tag{5.3}$$

where λ_i represents the Lagrange multipliers with respect to $f_i(x)$ and ω_i represents that with respect to $h_i(x)$.

5.1.2.4 Karush–Kuhn–Tucker Conditions

When the original problem is a convex problem, the point satisfying the Karush–Kuhn–Tucker (KKT) conditions is the optimal solution of the original and dual problems. If x^*, λ^*, and ω^* satisfy KKT conditions of (P5.1.1), x^* is the optimal solution to the original problem. For instance, the KKT conditions for (P5.1.1) are as follows:

$$\begin{cases} f_i(x^*) \leq 0, \\ h_i(x^*) = 0, \\ \lambda_i^* \geq 0, \\ \lambda_i^* f_i(x^*) = 0, \\ \nabla f_0(x^*) + \sum_{i=0}^{m} \lambda_i^* f_i(x^*) + \sum_{i=1}^{p} \omega_i^* h_i(x^*) = 0. \end{cases} \tag{5.4}$$

5.2 Multidimensional Resource Allocation in Full-Duplex Wireless Powered Communication Networks

In order to support information uploading of batteryless user equipments (UEs), a full-duplex (FD) hybrid access point (H-AP) is relied upon for downlink wireless power transfer (WPT) and uplink information collection, which yields a classic wireless powered communication network (WPCN). These UEs harvest energy from downlink radio frequency (RF) signals in order to power their own uplink wireless information transfer (WIT). However, the uplink information collection of the H-AP suffers from the self-interference (SI) induced by the downlink WPT. In this section, for the sake of maximizing the uplink fair-throughput of UEs, we jointly design the transmit beamforming and receive combining at the H-AP having multiple antennas in the spatial domain, the resource allocation in the time-domain, and the UEs' scheduling in the user domain. Since the resultant optimization problem is non-convex in nature, an iterative algorithm is developed by decoupling the original joint design into several subproblems. Our numerical results demonstrate the fast convergence, the uplink fair-throughput gain, and the great fairness of our multi-domain resource scheduling design.

5.2.1 Introduction

In the Internet of Things (IoT) era, the lifetime of massive miniature communication devices mainly depends on their limited battery capacity. Dead devices may significantly degrade the performance of these IoT applications. Moreover, frequently replacing batteries may result in significant growth of the maintenance expenditure. Therefore, dedicated wireless power transfer (WPT) has been widely recognized as a key technology for charging massively deployed low-power communication devices, which will realize self-sustainable IoT in the future era of 6G [2]. Inductive coupling and magnetic resonance coupling based WPT can only deliver energy to the near-field ($\leq$1 m) [3]. They are not suitable for charging remote miniature IoT devices [4, 5]. Hence, radio frequency (RF) signal based WPT is preferred, due to its controllability, flexibility, wide coverage ($\geq$10 m), and easy implementation [6]. With the aid of RF-based WPT, low-power user equipments (UEs) may be finally freed from the constraints of embedded batteries [7]. However, batteryless UEs have to gain sufficient energy from downlink RF signals emitted by energy transmitters before their own uplink wireless information transfer (WIT), which yields wireless powered communication networks (WPCNs) [8–12]. If both downlink WPT and uplink information reception are carried out by a single transceiver, it is normally known as a hybrid access point (H-AP) [13].

In order to improve the performance of WPCN, half-duplex (HD) H-APs are normally equipped with multiple antennas by introducing both multiplexing and diversity gains in the spatial domain [14–16]. Specifically, Pehlivan et al. [14] jointly optimized a hybrid transmit beamformer associated with a limited number of RF chains and the time-slot durations allocated for WPT toward individual users in order to minimize the total duration for the downlink WPT. Lim et al. [15] maximized the sum energy efficiency of UEs by jointly optimizing the WPT beamforming of the AP and the uplink transmit power of the UEs. Furthermore, Choi et al. [16] jointly designed downlink WPT beamformer and uplink receive combiner of the H-AP as well as the zero-forcing-based WIT precoders of UEs, which maximized the minimum uplink WIT throughput of UEs. Multi-domain resource scheduling in WPCNs was firstly considered in [17], where downlink/uplink resource allocation scheme was jointly designed in spatial, power-, and time-domains in an HD aided WPCN for maximizing both the sum-throughput and fair-throughput[1] of the uplink WIT.

Furthermore, H-APs may adopt the full-duplex (FD) technique for simultaneously operating the downlink WPT and uplink WIT in the same spectral band. However, its main limitation is the self-interference (SI) due to signal leakage from the downlink WPT of an H-AP to its uplink information reception. The large SI spans most of the dynamic range of the analog-to-digital converter at the receiver side of the H-AP, and thus its mitigation is critical for the implementation of the FD [18]. Kang et al. [19] designed the time allocation for maximizing the uplink sum-throughput and minimizing the frame duration, respectively. Hu et al. [20] jointly designed UEs' scheduling and time allocation for maximizing their uplink sum-throughput in an FD-WPCN. This is the first work for resource scheduling in both the time- and user domains. Unfortunately, they [19, 20] assumed perfect self-interference cancellation (SIC). Furthermore, Abd-Elmagid et al. [21] characterized an optimal policy in an FD-WPCN with two UEs for maximizing the long-term weighted throughput by considering perfect as well as imperfect SIC. The sum- and fair-throughput performances were demonstrated by modifying the weighted factor. However, they [19–21] all ignored potential performance gain from the spatial domain.

By implementing multiple antennas at H-APs, resource scheduling in the spatial domain may substantially improve the performance of FD-WPCNs [22, 23]. Specifically, Peng et al. [22] jointly designed optimal power allocation and time allocation for the uplink WIT in order to maximize the sum-throughput associated with finite alphabet. However, they ignored the downlink WPT beamforming design of H-AP, while they assumed perfect SIC. Moreover, Lee et al. [23] maximized the weighted sum-throughput of UEs against imperfect knowledge of channel state information (CSI) by jointly designing a robust WPT beamformer of the H-AP and the associated time allocation to UEs' uplink WIT, where residual SI is conceived for characterizing imperfect SIC.

Apart from [21] and [23], several works [24–28] all considered the impact of SI on FD-WPCNs. Specifically, Kim et al. [24] optimized subcarrier scheduling and associated power allocation by considering residual SI, which was derived by multiplying an attenuation factor on downlink WPT signals, in order to maximize the sum-throughput of uplink WIT. Similar model of residual SI was also conceived in [25], which maximized sum-throughput of the uplink WIT by jointly optimizing the power allocation of the H-AP for downlink WPT and the time allocation for UEs' uplink WIT. Since the SI can be modeled as a polynomial with respect to multiple orders of transmit signals [26], Nikoloska et al. [27] adopted first-order approximation for modeling SI as the worst case. They further derived the capacity of the uplink WIT channel of a single UE by considering the power consumption for various signal processing. Moreover, Asiedu et al. [28] jointly optimized channel assignment, time and power allocations to maximize sum-throughput by considering SI at H-AP and linear energy harvesting (EH) mode at users. Unfortunately, the authors of [21, 23–28] only modeled the SI and residual SI in a mathematical manner, but they did not consider a specific SIC technique in FD-WPCNs.

We compare the contributions of the existing works [8, 19–23, 25–28] to ours. In a nut shell, these existing works have the following drawbacks:

- They only considered resource scheduling in spatial and time-domains in multi-antenna aided FD-WPCNs. Although energy causality was conceived, which requires that energy consumption of UEs for uplink WIT has to be lower than total energy harvested, they do not consider optimal UEs' scheduling for maximizing their WIT performance.
- They only considered mathematical modeling of SI and residual SI, but they did not fully exploit benefit of multiple antennas at the H-AP for practical SIC and for WIT performance improvement by properly combining received uplink WIT signals.
- They only maximized fair-throughput for a pair of UEs or sum-throughput performance for multiple UEs in an FD-WPCN.

In order to overcome the abovementioned drawbacks, our novel contributions are summarized as follows:

- *We investigate an FD-WPCN having a single H-AP equipped with multiple antennas and many batteryless UEs equipped with a signal antenna. A nonlinear energy harvesting model is conceived at UEs, while the impact of SI imposed by downlink WPT on uplink WIT as well as the energy causality of UEs is considered in the system.*

[1] Due to doubly near–far effect of WPCNs, sum-throughput maximization of uplink WIT may result in unfair resource allocation among batteryless UEs, which sacrifices the quality of service (QoS) of UEs distant from the H-AP. Therefore, fair-throughput is proposed, which is defined as the minimum uplink WIT throughput among the UEs.

- *We maximize the fair-throughput of UEs' uplink WIT for overcoming doubly near–far effect. Therefore, every UE's QoS can be guaranteed in the FD-WPCN. By jointly designing the downlink WPT beamformer and the uplink receive combiner at the H-AP, the time allocation, and the access order of UEs' uplink WIT, we obtain a joint scheduling of resources in spatial, time-, and user domain.*
- *The original non-convex and multivariable optimization problem is decoupled into several subproblems having identical objective, which includes transceiver design of the H-AP operating in an FD mode by considering SIC, UEs' scheduling by considering energy causality as well as the time allocation of UEs' uplink WIT. Since every decoupled subproblem can be optimally solved, our resource scheduling design in multi-domain can be obtained by invoking an iterative algorithm.*
- *We demonstrate that our joint design substantially improves the performance of the FD-WPCN.*

5.2.2 System Model

5.2.2.1 Network Description

A general FD-WPCN is illustrated in Fig. 5.1, which is constituted by a single H-AP operating in an FD mode and a set of UEs $\mathcal{UE} = \{UE_i | i = 1, 2, \cdots, K\}$ operating in an HD mode.[2] In this FD-WPCN, we consider a data collection application, where the H-AP hires selfish UEs for collecting their sensing information. Specifically, the transceiver of the H-AP has the following pair of functions:

- It transmits RF signals in the downlink for remotely charging batteryless UEs by forming a WPT beam with its N_t transmit antennas.
- It receives information uploaded by UEs by properly combining WIT signals with its N_r receive antennas, while cancelling SI induced by its downlink WPT.

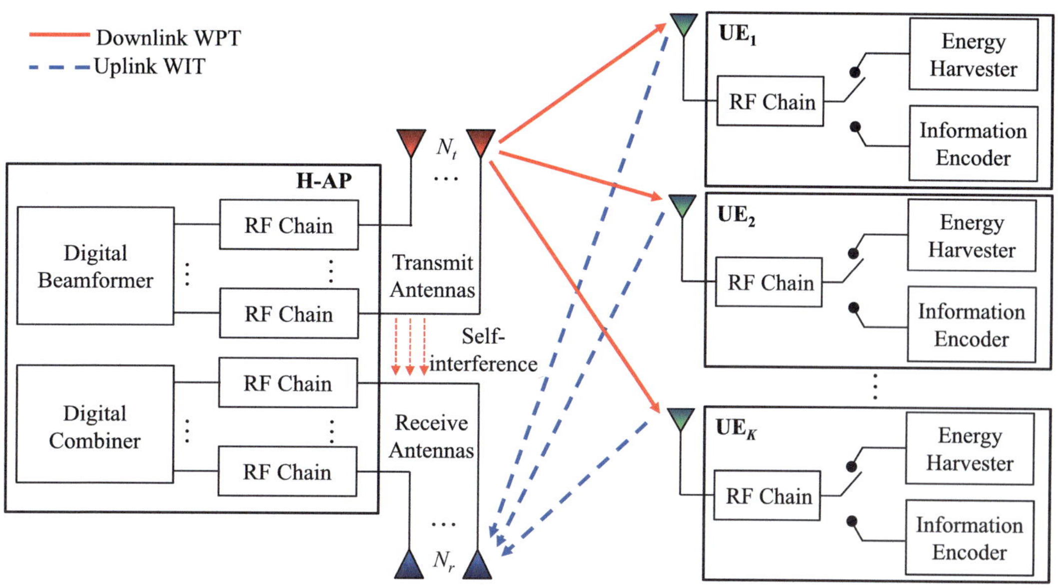

Fig. 5.1 An FD-WPCN having a single H-AP equipped with multiple antennas and K UEs equipped with a single antenna

[2] Note that any geographic distribution of UEs can be conceived, which may not make any difference in our resource scheduling design.

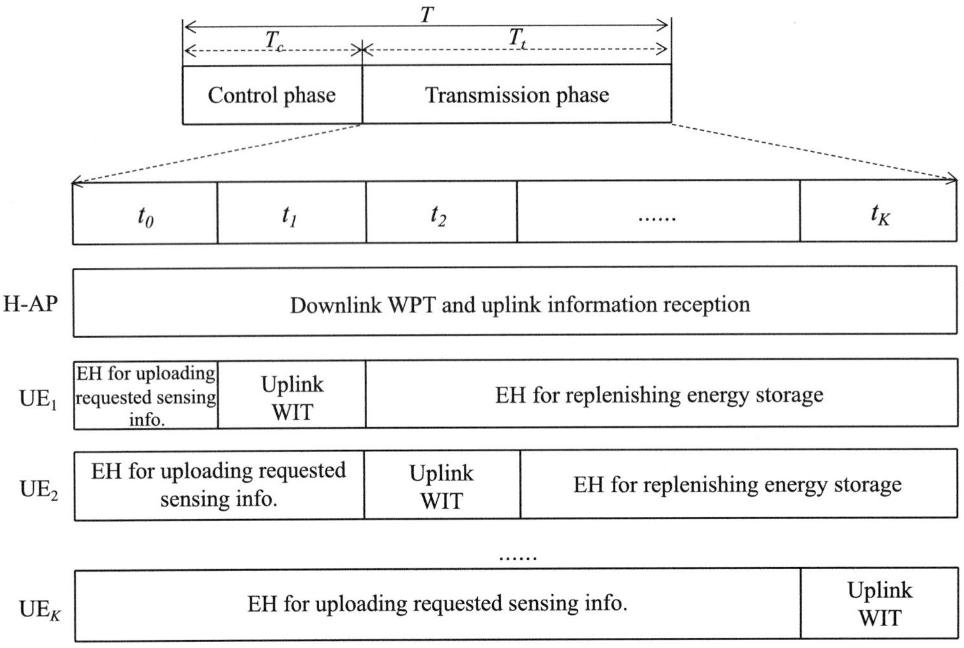

Fig. 5.2 Frame structure of the H-AP's and UEs' slotted operations

Moreover, the UEs in the FD-WPCN only have a single antenna for energy harvesting in the downlink and for WIT in the uplink, while they are all equipped with super-capacitors having a short discharging cycle.[3] Therefore, the UEs operate by obeying the following principles:

- They harvest energy from the H-AP and replenish the energy storage.
- These selfish UEs do not consume their energy storage for uploading the sensing information requested by the H-AP. Therefore, they harvest additional energy from the downlink WPT of the H-AP during the time slots before their information uploading. They then deplete this part of energy harvested for powering their information uploading.

As shown in Fig. 5.1, the UEs switch between the energy harvesting mode and information uploading mode in a time-division manner. We have to consider energy causality at the UEs [29].

5.2.2.2 MAC Layer Description

In order to support multiple UEs, our FD-WPCN adopts a classic time-division multiple access (TDMA) protocol in the medium access control (MAC) layer [30]. A complete transmission frame is characterized in Fig. 5.2, which includes a control phase having a duration of T_c and a transmission phase having a duration of T_t. Specifically, during the control phase, channel estimation [31], resource scheduling, and synchronisation[4] are carried out. UEs execute the corresponding tasks in the control phase by consuming their energy storage. Moreover, T_c is a deterministic duration to ensure the completion of the forementioned tasks, which is generally lower than 300 μs with a bandwidth of 15 kHz [34].

[3] We consider that UEs are miniature low-power devices, such as micro-sensors. As compared to the rechargeable lithium batteries, super-capacitors have very small sizes as well as fast recharging and discharging cycle. Moreover, RF signals received by UEs are too weak to recharge lithium batteries. Therefore, such UEs typically store the harvested energy in super-capacitors [19], which are generally equipped with single antennas in practical implementation.

[4] Since all the UEs are distributed in the coverage of the H-AP, they may readily be synchronized together by invoking the reference-based clock synchronisation approach [32]. Specifically, the H-AP broadcasts its locally recorded clock information to all the UEs during the control phase. Once the clock information is successfully received, the UEs may adjust their local clock in order to complete their synchronisation process. We assume perfect synchronisation since its accuracy may achieve 3.67 μs in the worst case, which is far lower than our time-slot length [33]. A large asynchronization rarely happens in a centralized communication system. If it really happens, it may result in UEs' energy error, bit error, inter-UEs interference, and even outage, which will be considered in our future work.

As shown in Fig. 5.2, a transmission phase is further divided into $(K + 1)$ time slots having durations of $\mathbf{t} = \{t_j | j = 0, 1, 2, \cdots, K\}$, which satisfy

$$\sum_{j=0}^{K} t_j \leq T_t. \tag{5.5}$$

The H-AP always transmits RF signals for the downlink WPT during the entire transmission phase, while each UE harvests energy during the whole transmission frame except the time slot scheduled for its uplink WIT. Specifically, UE_i $(\forall i = 1, \cdots, K)$ is scheduled to upload information on the j-th $(\forall j = 1, \cdots, K)$ time slot for a duration of t_j. No UE is allowed to upload information on the 0-th time slot having a duration of t_0, which is for the downlink WPT only. As a result, UE_i harvests energy from the downlink WPT of the H-AP within a duration of $\sum_{k=0}^{j-1} t_k$ for powering its uplink WIT in the j-th time slot in the current transmission frame. Moreover, after its uplink WIT, UE_i may still harvest energy from the rest of time slots with a duration of $\sum_{k=j+1}^{K} t_k$ in the current transmission frame, which is for its energy replenishment. The access pattern of UEs can be represented by a binary indicator matrix $\mathbf{G} = [g_{i,j}]_{K \times K}$ obeying the following constraints:

$$\sum_{i=1}^{K} g_{i,j} \leq 1, \forall j = 1, \cdots, K, \tag{5.6}$$

$$\sum_{j=1}^{K} g_{i,j} \leq 1, \forall i = 1, \cdots, K, \tag{5.7}$$

$$g_{i,j} = 0 \text{ or } 1, \tag{5.8}$$

where constraint (5.6) represents that a single time slot can only be accessed by a single UE, while constraint (5.7) represents that a single UE can only be scheduled on a single time slot for the uplink WIT. Furthermore, as shown in constraint (5.8), $g_{i,j} = 1$ represents UE_i is scheduled on the j-th time slot; otherwise, $g_{i,j} = 0$.

Therefore, the amount of energy harvested by UEs and their resultant uplink WIT throughput are then jointly determined by the access pattern matrix $\mathbf{G}$ and the time-slot durations $\mathbf{t}$.

5.2.2.3 Wireless Channel Description

The normalized fading coefficients of the downlink channel from the H-AP's N_t transmit antennas to UE_i are denoted as a $1 \times N_t$ vector $\mathbf{h}_{dl,i}$, while those from the H-AP's N_r receive antennas to UE_i are denoted as a $1 \times N_r$ vector $\mathbf{h}_{ul,i}$. By assuming channel reciprocity [35], the normalized fading coefficients of the actual uplink channel from UE_i to the H-AP's N_r receive antennas are represented by $\mathbf{h}_{ul,i}^{\dagger}$. Moreover, the normalized fading coefficients of the SI channel from N_t downlink transmit antennas to N_r receive antennas at the H-AP are denoted as an $N_r \times N_t$ matrix $\mathbf{H}_{si}$. Since the line-of-sight (LoS) path is considered, all the normalized fading coefficients $\mathbf{H}_x = \{\mathbf{h}_{dl,i}, \mathbf{h}_{ul,i}, \mathbf{H}_{si}\}$ follow Rician distribution, namely [36]

$$\mathbf{H}_x = \sqrt{\frac{K_r}{K_r + 1}}\mathbf{H}_0 + \sqrt{\frac{1}{K_r + 1}}\mathbf{H}_{\omega}, \tag{5.9}$$

where K_r is the Rician factor, and $\mathbf{H}_{\omega}$ is the channel matrix with entries following the Rayleigh distribution, while $\mathbf{H}_0$ is a full-rank deterministic matrix satisfying $||\mathbf{H}_0||^2 = N_t$, N_r and $N_t N_r$ for $\mathbf{h}_{dl,i}$, $\mathbf{h}_{ul,i}$ and $\mathbf{H}_{si}$, respectively.

By assuming uncorrelated block fading [37], the channel coefficients $\{(\mathbf{h}_{dl,i}, \mathbf{h}_{ul,i}^{\dagger}) | i = 1, \cdots, K\}$ as well as $\mathbf{H}_{si}$ keep unchanged during a transmission frame of Fig. 5.2 but vary from one transmission frame to another.

Furthermore, large-scale path-loss Ω_i between the H-AP and UE_i can be modeled as

$$\Omega_i = \begin{cases} \Omega_0, & d_i < d_0, \\ \Omega_0 \cdot \left(\frac{d_i}{d_0}\right)^{\alpha}, & d_i \geq d_0, \end{cases} \tag{5.10}$$

for $\forall i = 1, \cdots, K$, where α and d_i denote the path-loss exponent and the distance between the H-AP and UE_i. Ω_0 is the path-loss at the reference distance d_0, which can be calculated by the free-space path-loss equation $\Omega_0 = \left(\frac{4\pi f_c d_0}{c}\right)^2$ [38], where f_c and c represent the transmit frequency and the speed of light, respectively. Moreover, Ω_{si} represents the path-loss of the SI channel.

5.2.3 Problem Formulation

5.2.3.1 Downlink WPT

Without loss of generality, the H-AP transmits a Gaussian distributed complex signal x_{wpt} for the downlink WPT, which has a zero mean and a variance of $\mathbb{E}\left[x_{wpt}x_{wpt}^{\dagger}\right] = 1$. After being attenuated by wireless channel, the received WPT signal $y_{wpt,i}$ at UE_i is expressed as

$$y_{wpt,i} = \sqrt{P_{wpt}\Omega_i^{-1}}\mathbf{h}_{dl,i}\mathbf{w}_{wpt}x_{wpt} + n_0, \tag{5.11}$$

where P_{wpt} is the transmit power of the H-AP, $\mathbf{w}_{wpt}$ is an $N_t \times 1$ normalized WPT beamformer satisfying $\mathrm{Tr}\left(\mathbf{w}_{wpt}\mathbf{w}_{wpt}^{\dagger}\right) \leq 1$, and n_0 is a Gaussian distributed complex noise having a zero mean and a variance of σ_0^2.

The power of the direct current (DC) converted from the RF signal $y_{wpt,i}$ by a nonlinear energy harvester of UE_i is formulated as [39]

$$P_{dc,i} = \frac{\psi - M\Theta}{1 - \Theta}, \tag{5.12}$$

where we have $\Theta = \frac{1}{1+e^{ab}}$ and $\psi = \frac{M}{1+e^{-a\left(\mathbb{E}[y_{wpt,i}y_{wpt,i}^{\dagger}]-b\right)}}$. Note that the parameters M, a, and b are determined by the circuitry of the energy harvester [39]. Moreover, we have $\mathbb{E}\left[y_{wpt,i}y_{wpt,i}^{\dagger}\right] \approx P_{wpt}\Omega_i^{-1} \cdot \mathrm{Tr}\left(\mathbf{h}_{dl,i}\mathbf{w}_{wpt}\mathbf{w}_{wpt}^{\dagger}\mathbf{h}_{dl,i}^{\dagger}\right)$ by ignoring the negligible noise power σ_0^2. According to the MAC layer mechanism of Sect. 5.2.2.2, if the uplink WIT of UE_i is scheduled on the j-th time slot, which corresponds to $g_{i,j} = 1$, the total energy harvested by UE_i during a transmission frame before its uplink WIT is derived as

$$E_{i,j} = \sum_{j'=0}^{j-1} t_{j'}P_{dc,i}, \ \forall i, j = 1, \cdots, K. \tag{5.13}$$

Note that the beamformer $\mathbf{w}_{wpt}$ is unchanged during a single transmission frame, but it has to be redesigned when the channel coefficients change from one transmission frame to another.

5.2.3.2 Uplink WIT

Batteryless UEs deplete all the energy harvested during a transmission frame for its uplink WIT on the j-th time slot. The corresponding transmit power is obtained as $P_i = E_{i,j}/t_j$, where t_j is the duration of the j-th time slot. The RF signal $\mathbf{y}_i$ received by the N_r receive antennas of the H-AP can be expressed as

$$\mathbf{y}_i = \underbrace{\sqrt{P_i\Omega_i^{-1}}\mathbf{h}_{ul,i}^{\dagger}x_i}_{\text{Desired WIT signal}} + \underbrace{\sqrt{P_{wpt}\Omega_{si}^{-1}}\mathbf{H}_{si}\mathbf{w}_{wpt}x_{wpt}}_{\text{SI}} + \mathbf{n}_0, \tag{5.14}$$

where x_i is a Gaussian distributed complex signal having a zero mean and $\mathbb{E}\left[x_ix_i^{\dagger}\right] = 1$. Moreover, in Eq. (5.14), the $N_r \times 1$ additive noise vector $\mathbf{n}_0$ has all its elements independently obeying a complex Gaussian distribution having a zero mean and a variance of σ_0^2. If the H-AP adopts an $N_r \times 1$ normalized combiner $\mathbf{w}_{wit,i}$ satisfying $\text{Tr}\left(\mathbf{w}_{wit,i}^\dagger \mathbf{w}_{wit,i}\right) \leq 1$ for combining the signal $\mathbf{y}_i$ received by N_r antennas, the equivalent signal $\hat{y}_i$ is formulated as

$$\hat{y}_i = \mathbf{w}_{wit,i}^\dagger \mathbf{y}_i = \underbrace{\sqrt{P_i \Omega_i^{-1}} \mathbf{w}_{wit,i}^\dagger \mathbf{h}_{ul,i}^\dagger x_i}_{\text{Combined WIT signal}} + \underbrace{\mathbf{w}_{wit,i}^\dagger \mathbf{n}_0}_{\text{Combined noise}} + \underbrace{\sqrt{P_{wpt} \Omega_{si}^{-1}} \mathbf{w}_{wit,i}^\dagger \mathbf{H}_{si} \mathbf{w}_{wpt} x_{wpt}}_{\text{Combined SI}}. \tag{5.15}$$

According to Eq. (5.15), the uplink WIT throughput of UE_i can be formulated as Eq. (5.16) on the next page, where $E_{i,j}$ is given by Eq. (5.13). The power P_{si} of the SI is also defined in Eq. (5.16). Since the fair-throughput R_{fair} is defined as the minimum of all the UEs' uplink WIT throughput, it is formulated as $R_{\text{fair}} = \min\{R_i\}$, for $\forall i = 1, \cdots, K$.

$$R_i = \sum_{j=1}^{K} g_{i,j} \cdot \underbrace{t_j \log\left(1 + \frac{E_{i,j}}{t_j \Omega_i} \cdot \frac{\mathbf{w}_{wit,i}^\dagger \mathbf{h}_{ul,i}^\dagger \mathbf{h}_{ul,i} \mathbf{w}_{wit,i}}{\underbrace{P_{wpt} \Omega_{si}^{-1} \mathbf{w}_{wit,i}^\dagger \mathbf{H}_{si} \mathbf{w}_{wpt} \mathbf{w}_{wpt}^\dagger \mathbf{H}_{si}^\dagger \mathbf{w}_{wit,i}}_{\text{SI power } P_{si}} + N_r \sigma_0^2}\right)}_{\text{Uplink WIT throughput } R_{i,j}\text{, when } UE_i \text{ is scheduled on the } j\text{-th time slot}} \times [\text{bit/Hz}], \ \forall i = 1, \cdots, K. \tag{5.16}$$

5.2.3.3 Uplink Fair-Throughput Maximization

Our multi-domain resource scheduling design can be formulated as the following fairness optimization problem:

$$\text{(P5.3.1):} \quad \max_{\mathbf{w}_{wpt}, \mathbf{w}_{wit,i}, \mathbf{G}, \mathbf{t}} \quad R_{\text{fair}}, \tag{5.17}$$

$$\text{s. t. } R_i \geq R_{\text{fair}}, \forall i = 1, \cdots, K, \tag{5.17a}$$

$$\text{Tr}\left(\mathbf{w}_{wpt} \mathbf{w}_{wpt}^\dagger\right) \leq 1, \ \text{Tr}\left(\mathbf{w}_{wit,i} \mathbf{w}_{wit,i}^\dagger\right) \leq 1, \forall i = 1, \cdots, K, \tag{5.17b}$$

$$\text{(5.5), (5.6), (5.7) and (5.8).}$$

(P5.3.1) aims for maximizing the fair-throughput R_{fair} by optimizing the downlink WPT beamformer $\mathbf{w}_{wpt}$ and the uplink WIT combiner $\mathbf{w}_{wit,i}$ ($\forall i = 1, \cdots, K$) of the H-AP in the spatial domain, the UEs' binary scheduling indicator matrix $\mathbf{G}$ in the user domain, and the durations $\mathbf{t}$ of the $(K+1)$ time slots in the time-domain. Constraint (5.17a) ensures that the fair-throughput R_{fair} is equivalent to its definition $R_{\text{fair}} = \min\{R_i\}$, for $\forall i = 1, \cdots, K$. Constraint (5.17b) ensures normalized $\mathbf{w}_{wpt}$ and $\mathbf{w}_{wit,i}$.

5.2.4 *Joint Design Iterative Algorithm*

Due to the coupled optimization variables $\{\mathbf{w}_{wpt}, \mathbf{w}_{wit,i}, \mathbf{G}, \mathbf{t}\}$ as well as the discrete matrix $\mathbf{G}$, (P5.3.1) is non-convex. In this section, we decouple (P5.3.1) into four subproblems, which aim for optimizing the spatial domain design $\{\mathbf{w}_{wpt}, \mathbf{w}_{wit,i}\}$, the user-domain design $\mathbf{G}$, and the time-domain design $\mathbf{t}$, respectively, by considering that the optimization variables in other domains are fixed. Finally, the joint design is obtained by iteratively solving these subproblems until it converges.

5.2.4.1 Spatial Domain Design

When fixed $\{\mathbf{G}, \mathbf{t}\}$ is given, by letting $\mathbf{W}_{wpt} = \mathbf{w}_{wpt}\mathbf{w}_{wpt}^{\dagger}$ and $\mathbf{W}_{wit,i} = \mathbf{w}_{wit,i}\mathbf{w}_{wit,i}^{\dagger}$ as well as by relaxing the rank-one constraint on the positive semi-definite covariance matrices $\mathbf{W}_{wpt}$ and $\mathbf{W}_{wit,i}$, (P5.3.1) can be reformulated as the following covariance matrices design in the spatial domain:

$$\text{(P5.3.2):} \quad \max_{\mathbf{W}_{wpt}, \mathbb{W}_{wit}} \quad R_{\text{fair}}, \tag{5.18}$$

$$\text{s. t. } \operatorname{Tr}(\mathbf{W}_{wpt}), \operatorname{Tr}(\mathbf{W}_{wit,i}) \le 1 \text{ and (5.17a)},$$

where $\mathbb{W}_{wit} = \{\mathbf{W}_{wit,i}, \forall i = 1, \cdots, K\}$. However, due to the coupled $\mathbf{W}_{wpt}$ and $\mathbf{W}_{wit,i}$, (P5.3.2) is still a non-convex problem.

Observe from Eq. (5.15) that the combined WIT signal has a far lower power than the SI for the following reasons:

- The combined WIT signal experiences double path-loss Ω_i^2 during the round-trip propagation between the H-AP and UE_i, according to Eqs. (5.11)–(5.13) and (5.15).
- The SI signal's propagation distance d_{si} from the N_t transmit antennas to the N_r receive antennas of the H-AP is far shorter than d_i from the H-AP to UE_i, which results in $\Omega_{si} \ll \Omega_i$.

Therefore, cancelling the SI is of paramount importance for maximizing the fair-throughput R_{fair}, and we then impose a zero-forcing constraint $P_{si} = 0$ on the covariance matrices design, which can make the following subproblems (P5.3.3) and (P5.3.4) solvable. Although this zero-forcing method normally results in sub-optimal solutions, its performance is demonstrated to be nearly optimal in the high SNR regime [18]. Therefore, given a fixed receive combining covariance matrix set $\mathbb{W}_{wit}$, the transmit beamforming covariance matrix design can be formulated as

$$\text{(P5.3.3):} \quad \max_{\mathbf{W}_{wpt}} R_{\text{fair}}, \tag{5.19}$$

$$\text{s. t. } P_{si} = 0, \operatorname{Tr}(\mathbf{W}_{wpt}) \le 1 \text{ and (5.17a)}.$$

However, (P5.3.3) is a non-convex problem owing to Eq. (5.17a), which can be equivalently transformed into

$$\operatorname{Tr}\left(\mathbf{h}_{dl,i}\mathbf{W}_{wpt}\mathbf{h}_{dl,i}^{\dagger}\right) \ge \frac{\Omega_i}{P_{wpt}}\left(b - \frac{1}{a}\ln\left(\frac{M\Lambda_i}{2^{\frac{R_{\text{fair}}}{t_j}} + M\Theta\Lambda_i - 1} - 1\right)\right), \tag{5.20}$$

for $R_{\text{fair}} \le t_j \log(1 + M\Lambda_i(1-\Theta))$, $\forall i = 1, \cdots, K$, where we have $\Lambda_i = \frac{\sum_{j'=0}^{j-1} t_{j'}\mathbf{h}_{ul,i}\mathbf{W}_{wit,i}\mathbf{h}_{ul,i}^{\dagger}}{t_j\Omega_i(1-\Theta)(P_{si}+N_r\sigma_0^2)}$. Note that the right-hand side of Eq. (5.20) is a monotonously increasing function with respect to R_{fair}, and the left-hand side of Eq. (5.20) is a linear function with respect to each entry of $\mathbf{W}_{wpt}$. Therefore, given a fixed R_{fair}, we can obtain the feasible region for the downlink transmit beamforming covariance matrix $\mathbf{W}_{wpt}$ by invoking CVX toolbox [40]. A bisection method is adopted here for searching the maximum value R_{fair}^* of the fair-throughput, which is detailed in the pseudocode of Algorithm 5.1. The main steps of this algorithm are summarized as below:

- *Step 1*: Initialize the maximum $R_{\max}$ and minimum $R_{\min}$ of the bisection method, as shown in Line 1 of Algorithm 5.1.
- *Step 2*: Update the current fair-throughput R_{fair}. If there is a feasible region for $\mathbf{W}_{wpt}$ in (P5.3.3), the optimal fair-throughput R_{fair}^* satisfies $R_{\text{fair}}^* \ge R_{\text{fair}}$. Therefore, we update $R_{\min} \leftarrow R_{\text{fair}}$. Otherwise, R_{fair}^* satisfies $R_{\text{fair}}^* < R_{\text{fair}}$. We then update $R_{\max} \leftarrow R_{\text{fair}}$. This step is characterized in Lines 3–9 of Algorithm 5.1, where F is a binary indicator to record if there is a feasible region for $\mathbf{W}_{wpt}$.
- *Step 3*: If the difference between $R_{\max}$ and $R_{\min}$ is higher than the threshold ς, repeat Step 2, as shown in Line 3 of Algorithm 5.1. Otherwise, let $R_{\text{fair}}^* \leftarrow R_{\min}$, since R_{fair} output by the bisection method may result in the situation that there is not a feasible region for $\mathbf{W}_{wpt}$. We then find the solution $\mathbf{W}_{wpt}^*$ to (P5.3.3) by substituting R_{fair} into R_{fair}^*, as shown in Lines 11–13 of Algorithm 5.1.

Algorithm 5.1 Transmit beamforming design for solving (P5.3.3)

Require: Fixed resource scheduling scheme $\{\mathbf{G}, \mathbb{W}_{wit}, \mathbf{t}\}$ in user-, spatial- and time-domain. Difference threshold ς, Other parameters.
Ensure: Transmit beamforming $\mathbf{W}^*_{wpt}$; Fair-throughput R^*_{fair}

```
Initialise R_min ← 0 and R_max ← min_i (t_j log(1 + M Λ_i (1 − Θ)));
while R_max − R_min > ς do
    Update R_fair ← (R_max + R_min)/2;
    Find whether there is a feasible region for W_wpt in (P5.3.3) by using CVX toolbox [40]. If there is a feasible solution, let F ← 1; otherwise, F ← 0;
    if F = 1 then
        Update R_min ← R_fair;
    else
        Update R_max ← R_fair;
    end if
end while
Let R*_fair ← R_min;
Find the solution W*_wpt to (P5.3.3);
return The optimal transmit beamforming W*_wpt and the corresponding fair-throughput R*_fair.
```

Complexity analysis: The complexity of solving the feasibility problem for $\mathbf{W}_{wpt}$ and that of the bisection method are $\mathcal{O}\left(\left(\frac{N_t^2+N_t}{2}\right)^{3.5}\right)$ [41] and $\mathcal{O}\left(\log\left(\frac{R_{\max}-R_{\min}}{\varsigma}\right)\right)$, respectively. Therefore, the complexity of Algorithm 5.1 is $\mathcal{O}\left(\left(\frac{N_t^2+N_t}{2}\right)^{3.5}\log\left(\frac{R_{\max}-R_{\min}}{\varsigma}\right)\right)$.

Performance analysis: The optimal solution to (P5.3.3) can be obtained when the difference threshold ς is sufficiently small, since any R_{fair} satisfying $R_{\text{fair}} > R^*_{\text{fair}}$ may result in a situation of no feasible solution for $\mathbf{W}_{wpt}$.

Given a fixed $\mathbf{W}_{wpt}$, the receive combining covariance matrices design can be formulated as

$$(\text{P5.3.4}): \max_{\mathbb{W}_{wit}} R_{\text{fair}}, \tag{5.21}$$
$$\text{s. t. } P_{si} = 0, \text{Tr}(\mathbf{W}_{wit,i}) \leq 1 \text{ and (5.17a)}.$$

The objective of (P5.3.4) is an affine function, while its corresponding constraints are either affine or concave. Therefore, (P5.3.4) is a semi-definite programming problem, which can be solved by CVX toolbox [40].

5.2.4.2 User-Domain Design

Given fixed $\{\mathbf{W}_{wpt}, \mathbb{W}_{wit}, \mathbf{t}\}$ in both the spatial and time-domain, (P5.3.1) can be reformulated as

$$(\text{P5.3.5}): \max_{\mathbf{G}} R_{\text{fair}}, \tag{5.22}$$
$$\text{s. t. (5.17a), (5.6), (5.7) and (5.8).}$$

In order to solve the non-convex (P5.3.5), we first define an uplink WIT throughput matrix $\mathbf{R} = \left[R_{i,j}\right]_{K\times K}$. The entry $R_{i,j}$ is the uplink WIT throughput of UE_i scheduled on the j-th time slot, which is defined in Eq. (5.16). The objective of (P5.3.5) is to select K different entries from the matrix $\mathbf{R}$ for maximizing the minimum among them. Any pair of these selected entries is neither allowed in the same column nor in the same row. Normally, (P5.3.5) can be solved by exploiting exhaustive search having a complexity of $\mathcal{O}(K!)$, which becomes impractical when we have a large K.

By exploiting the constraints (5.6)–(5.8) and the objective of fair-throughput maximization, a Hungarian method[5] aided algorithm is developed for obtaining the optimal solution to (P5.3.5), which is detailed in the pseudocode of Algorithm 5.2. The main steps of this algorithm are summarized as below:

[5] Hungarian method [42] is a classic one-to-one task allocation algorithm in graph theory, which can search the optimal match in undirected and unweighted bipartite graph $G(V, E)$ with vertex set V and edge set E. We have $V = \{V_1, V_2\}$, where V_1 and V_2 are disjoint subsets. When the number of matches M satisfies $M = |V_1| = |V_2|$, it is a perfect matching.

Algorithm 5.2 UEs' scheduling design for solving (P5.3.5)

Require: Fixed resource scheduling scheme $\{\mathbf{W}_{wpt}, \mathbb{W}_{wit}, \mathbf{t}\}$ in spatial- and time-domain. Difference threshold η, Other parameters.
Ensure: Binary scheduling indicator matrix $\mathbf{G}^*$; Fair-throughput R_{fair}^*

1: Initialise matrices $\mathbf{R} \leftarrow [R_{i,j}]_{K\times K}$ by Eq. (5.16), $\hat{\mathbf{R}} \leftarrow [\hat{R}_i]_{1\times K}$, $\mathbf{U} \leftarrow [U_{i,j}]_{K\times K}$ and $\hat{\mathbf{G}} \leftarrow [\hat{g}_{i,j}]_{K\times K}$ with all zeros;
2: Initialise $R'_{\text{min}} \leftarrow 0$ and $R'_{\text{max}} \leftarrow \max_{(i,j)} R_{i,j}$;
3: **while** $R'_{\text{max}} - R'_{\text{min}} > \eta$ **do**
4: $\quad R_{\text{fair}} \leftarrow \frac{R'_{\text{max}}+R'_{\text{min}}}{2}$;
5: $\quad$ Find all (i', j') associated with $R_{i,j} < R_{\text{fair}}$ for $i, j = 1, \cdots, K$, let $U_{i',j'} \leftarrow 0$;
6: $\quad$ Find all (i', j') associated with $R_{i,j} \geq R_{\text{fair}}$ for $i, j = 1, \cdots, K$, let $U_{i',j'} \leftarrow 1$;
7: $\quad$ By exploiting Hungarian method in matrix $\mathbf{U}$, obtain the searched scheduling indicator matrix $\hat{\mathbf{G}}$;
8: $\quad$ **if** $\sum_{i=1}^{K}\sum_{j=1}^{K} \hat{g}_{i,j} = K$ **then**
9: $\quad\quad$ Update $R'_{\text{min}} \leftarrow R_{\text{fair}}$;
10: $\quad$ **else**
11: $\quad\quad$ Update $R'_{\text{max}} \leftarrow R_{\text{fair}}$;
12: $\quad$ **end if**
13: **end while**
14: Find all (i', j') associated with $R_{i,j} < R'_{\text{min}}$ for $i, j = 1, \cdots, K$, let $U_{i',j'} \leftarrow 0$;
15: Find all (i', j') associated with $R_{i,j} \geq R'_{\text{min}}$ for $i, j = 1, \cdots, K$, let $U_{i',j'} \leftarrow 1$;
16: By exploiting Hungarian method in matrix $\mathbf{U}$, obtain the searched scheduling indicator matrix $\hat{\mathbf{G}}$;
17: **for** $i = 1 : K$ **do**
18: $\quad$ Find $\hat{j} \leftarrow \arg\max_j [\mathbf{R}.\times\hat{\mathbf{G}}]_{i,j}$, let $\hat{R}_i \leftarrow [\mathbf{R}.\times\hat{\mathbf{G}}]_{i,\hat{j}}$;
19: **end for**
20: **return** $R_{\text{fair}}^* \leftarrow \min_i \hat{\mathbf{R}}$ and $\mathbf{G}^* \leftarrow \hat{\mathbf{G}}$.

- *Step 1*: Initialize the throughput matrix $\mathbf{R}$, the optimal actual uplink throughput vector $\hat{\mathbf{R}}$, the bipartite-graph indicator matrix $\mathbf{U}$, and the searched scheduling indicator matrix $\hat{\mathbf{G}}$. Initialize the searching maximum R'_{max} and minimum R'_{min}, as shown in Lines 1 and 2 of Algorithm 5.2.
- *Step 2*: Update the current fair-throughput R_{fair}. For all (i', j') satisfying $R_{i',j'} < R_{\text{fair}}$, set $U_{i',j'}$ to 0; otherwise, $U_{i',j'} = 1$, where $\mathbf{U}$ represents a bipartite graph. Specifically, $U_{i,j} = 1$ indicates that there is a link between point i and point j in the graph. Therefore, the scheduling indicator matrix $\hat{\mathbf{G}}$ can be obtained by exploiting Hungarian method [42] with an input $\mathbf{U}$, where $\hat{g}_{i,j} = 1$ represents that the link between point i and point j in $\mathbf{U}$ is chosen. If $\mathbf{U}$ has a perfect matching, i.e., $\sum_{i=1}^{K}\sum_{j=1}^{K} \hat{g}_{i,j} = K$, it indicates that there is a feasible solution of UE scheduling, so that the actual throughput of all the UEs is higher than or equal to the current fair-throughput R_{fair}, i.e., $R_{\text{fair}}^* \geq R_{\text{fair}}$. It means that the current fair-throughput R_{fair} can be further increased. Therefore, update $R'_{\text{min}} \leftarrow R_{\text{fair}}$. Otherwise, we need to reduce R_{fair}. Therefore, update $R'_{\text{max}} \leftarrow R_{\text{fair}}$. This step is characterized in Lines 4–12 of Algorithm 5.2.
- *Step 3*: If the difference between R'_{max} and R'_{min} is higher than the threshold η, repeat Step 2, as shown in Line 3 of Algorithm 5.2. Otherwise, the similar operation to Lines 5–7 is carried out by replacing the final R_{fair} with the final R'_{min} to obtain the optimal scheduling indicator matrix $\mathbf{G}^*$, since this R_{fair} may not satisfy $R_{\text{fair}} \leq R_{\text{fair}}^*$. After obtaining the optimal actual uplink throughput vector $\hat{\mathbf{R}}$, the maximum fair-throughput R_{fair}^* is the minimum $\hat{R}_i$ and the optimal $\mathbf{G}^*$ is $\hat{\mathbf{G}}$. This step is characterized in Lines 14–20 of Algorithm 5.2.

Complexity Analysis The complexity of Algorithm 5.2 is $\mathcal{O}\left(K^3\log(\frac{R'_{\text{max}}-R'_{\text{min}}}{\eta}) + K^3 + K\right)$, since that of the Hungarian method is $\mathcal{O}(K^3)$ and that of the bisection method is $\mathcal{O}\left(\log(\frac{R'_{\text{max}}-R'_{\text{min}}}{\eta})\right)$.

Performance Analysis Since Hungarian method is capable of searching the optimal match of the bipartite graph represented by the matrix $\mathbf{U}$ [42]. When the difference threshold $\eta \to 0$, the searched R_{fair}^* associated with the final R'_{min} is optimal, since for $\forall R_{\text{fair}} > R'_{\text{min}}$, the corresponding $\mathbf{U}$ does not have a perfect match, i.e., $\sum_{i=1}^{K}\sum_{j=1}^{K} \hat{g}_{i,j} < K$. Therefore, Algorithm 5.2 can obtain the optimal solution to (P5.3.5).

5.2.4.3 Time-Domain Design

Given fixed $\{\mathbf{W}_{wpt}, \mathbb{W}_{wit}, \mathbf{G}\}$, the time-domain design can be formulated as

$$\text{(P5.3.6)}: \max_{\mathbf{t}} R_{\text{fair}}, \tag{5.23}$$
$$\text{s. t. (5.17a) and (5.5).}$$

The scheduling indicator matrix $\mathbf{G}$ can be interpreted as a sequence of $\{UE_{(j)}|\forall j = 1, \cdots, K\}$, where $UE_{(j)}$ represents the UE scheduled on the j-th time slot for uplink WIT.

According to Eqs. (5.13) and (5.16), we obtain a variable $\gamma_{(j)}$ of $UE_{(j)}$ as

$$\gamma_{(j)} = \frac{P_{dc,(j)}}{\Omega_{(j)}} \cdot \frac{\mathbf{w}_{wit,(j)}^{\dagger}\mathbf{h}_{ul,(j)}^{\dagger}\mathbf{h}_{ul,(j)}\mathbf{w}_{wit,(j)}}{P_{wpt}\Omega_{si}^{-1}\mathbf{w}_{wit,(j)}^{\dagger}\mathbf{H}_{si}\mathbf{w}_{wpt}\mathbf{w}_{wpt}^{\dagger}\mathbf{H}_{si}^{\dagger}\mathbf{w}_{wit,(j)} + N_r\sigma_0^2}, \tag{5.24}$$

for $\forall j = 1, \cdots, K$, which only depends on wireless channels and other physical layer parameters of $UE_{(j)}$. Moreover, the SINR $z_{(j)}$ of $UE_{(j)}$'s uplink transmission can be formulated as

$$z_{(j)} = \frac{\gamma_{(j)}\sum_{j'=0}^{j-1} t_{j'}}{t_j}, \forall j = 1, \cdots, K, \tag{5.25}$$

which also depends on a specific time-domain design $\mathbf{t}$. We then aim for finding the optimal $\{z_{(j)}^*|j = 1, \cdots, K\}$ to maximize R_{fair}. Furthermore, we define the operational period of all UEs in the first $(j+1)$ time slots as

$$\tau_j = \sum_{j'=0}^{j} t_{j'}. \tag{5.26}$$

Since (P5.3.6) is a convex optimization problem, its Lagrangian function is obtained as

$$\mathcal{L}(\mathbf{t}, R_{\text{fair}}, \boldsymbol{\lambda}) = R_{\text{fair}} + \lambda_0(T_t - \sum_{j=0}^{K} t_j) + \sum_{i=1}^{K} \lambda_i(R_i - R_{\text{fair}}),$$

where $\boldsymbol{\lambda}$ is a vector of all the Lagrangian multipliers. The Karush–Kuhn–Tucker (KKT) conditions are then derived as

$$\begin{cases} -\lambda_0 + \sum_{i=1}^{K} \lambda_i \frac{\gamma_{(i)}}{(1+z_{(i)})\ln 2} = 0, \\ -\lambda_0 + \lambda_j \left(\log(1+z_{(j)}) - \frac{z_{(j)}}{(1+z_{(j)})\ln 2}\right) + \sum_{i=j+1}^{K} \lambda_i \frac{\gamma_{(i)}}{(1+z_{(i)})\ln 2} = 0, \\ -\lambda_0 + \lambda_K \left(\log(1+z_{(K)}) - \frac{z_{(K)}}{(1+z_{(K)})\ln 2}\right) = 0, \\ 1 - \sum_{i=1}^{K} \lambda_i = 0, \\ \lambda_0(T_t - \sum_{i=0}^{K} t_i) = 0, \\ \lambda_i(R_i - R_{\text{fair}}) = 0, \forall i = 1, \cdots, K, \end{cases} \tag{5.27}$$

for $\forall j = 1, \cdots, K-1$. The optimal solution $\mathbf{t}^*$ cannot be obtained directly from the above KKT conditions. Therefore, it is solved by the following lemmas and theorem.

Lemma 5.1 *For $UE_{(1)}$, the optimal SINR $z_{(1)}^*$ and the optimal equivalent fair rate r_1^**[6] *during its operational period τ_1 are derived as*

$$(1+z_{(1)}^*)\ln(1+z_{(1)}^*) - z_{(1)}^* = \gamma_{(1)}, \tag{5.28}$$

$$r_1^* = \frac{\gamma_{(1)}}{z_{(1)}^* + \gamma_{(1)}}\log(1+z_{(1)}^*), \tag{5.29}$$

[6] According to $R_{\text{fair}} = \tau_j \cdot r_j^*$ for $j = 1, 2, \cdots, K$ in Lemma 5.1, Lemma 5.2 and Theorem 5.1, r_j^* can be regarded as the equivalent fair rate within τ_j, which cannot represent any practical WIT rate.

respectively, where Eq. (5.28) *can be solved by a bisection method. Given a fixed* τ_1*, the maximum throughput of* $UE_{(1)}$ *is then obtained as* $R_1 = \tau_1 \cdot r_1^*$.

Proof Observe from the KKT conditions of Eq. (5.27) that if $\lambda_0 = 0$, we then have $\lambda_1 = \lambda_2 = \cdots = \lambda_K = 0$ owing to the fact $f(x) = \log(1+x) - \frac{x}{(1+x)\ln 2} > 0$ for $\forall x > 0$. However, the equality in the fourth line of Eq. (5.27) is violated. Therefore, we have $\lambda_0 \neq 0$. We then have $\lambda_K \neq 0$, according to the third line. Furthermore, we find $\sum_{i=j+1}^{K} \lambda_i \frac{\gamma_{(i)}}{(1+z_{(i)})\ln 2} - \lambda_0 \neq 0$, according to the first line of Eq. (5.27). Hence, we also have $\lambda_i \neq 0$ for $\forall i = 1, \cdots, K-1$, based on the second line. As a result, we have $\lambda_i \neq 0$ for $\forall i = 1, \cdots, K$.

According to the first and second lines of Eq. (5.27), we have

$$\lambda_1 \frac{\gamma_{(1)}}{(1+z_{(1)})\ln 2} - \lambda_1 \left(\log(1+z_{(1)}) - \frac{z_{(1)}}{(1+z_{(1)})\ln 2} \right) = 0. \tag{5.30}$$

Since we have $\lambda_1 \neq 0$, Eq. (5.30) can be transformed into Eq. (5.28). Since the left-hand side of Eq. (5.28) is a monotonously increasing function with respect to $z_{(1)}^*$, a bisection method can be exploited for obtaining $z_{(1)}^*$. By jointly considering Eqs. (5.25) and (5.26), the optimal time slots t_0^* and t_1^* are expressed as $t_0^* = \frac{z_{(1)}^* \tau_1}{z_{(1)}^* + \gamma_{(1)}}$, $t_1^* = \frac{\gamma_{(1)} \tau_1}{z_{(1)}^* + \gamma_{(1)}}$. As a result, given a fixed τ_1, the maximum throughput of $UE_{(1)}$ is derived as $R_1 = \frac{\gamma_{(1)} \tau_1}{z_{(1)}^* + \gamma_{(1)}} \log(1+z_{(1)}^*) = \tau_1 \cdot r_1^*$. Obviously, $z_{(1)}^*$ is a constant and R_1 is only related to τ_1. Therefore, given a fixed τ_1, the optimal time slots t_0^* and t_1^* are unchanged. Lemma 5.1 is proved.

According to the conclusion in Lemma 5.1, the original subproblem (P5.3.6) can be reformulated as

$$\text{(P5.3.7):} \quad \max_{\tau_1, t_2, \cdots, t_K, R_{\text{fair}}} \quad R_{\text{fair}}, \tag{5.31}$$

$$\text{s. t. } r_1^* \tau_1 \geq R_{\text{fair}}, \tag{5.31a}$$

$$t_2 \log\left(1 + \frac{\gamma_{(2)} \tau_1}{t_2}\right) \geq R_{\text{fair}}, \tag{5.31b}$$

$$t_i \log\left(1 + \frac{\gamma_{(i)} (\tau_1 + \sum_{j=2}^{i-1} t_j)}{t_i}\right) \geq R_{\text{fair}}, \forall i = 3, \cdots, K, \tag{5.31c}$$

$$\tau_1 + \sum_{j=2}^{K} t_j \leq T_t, \tag{5.31d}$$

where $r_1^* \tau_1$ represents $UE_{(1)}$'s uplink WIT throughput R_1. Moreover, the left-hand side of Eq. (5.31b) and that of Eq. (5.31c) represent the uplink WIT throughput R_2 of $UE_{(2)}$ and R_i of $UE_{(i)}$ for $\forall i = 3, 4, \cdots, K$, respectively. Since (P5.3.7) is also convex, Lemma 5.2 can be obtained by invoking the traditional Lagrange multiplier method.

Lemma 5.2 *The optimal SINR* $z_{(2)}^*$ *of* $UE_{(2)}$ *and the optimal equivalent fair rate* r_2^* *of* $UE_{(1)}$ *and* $UE_{(2)}$ *during their operational period* τ_2 *are derived as*

$$\frac{\log(1+z_{(2)}^*)}{z_{(2)}^*} = \frac{r_1^*}{\gamma_{(2)}}, \tag{5.32}$$

$$r_2^* = \frac{r_1^* z_{(2)}^*}{z_{(2)}^* + \gamma_{(2)}}, \tag{5.33}$$

respectively, where r_1^* *is obtained in Lemma 5.1. Given a fixed* τ_2*, the maximum fair-throughput of* $UE_{(1)}$ *and* $UE_{(2)}$ *is then obtained as* $R_{fair} = \tau_2 \cdot r_2^*$.

Proof Since (P5.3.7) is also convex, its Lagrangian function is obtained as

$$\mathcal{L}(\tau_1, t_2, \cdots, t_K, R_{\text{fair}}, \boldsymbol{\lambda}) = R_{\text{fair}} + \lambda_0 \left(T_t - \tau_1 - \sum_{j=2}^{K} t_j \right)$$

$$+\lambda_1(r_1^*\tau_1 - R_{\text{fair}}) + \lambda_2(R_2 - R_{\text{fair}})$$
$$+\sum_{i=3}^{K}\lambda_i(R_i - R_{\text{fair}}),$$

where $\boldsymbol{\lambda}$ is a vector of all the Lagrangian multipliers. The KKT conditions are then derived as

$$\begin{cases} -\lambda_0 + \lambda_1 r_1^* + \sum_{i=2}^{K}\lambda_i \frac{\gamma_{(i)}}{(1+z_{(i)})\ln 2} = 0, \\ -\lambda_0 + \lambda_j\left(\log(1+z_{(j)}) - \frac{z_{(j)}}{(1+z_{(j)})\ln 2}\right) + \sum_{i=j+1}^{K}\lambda_i\frac{\gamma_{(i)}}{(1+z_{(i)})\ln 2} = 0, \\ -\lambda_0 + \lambda_K\left(\log(1+z_K) - \frac{z_K}{(1+z_K)\ln 2}\right) = 0, \\ 1 - \sum_{i=1}^{K}\lambda_i = 0, \\ \lambda_0(T_t - \tau_1 - \sum_{i=2}^{K}t_i) = 0, \\ \lambda_1(r_1^*\tau_1 - R_{\text{fair}}) = 0, \\ \lambda_i(t_i\log(1+z_{(i)}) - R_{\text{fair}}) = 0, \forall i = 2, \cdots, K, \end{cases} \tag{5.34}$$

for $\forall j = 2, \cdots, K-1$. Similar to Eq. (5.27), we can obtain that $\lambda_i \neq 0$ for $\forall i = 0, \cdots, K$. Therefore, according to the 6-th and 7-th lines of Eq. (5.34), we have $r_1^* z_{(2)} = \gamma_{(2)}\log(1+z_{(2)})$. It can be further transformed into Eq. (5.32), which can be solved by a bisection method. By jointly considering Eqs. (5.25) and (5.26), the optimal time slots τ_1^* and t_2^* are expressed as $\tau_1^* = \frac{z_{(2)}^*\tau_2}{z_{(2)}^*+\gamma_{(2)}}$, $t_2^* = \frac{\gamma_{(2)}\tau_2}{z_{(2)}^*+\gamma_{(2)}}$. Moreover, given a fixed τ_2, the maximum fair-throughput of $UE_{(1)}$ and $UE_{(2)}$ is obtained as $R_{\text{fair}} = \frac{r_1^* z_{(2)}^*}{z_{(2)}^*+\gamma_2}\tau_2 = r_2^*\tau_2$, where r_2^* is expressed as Eq. (5.33). Similar to Lemma 5.1, given a fixed τ_2, the optimal τ_1^* and t_2^* are unchanged. Lemma 5.2 is proved.

We then assume that the optimal $z_{(j-1)}^*$ and r_{j-1}^* have already been obtained. Therefore, (P5.3.6) can be reformulated as

$$\text{(P5.3.8):} \quad \max_{\tau_{j-1}, t_j, \cdots, t_K, R_{\text{fair}}} R_{\text{fair}}, \tag{5.35}$$

$$\text{s. t. } r_{j-1}^*\tau_{j-1} \geq R_{\text{fair}}, \tag{5.35a}$$

$$t_j\log\left(1 + \frac{\gamma_{(j)}\tau_{j-1}}{t_j}\right) \geq R_{\text{fair}}, \tag{5.35b}$$

$$t_i\log\left(1 + \frac{\gamma_{(i)}(\tau_{j-1} + \sum_{k=j}^{i-1}t_k)}{t_i}\right) \geq R_{\text{fair}}, \forall i = (j+1), \cdots, K, \tag{5.35c}$$

$$\tau_{j-1} + \sum_{k=j}^{K}t_k \leq T_t. \tag{5.35d}$$

According to Lemma 5.2, we can obtain the following Theorem 5.1.

Theorem 5.1 *For $\{UE_{(1)}, UE_{(2)}, \cdots, UE_{(j)}\}$, the optimal SINR $z_{(j)}^*$ of $UE_{(j)}$ and the optimal equivalent fair rate r_j^* of these UEs during their operational period τ_j are derived as*

$$\frac{\log(1+z_{(j)}^*)}{z_{(j)}^*} = \frac{r_{j-1}^*}{\gamma_{(j)}}, \tag{5.36}$$

$$r_j^* = \frac{r_{j-1}^* z_{(j)}^*}{z_{(j)}^* + \gamma_{(j)}}. \tag{5.37}$$

Given a fixed τ_j, the maximum fair-throughput of $\{UE_{(1)}, UE_{(2)}, \cdots, UE_{(j)}\}$ is then obtained as $R_{fair} = \tau_j \cdot r_j^*$.

Algorithm 5.3 Optimal time allocation algorithm for solving (P5.3.6)

Require: Fixed resource scheduling schemes $\{\mathbf{W}_{wpt}, \mathbb{W}_{wit}, \mathbf{G}\}$ in both spatial- and user-domain; Other parameters.
Ensure: Optimal fair-throughput R^*_{fair}; Optimal time allocation scheme $\mathbf{t}^*$.

```
Initialise γ_(j) | j = 1, ··· , K by Eq. (5.24);
Obtain z*_(1) and r*_1 by Lemma 5.1;
Obtain z*_(j) and r*_j for ∀j = 2, ··· , K by Theorem 5.1 sequentially;
Obtain t* by Eq. (5.39);
Obtain R*_fair ← T_t r*_K according to Theorem 5.1;
return t* and R*_fair.
```

Proof Observe from both (P5.3.7) and (P5.3.8) that they have exactly the same form. By substituting $j = 2$ into Eqs. (5.36) and (5.37), according to Lemma 5.2, this theorem holds. As a result, the optimal $z^*_{(j)}$ and $r^*_{(j)}$ have the same form with Eqs. (5.32) and (5.33), which returns us Eqs. (5.36) and (5.37) in Theorem 5.1. According to the mathematical induction, Theorem 5.1 is proved.

Remark 1 According to the last line of Eq. (5.27), we have $R_i = R_{\text{fair}}$, for $\forall i = 1, \cdots, K$. Therefore, each UE achieves the same uplink throughput with each other.

Remark 2 According to Eqs. (5.25) and (5.26), we have $\frac{\sum_{j'=0}^{j-1} t_{j'}}{t_j} = \frac{z^*_{(j)}}{\gamma_{(j)}}$. Therefore, the time allocation vector $\mathbf{t}$ satisfies

$$
\begin{aligned}
&t_0^* : t_1^* : \cdots : t_j^* : \cdots : t_K^* = \\
&1 : \frac{\gamma_{(1)}}{z^*_{(1)}} : \cdots : \frac{\gamma_{(j)}}{z^*_{(j)}} \prod_{j'=1}^{j-1} \frac{z^*_{(j')} + \gamma_{(j')}}{z^*_{(j')}} : \cdots : \frac{\gamma_{(K)}}{z^*_{(K)}} \prod_{j'=1}^{K-1} \frac{z^*_{(j')} + \gamma_{(j')}}{z^*_{(j')}}.
\end{aligned} \tag{5.38}
$$

Since $z_{(j)}$ is only related to $\gamma_{(j)}$, for $\forall j = 1, \cdots, K$, Eq. (5.38) is a function with respect to $\gamma_{(j)}$. As a result, given a fixed set of UEs SNR $\{\gamma_{(j)} | \forall j = 1, \cdots, K\}$, the ratio of time-slot duration is unchanged.

Given a specific UE scheduling scheme $\{UE_{(1)}, UE_{(2)}, \cdots, UE_{(K)}\}$, the optimal time allocation $\mathbf{t}^*$ for maximizing R_{fair} is obtained in the following steps:

- *Step 1*: Find the optimal $z^*_{(1)}$ and r^*_1 by exploiting Lemma 5.1.
- *Step 2*: Sequentially obtain $z^*_{(j)}$ and r^*_j for $j = 2, 3, \cdots, K$ by exploiting Theorem 5.1, while the optimal fair-throughput is obtained as $R^*_{\text{fair}} = T_t \cdot r^*_K$.
- *Step 3*: According to Eq. (5.38) and $\sum_{j=0}^{K} t_j = T_t$, the optimal $\mathbf{t}^*$ is calculated as

$$
\begin{cases}
t_0^* = \dfrac{T_t}{1 + \frac{\gamma_{(1)}}{z_1^*} + \sum_{j=2}^{K} \frac{\gamma_{(j)}}{z^*_{(j)}} \prod_{j'=1}^{j-1} \frac{z^*_{(j')} + \gamma_{(j')}}{z^*_{(j')}}}, \\
t_1^* = \dfrac{\gamma_{(1)}}{z^*_{(1)}} t_0^*, \\
t_j^* = \dfrac{\gamma_{(j)}}{z^*_{(j)}} \displaystyle\prod_{j'=1}^{j-1} \frac{z^*_{(j')} + \gamma_{(j')}}{z^*_{(j')}} t_0^*, \ \forall j = 2, \cdots, K.
\end{cases} \tag{5.39}
$$

The pseudocode for the optimal time allocation is provided in Algorithm 5.3.

Complexity Analysis The complexity of the bisection method for $z^*_{(j)}$ is $\mathcal{O}\left(\log\left(\frac{z_{\max}}{\epsilon}\right)\right)$, where $z_{\max}$ and ϵ represent the upper bound of $z_{(j)}$ and the convergence threshold of the bisection method, respectively. Since we have K UEs, the complexity of Algorithm 5.3 is $\mathcal{O}\left(K\left(\log\left(\frac{z_{\max}}{\epsilon}\right) + 1\right)\right)$, which is far lower than $\mathcal{O}(K+1)^{3.5}$ [41] of the interior point method based convex optimization tool.

Algorithm 5.4 An iterative algorithm for solving (P5.3.1)

Require: All the physical/MAC layer parameters, such as K, T_t, N_t, N_r, P_t, $\{\Omega_i | i = 1, \cdots, K\}$, Ω_{si}, $\{(\mathbf{h}_{dl,i}, \mathbf{h}_{ul,i}) | i = 1, \cdots, K\}$, $\mathbf{H}_{si}$ and σ_0^2. Difference threshold ξ.
Ensure: Solution $\{\mathbf{W}_{wpt}^*, \mathbb{W}_{wit}^*, \mathbf{G}^*, \mathbf{t}^*\}$ in multi-domain. Fair-throughput R_{fair}^*.
1: Initialise $\mathbf{W}_{wit,i,(0)} \leftarrow \frac{1}{\sqrt{N_r}} \mathbf{I}_{N_r \times N_r}$, $\mathbf{G}_{(0)} \leftarrow \mathbf{I}_{K \times K}$ and $\mathbf{t}_{(0)} \leftarrow [t_j \leftarrow \frac{T_t}{K+1}]_{1 \times K}$;
2: Initialise $R_{\text{fair},pre} \leftarrow -2\xi$ and $R_{\text{fair},cur} \leftarrow 0$ and iteration number $k \leftarrow 0$;
3: **while** $|R_{\text{fair},cur} - R_{\text{fair},pre}| > \xi$ **do**
4: $R_{\text{fair},pre} \leftarrow R_{\text{fair},cur}$;
5: Update $\mathbf{W}_{wpt,(k+1)} \leftarrow \mathbf{W}_{wpt}^*$ by carrying out Algorithm 5.1 with input of $\{\mathbb{W}_{wit,(k)}, \mathbf{G}_{(k)}, \mathbf{t}_{(k)}\}$ into it;
6: Update $\mathbb{W}_{wit,(k+1)} \leftarrow \mathbb{W}_{wit}^*$ by solving (P5.3.4) after substituting $\{\mathbf{W}_{wpt,(k+1)}, \mathbf{G}_{(k)}, \mathbf{t}_{(k)}\}$ into it;
7: Update $\mathbf{G}_{(k+1)} \leftarrow \mathbf{G}^*$ by carrying out Algorithm 5.2 with input of $\{\mathbf{W}_{wpt,(k+1)}, \mathbb{W}_{wit,(k+1)}, \mathbf{t}_{(k)}\}$;
8: Update $\mathbf{t}_{k+1} \leftarrow \mathbf{t}^*$ and $R_{\text{fair},cur} \leftarrow R_{\text{fair}}^*$ by carrying out Algorithm 5.3 with input of $\{\mathbf{W}_{wpt,(k+1)}, \mathbb{W}_{wit,(k+1)}, \mathbf{G}_{(k+1)}\}$;
9: Update $k \leftarrow k + 1$;
10: **end while**
11: **return** $\{\mathbf{W}_{wpt}^*, \mathbb{W}_{wit}^*, \mathbf{G}^*, \mathbf{t}^*\} \leftarrow \{\mathbf{W}_{wpt,(k)}, \mathbb{W}_{wit,(k)}, \mathbf{G}_{(k)}, \mathbf{t}_{(k)}\}$ and $R_{\text{fair}}^* \leftarrow R_{\text{fair},cur}$.

5.2.4.4 Joint Design in Multi-Domain

Although (P5.3.1) is non-convex, all its subproblems can be optimally solved. The subproblem (P5.3.3) of the transmit beamforming design in the spatial domain can be optimally solved by carrying out Algorithm 5.1. The subproblem (P5.3.4) of receive combining design in the spatial domain can obtain its optimal solution $\mathbb{W}_{wit}^*$ since (P5.3.4) is a convex problem. Then, a Hungarian method aided Algorithm 5.2 can guarantee the optimal solution $\mathbf{G}^*$ to (P5.3.5) in the user domain. Finally, Algorithm 5.3 for solving the subproblem (P5.3.6) is proved to obtain the optimal time allocation $\mathbf{t}^*$. Therefore, we develop an iterative algorithm to solve the mixed-integer nonlinear programming problem (P5.3.1).

The pseudocode of the joint design is detailed in Algorithm 5.4, which has the following main steps:

- *Step 1*: Initialize $\mathbf{t}_{(0)}$ by equally allocating time to every time slot. Initialize receive combining covariance matrices $\mathbf{W}_{wit,i,(0)}$ ($\forall i = 1, \cdots, K$) as a diagonal matrix $diag(1/\sqrt{N_r}, \cdots, 1/\sqrt{N_r})$, while initializing the scheduling indicator matrix $\mathbf{G}_{(0)}$ as an identity matrix $\mathbf{I}_{K \times K}$. This step is characterized in Line 1 of Algorithm 5.4.
- *Step 2*: By exploiting the solution $\{\mathbf{W}_{wpt,(k)}, \mathbb{W}_{wit,(k)}, \mathbf{G}_{(k)}, \mathbf{t}_{(k)}\}$ obtained in the last iteration, we can sequentially update them as $\{\mathbf{W}_{wpt,(k+1)}, \mathbb{W}_{wit,(k+1)}, \mathbf{G}_{(k+1)}, \mathbf{t}_{(k+1)}\}$ by solving (P5.3.3) with Algorithm 5.1, (P5.3.4) with CVX toolbox and (P5.3.5) with Algorithm 5.2 as well as (P5.3.6) with Algorithm 5.3, respectively, as characterized in Lines 5–8 of Algorithm 5.4.
- *Step 3*: If the difference between the fair-throughput $R_{\text{fair},cur}$ obtained in the current iteration and $R_{\text{fair},pre}$ obtained in the last one is higher than a predefined threshold ξ, we start another iteration by repeating Step 2, as shown in Line 3 of Algorithm 5.4. Otherwise, $\{\mathbf{W}_{wpt,(k+1)}, \mathbb{W}_{wit,(k+1)}, \mathbf{G}_{(k+1)}, \mathbf{t}_{(k+1)}\}$ is output as the solution, and $R_{\text{fair},cur}$ is output as the fair-throughput, as shown in Line 11 of Algorithm 5.4.

Convergence Analysis The objective function of (P5.3.1) is nondecreasing after each iteration in Algorithm 5.4. Since a continuous function has an upper bound in a closed interval, the fair-throughput R_{fair} has upper bounds with respect to the continuous inputs $\mathbf{W}_{wpt}$, $\mathbb{W}_{wit}$, and $\mathbf{t}$. Moreover, since all entries in the uplink WIT throughput matrix $\mathbf{R} = [R_{i,j}]_{K \times K}$ are finite, the optimal fair-throughput R_{fair}^* in (P5.3.5) is upper-bounded by the maximum $R_{i,j}$. Therefore, R_{fair} is upper-bounded with respect to the binary matrix $\mathbf{G}$. As a result, the proposed Algorithm 5.4 is guaranteed to converge [43].

Complexity Analysis Since Algorithm 5.4 definitely converges after a finite number of iterations, the complexity of Algorithm 5.4 is denoted by $\mathcal{C}$ in Eq. (5.40) on the next page, where I represents the number of iterations.

$$\mathcal{C} = \mathcal{O}\Bigg(I\Bigg(\left(\frac{N_t^2 + N_t}{2}\right)^{3.5} \log\left(\frac{R_{\max} - R_{\min}}{\varsigma}\right) + \left(K \cdot \frac{N_r^2 + N_r}{2}\right)^{3.5}$$
$$+ K^3\left(\log\left(\frac{R'_{\max} - R'_{\min}}{\eta}\right) + 1\right) + K\left(\log\left(\frac{z_{\max}}{\epsilon}\right) + 2\right)\Bigg)\Bigg). \tag{5.40}$$

Since rank-1 relaxations are applied to (P5.3.2), the rank-1 solutions need to be recovered after the convergence of Algorithm 5.4. The vectors $\mathbf{w}_{wpt}^*$ and $\mathbf{w}_{wit,i}^*$ can be approximated by a Gaussian randomization procedure [44], which is detailed in Algorithm 5.5.

Algorithm 5.5 Recover the rank-1 solutions from $\mathbf{W}^*_{wpt}$ and $\mathbb{W}^*_{wit}$

Require: Optimal solutions $\mathbf{W}^*_{wpt}$ and $\mathbb{W}^*_{wit}$ to (P5.3.3) and (P5.3.4), number of randomisations $L \geq 1$
Ensure: $\mathbf{w}_{wpt,j^*}$ and $\mathbf{w}_{wit,i,j^*}$
1: Generate independent random vectors $\{\mathbf{w}_{wpt,j} \sim \mathcal{CN}(0, \mathbf{W}^*_{wpt})|\forall j = 1, \cdots, L\}$ and $\{\mathbf{w}_{wit,i,j} \sim \mathcal{CN}(0, \mathbf{W}^*_{wit,i})|\forall j = 1, \cdots, L\}$ for $\forall i = 1, \cdots, K$;
2: Let $j^* = \arg\max_{j=1,\ldots,L} \min_{\mathbf{w}_{wpt,j} \text{ or } \mathbf{w}_{wit,i,j}} R_{\text{fair}}$.

Table 5.1 Parameter Setting

Parameter	Value
Transmit power of H-AP	45 dBm
Number of transmit antennas of H-AP	10
Number of receive antennas of H-AP	10
Additive noise power	−70 dBm
Transmit frequency	900 MHz
Path-loss reference distance	1 m
Path-loss exponent	3
Number of UEs	5
Maximum distance between H-AP and UEs	5 m
Distance from H-AP's tx to rx (d_{si})	1 m
Rician factor	$3 + \sqrt{12}$

5.2.5 *Numerical Results*

Simulation parameters in both physical and MAC layers are in line with Table 5.1, unless some parameter changes are particularly mentioned. The parameters $a = 132.8$, $b = 0.01181$, and $M = 0.02337$ of the nonlinear energy harvester are obtained by fitting the measurement data [45]. UEs are randomly distributed within the maximum distance. All the results are obtained by averaging the randomness incurred by the Rician fading.

5.2.5.1 Convergence of the Iterative Algorithm

Observe from Fig. 5.3 that our joint design can converge after several iterations, when different downlink WPT transmit powers $P_{wpt} = \{40, 45, 47\}$ dBm of the H-AP are considered. Specifically, in Fig. 5.3, when we have $P_{wpt} = 45$ dBm, the uplink fair-throughput converges to 1.43 bit/Hz after 3 iterations. In the case of $P_{wpt} = 40$ dBm, the fair-throughput converges to 1.11 bit/Hz after 9 iterations.

5.2.5.2 Individual UEs Performance

We plot the individual UEs uplink throughput and the amount of energy harvested. In order to demonstrate the fairness performance of our joint design, we set the distance from UE_i to the H-AP as $d_i = i/K \cdot d_{\max}$, where we have $d_{\max} = 3$ m and $K = 5$. Observe from Fig. 5.4 that each UE has the same individual uplink throughput. Therefore, our joint design achieves great fairness, which verifies Remark 1 in Sect. 5.2.4.3. Moreover, observe from Fig. 5.4 that the UE farther from the H-AP harvests more energy. This is because its information signal suffers from higher path-loss. As a result, in order to guarantee its QoS, it has to harvest more energy for powering its information uploading.

5.2.5.3 Multiple Antennas

We then evaluate the uplink fair-throughput with a different number of antennas at the H-AP in the following three WPCN systems:

- *HD aided WPCN*: A transmission frame is generally partitioned into two groups of time slots. The first group is adopted by the H-AP for downlink WPT, while the second group is adopted by the UEs for uplink WIT. The H-AP's transmit

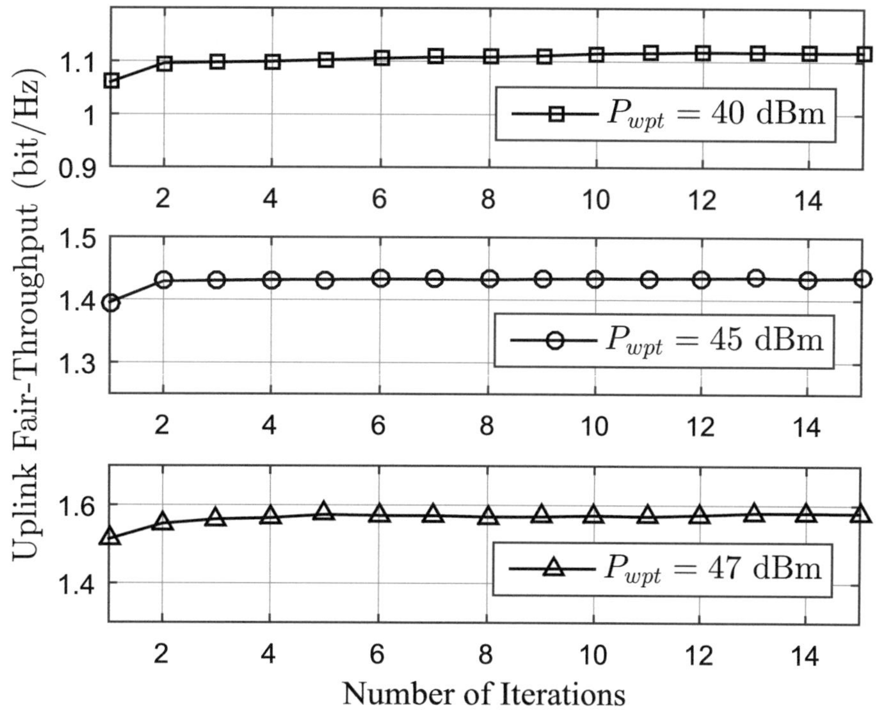

Fig. 5.3 Convergence of the iterative algorithm aided resource scheduling design in multi-domain

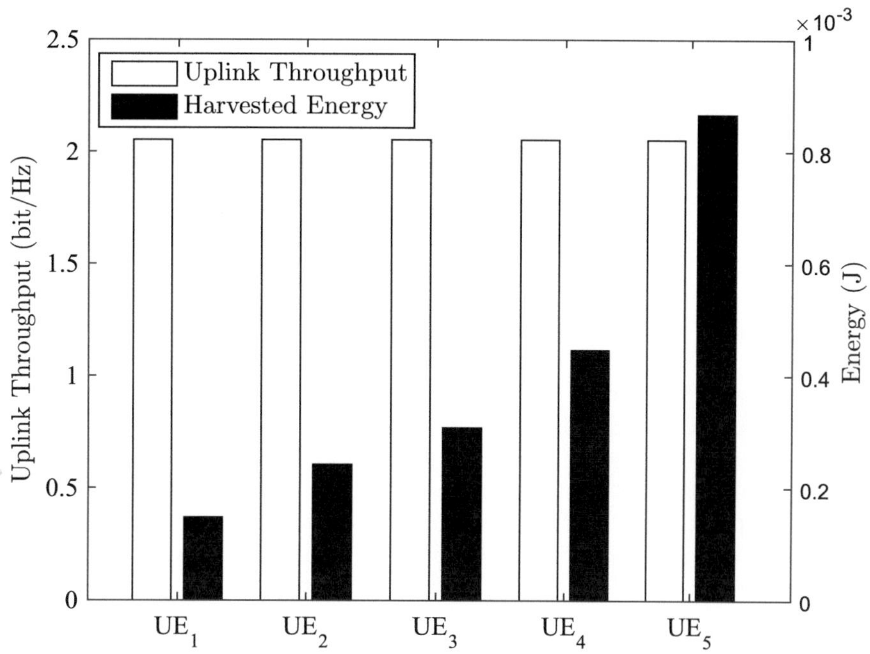

Fig. 5.4 Individual UEs uplink throughput and energy harvested

beamformer and receive combiner as well as the duration of downlink WPT and the time allocation for the UEs' uplink WIT are optimized for maximizing the uplink fair-throughput. Please refer to [17] for more details.

- *Ideal FD aided WPCN*: The SI of the H-AP can be perfectly cancelled by circuitry-level signal processing. Therefore, the self-interference part in the calculation of R_i in Eq. (5.16) can be ignored in our joint design.
- *Actual FD aided WPCN*: We consider the adverse effect of the H-AP's self-interference in our joint design.

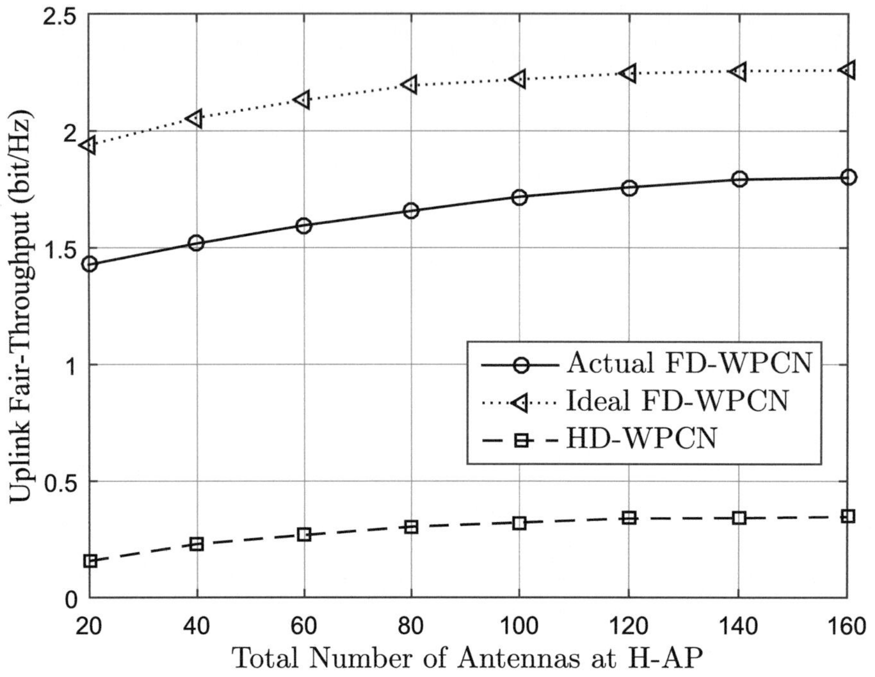

Fig. 5.5 Number of transmit and receive antennas versus uplink fair-throughput

In the HD aided WPCN, the H-AP utilizes all antennas for WPT and WIT, while in the other two WPCNs, the H-AP adopts half of the number of antennas for WPT and the other half for WIT.

Observe from Fig. 5.5 that when we increase the number of transmit antennas and that of receive antennas, the uplink fair-throughputs of all these three WPCN systems are substantially increased. This is because having a higher number of Tx/Rx antennas [46], we are able to focus more energy toward the batteryless UEs, to constructively combine the received signal for improving the uplink WIT performance and to efficiently cancel the adverse SI at the FD aided H-AP. Furthermore, as shown in Fig. 5.5, both ideal and actual FD aided WPCNs outperform the HD aided counterpart, since downlink WPT and uplink WIT can be carried out simultaneously.

5.2.5.4 UEs' Scheduling and Time Allocation

Observe from Fig. 5.6 that the uplink fair-throughput of the UEs reduces, as we increase the maximum distance $d_{\max}$. This is because increasing path-loss reduces the amount of energy harvested by the batteryless UEs, while it also degrades the uplink WIT. Furthermore, we also compare our optimal UE scheduling design (US) of Algorithm 5.2 to a method without UE scheduling in Fig. 5.6a. Resource scheduling in other domains is obtained by exploiting the iterative Algorithm 5.4. Observe from Fig. 5.6a that our optimal UE scheduling design substantially outperforms the counterpart. Similarly, as shown in Fig. 5.6b, we demonstrate the performance gain of our optimal time allocation (TA) of Algorithm 5.3 over the uniform counterpart, where all UEs' uplink time slots have equal durations.

5.2.5.5 Transmit Power

We then compare the performance of practical nonlinear energy harvesting (EH) model to that of the conventional linear counterpart. Observe from Fig. 5.7 that as the H-AP's transmit power increases, the fair-throughput generally increases at first. This is because, as we initially increase the H-AP's transmit power, the batteryless UEs may gain more energy from the downlink WPT. Therefore, they have a higher amount of energy for powering their own uplink WIT, which results in an increasing fair-throughput. However, as we continuously increase the transmit power, the fair-throughput of the nonlinear EH model keeps unchanged, while that of the linear one reduces. This is because a very high transmit power of the H-AP may

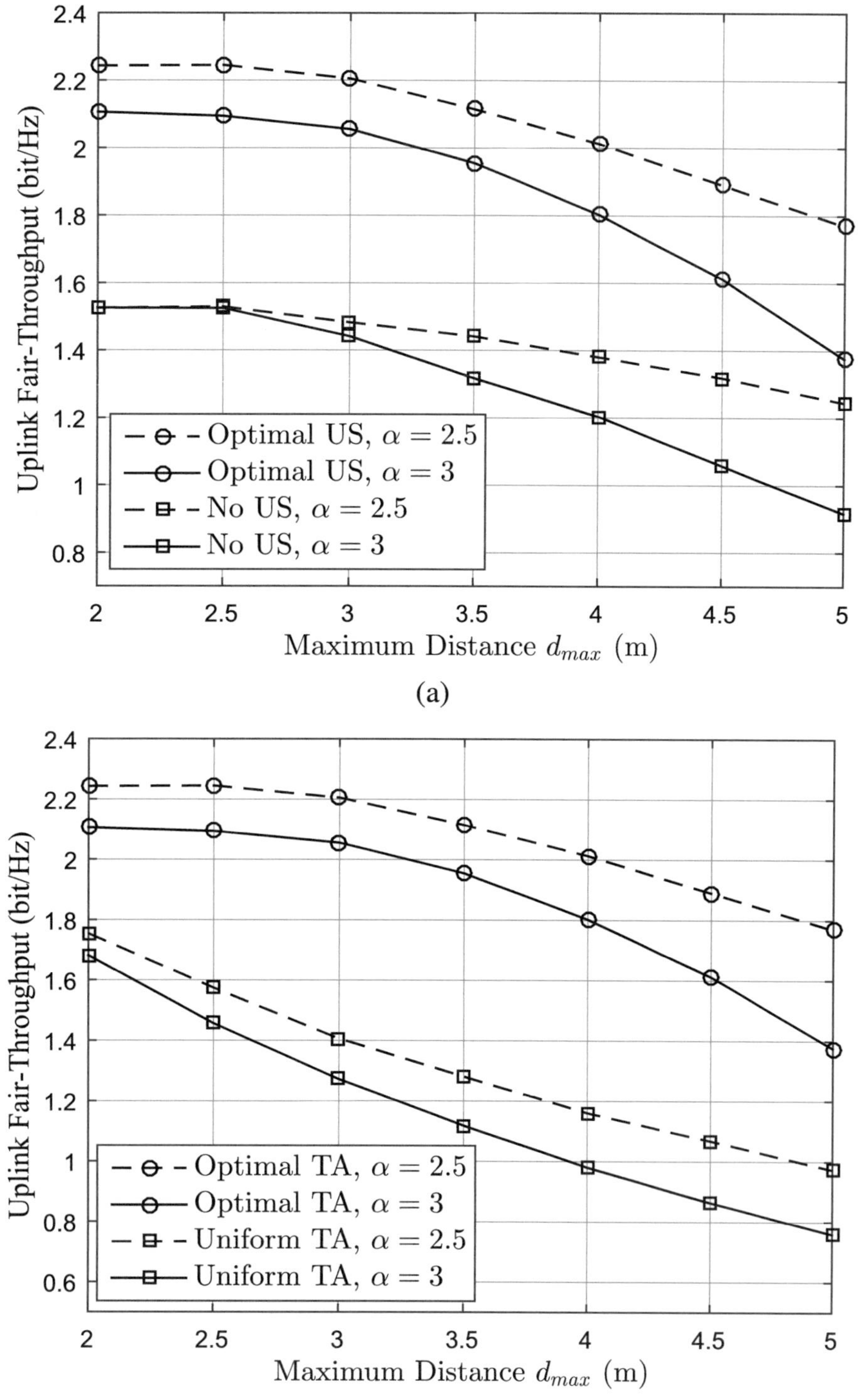

Fig. 5.6 Impact of maximum distance $d_{\max}$ on uplink fair-throughput (**a**) with different UE scheduling schemes and (**b**) with different time allocation schemes

severely interfere its own uplink information reception. This observation tells us that our nonlinear EH model is more robust than the linear counterpart. Furthermore, observe from Fig. 5.7 that our nonlinear EH model achieves higher fair-throughput than the linear one. Moreover, when we have more UEs, the fair-throughput becomes lower.

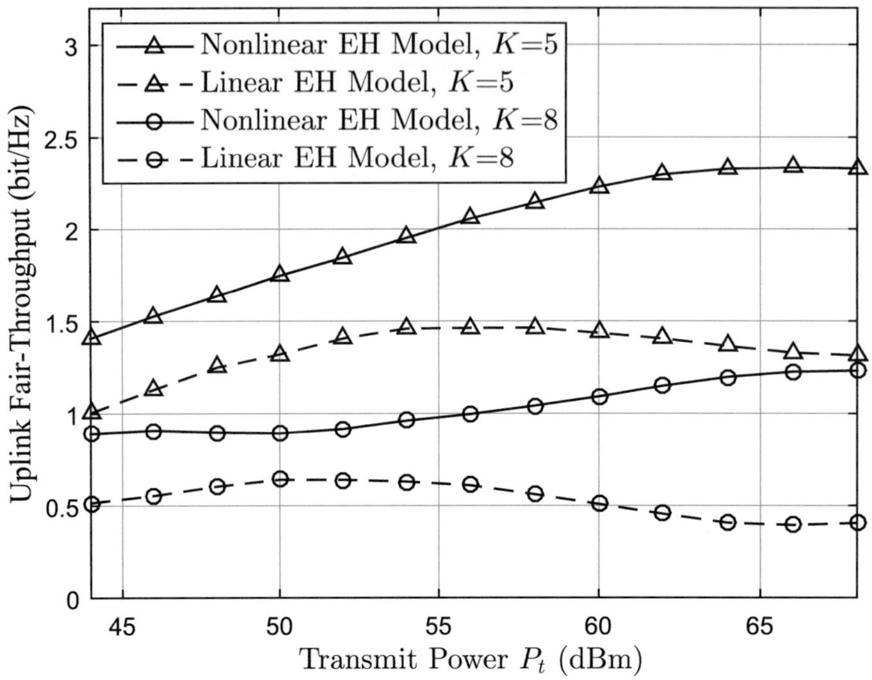

Fig. 5.7 Impact of the transmit power on uplink fair-throughput

5.3 Exercises

1. Please describe the Lagrangian duality method for the classic convex maximization and the minimization problems and compare the differences between them.
2. What imposes the self-interference (SI) of devices in the full-duplex system? List some methods for the SI cancellation.
3. Consider M users with the powers $P(1), \cdots, P(M)$ are to be allocated subject to the constraint $\sum_{i=1}^{M} P(i) \leq MP$. When we assume that they can upload their information simultaneously with the same resources allocated to them, determine the optimal power allocation $\mathbf{P}$ to maximize the throughput in the whole networks.
4. Please describe the optimal beamforming design in a point-to-point system.
5. Consider a network with M users and a single BS. The users operating in TDMA transmit their information sequentially to the BS within the uploading duration of T. Design a time allocation to maximize the uplink total throughput.
6. Consider the same scenario as 3. Please design users uploading sequence for the uplink total throughput maximization.

References

1. Boyd, S., & Vandenberghe, L. (2009). Convex Optimization. In *Algorithms* (pp. 466–475).
2. Saad, W., Bennis, M., & Chen, M. (2019). A vision of 6g wireless systems: Applications, trends, technologies, and open research problems. *IEEE Network, 34*(3), 1–9. ISSN: 1558–156X.
3. Krikidis, I. et al. (2014). Simultaneous wireless information and power transfer in modern communication systems. *IEEE Communications Magazine, 52*(11), 104–110. ISSN: 0163–6804.
4. Matias, R., Cunha, B., & Martins, R. (2013). Modeling inductive coupling for wireless power transfer to integrated circuits. In *2013 IEEE Wireless Power Transfer (WPT)* (pp. 198–201).
5. Qi, H., Chen, W., Sha, Y., Han, Y., Li, H., & Yang, X. (2016). A study of electromagnetic emission from two-coil WPT system using resonant magnetic field coupling. In *2016 Asia-Pacific international symposium on electromagnetic compatibility (APEMC)* (vol. 01, pp. 484–487).
6. Hu, J., Kun, Y., Guangjun, W., & Lajos, H. (2018). Integrated data and energy communication network: A comprehensive survey. *IEEE Communications Surveys Tutorials, 20*(4), 3169–3219.
7. Zhao, Y., Hu, J., Diao, Y., Yu, Q., & Yang, K. (2018). Modelling and performance analysis of wireless LAN enabled by RF energy transfer. *IEEE Transactions on Communications, 66*(11), 5756–5772. ISSN: 0090–6778.
8. Guo, C., Liao, B., & Huang, L. (2016). Time allocation and load balancing in multi-cell wireless powered communication networks. *IEEE Access, 4*, 7795–7805.

9. Wu, Q., Tao, M., Kwan Ng, D. W., Chen, W., & Schober, R. (2016). Energy-efficient resource allocation for wireless powered communication networks. *IEEE Transactions on Wireless Communications, 15*(3), 2312–2327. ISSN: 1536–1276.
10. Lei, M., Zhang, X., Yu, B., & Dong, X. (2018). Power and discrete rate adaptation in BER constrained wireless powered communication networks. *IET Communications, 12*(18), 2213–2221, .
11. Ju, H. & Zhang, R. (2014). Throughput maximization in wireless powered communication networks. *IEEE Transactions on Wireless Communications, 13*(1), 418–428. ISSN: 1536–1276.
12. Mao, S., Leng, S., Hu, J., & Yang, K. (2019). Energy-efficient transmission schemes for cooperative wireless powered cellular networks. *IEEE Transactions on Green Communications and Networking, 3*(2), 494–504. ISSN: 2473–2400.
13. Zhao, Y., Wang, D., Hu, J., & Yang, K. (2018). H-ap deployment for joint wireless information and energy transfer in smart cities. *IEEE Transactions on Vehicular Technology, 67*(8), 7485–7496. ISSN: 1939–9359.
14. Pehlivan, I. & Ergen, S. C. (2019). Scheduling of energy harvesting for MIMO wireless powered communication networks. *IEEE Communications Letters, 23*(1), 152–155.
15. Lim, H. & Hwang, T. (2019). User-centric energy efficiency optimization for MISO wireless powered communications. *IEEE Transactions on Wireless Communications, 18*(2), 864–878. ISSN: 1558–2248.
16. Choi, J., Song, C., & Joung, J. (2018). Wireless powered information transfer based on zero-forcing for multiuser MIMO systems. *IEEE Transactions on Vehicular Technology, 67*(9), 8561–8570. ISSN: 1939–9359.
17. Yang, J., Hu, J., Lv, K., Yu, Q., & Yang, K. (2018). Multi-dimensional resource allocation for uplink throughput maximisation in integrated data and energy communication networks. *IEEE Access, 6*, 47163–47180. ISSN: 2169–3536.
18. Suraweera, H. A., Krikidis, I., Zheng, G., Yuen, C., & Smith, P. J. (2014). Low-complexity end-to-end performance optimization in MIMO full-duplex relay systems. *IEEE Transactions on Wireless Communications, 13*(2), 913–927.
19. Kang, X., Ho, C. K., & Sun, S. (2015). Full-duplex wireless-powered communication network with energy causality. *IEEE Transactions on Wireless Communications, 14*(10), 5539–5551. ISSN: 1558–2248.
20. Hu, J., Xue, Y., Yu, Q., & Yang, K. (2017). A joint time allocation and UE scheduling algorithm for full-duplex wireless powered communication networks. In *2017 IEEE 86th vehicular technology conference (VTC-Fall)* (pp. 1–5).
21. Abd-Elmagid, M. A., Biason, A., ElBatt, T., Seddik, K. G., & Zorzi, M. (2016). On optimal policies in full-duplex wireless powered communication networks. In *2016 14th International symposium on modeling and optimization in mobile, Ad Hoc, and Wireless Networks (WiOpt)* (pp. 1–7).
22. Peng, Y. & Ke, F., Sum-throughput maximization for MIMO full-duplex wireless powered communication network with finite alphabet inputs. In *2018 IEEE 88th Vehicular Technology Conference (VTC-Fall)* (pp. 1–5).
23. Lee, J., Cho, Y., Park, D., & Chang, D. E. (2019). Robust beamforming and time allocation for full-duplex wireless-powered communication networks. *IEEE Communications Letters, 23*(9), 1665–1669.
24. Kim, H., Lee, H., Ahn, M., Kong, H. B., & Lee, I. (2016). Joint subcarrier and power allocation methods in full duplex wireless powered communication networks for OFDM systems. *IEEE Transactions on Wireless Communications, 15*(7), 4745–4753. ISSN: 1536–1276.
25. Do, T. P. & Kim, Y. H. (2016). Resource allocation for a full-duplex wireless-powered communication network with imperfect self-interference cancelation. *IEEE Communications Letters, 20*(12), 2482–2485. ISSN: 2373–7891.
26. Zeng, Y. & Zhang, R. (2015). Full-duplex wireless-powered relay with self-energy recycling. *IEEE Wireless Communications Letters, 4*(2), 201–204. ISSN: 2162–2345.
27. Nikoloska, I., Zlatanov, N., & Hadzi-Velkov, Z. (2018). Capacity of a full-duplex wirelessly powered communication system with self-interference and processing cost. *IEEE Transactions on Wireless Communications, 17*(11), 7648–7660. ISSN: 1558–2248.
28. Asiedu, D. K. P., Mahama, S., Song, C., Kim, D., & Lee, K.-J. (2020). Beamforming and resource allocation for multiuser full-duplex wireless-powered communications in IoT networks. *IEEE Internet of Things Journal, 7*(12), 11355–11370.
29. Jia, X., Zhang, C., & Kim, I. (2019). Worst-case robust beamforming design for wireless powered multirelay multiuser network with a nonlinear EH model. *IEEE Transactions on Vehicular Technology, 68*(3), 3038–3042. ISSN: 1939–9359.
30. Chi, K., Chen, Z., Zheng, K., Zhu, Y., & Liu, J. (2019). Energy provision minimization in wireless powered communication networks with network throughput demand: TDMA or NOMA?. *IEEE Transactions on Communications, 67*(9), 6401–6414. ISSN: 1558–0857.
31. Huang, W., Chen, H., Li, Y., & Vucetic, B. (2016). On the performance of multi-antenna wireless-powered communications with energy beamforming. *IEEE Transactions on Vehicular Technology, 65*(3), 1801–1808.
32. Sun, W., Ström, E. G., Brännström, F., & Gholami, M. R. (2015). Random broadcast based distributed consensus clock synchronization for mobile networks. *IEEE Transactions on Wireless Communications, 14*(6), 3378–3389. ISSN: 1536–1276.
33. Mahmood, A. & Jantti, R. (2009). Time synchronization accuracy in real-time wireless sensor networks. In *2009 IEEE 9th Malaysia International Conference on Communications (MICC)* (pp. 652–657).
34. 3gpp ts 38.213 v16.8.0 (2021-12).
35. Lee, K., Lim, J., & Choi, H. (2020). Impact of outdated CSI on the secrecy performance of wireless-powered untrusted relay networks. *IEEE Transactions on Information Forensics and Security, 15*, 1423–1433. ISSN: 1556–6021.
36. Zhao, F., Lin, H., Zhong, C., Hadzi-Velkov, Z., Karagiannidis, G. K., & Zhang, Z. (2018). On the capacity of wireless powered communication systems over Rician fading channels. *IEEE Transactions on Communications, 66*(1), 404–417.
37. Lv, K., Hu, J., Yu, Q., & Yang, K. (2017). Throughput maximization and fairness assurance in data and energy integrated communication networks. *IEEE Internet of Things Journal, 5*(2), 636–644. ISSN: 2372–2541.
38. Hu, J., Yang, L. L., & Hanzo, L. (2016). Delay analysis of social group multicast-aided content dissemination in cellular system. *IEEE Transactions on Communications, 64*(4), 1660–1673. ISSN: 0090–6778.
39. Boshkovska, E., Ng, D. W. K., Zlatanov, N., & Schober, R. (2015). Practical non-linear energy harvesting model and resource allocation for SWIPT systems. *IEEE Communications Letters, 19*(12), 2082–2085.
40. CVX Research, Inc. (2012). CVX: Matlab software for disciplined convex programming, version 2.0. http://cvxr.com/cvx.
41. Zhang, G., Wu, Q., Cui, M., & Zhang, R. (2019). Securing UAV communications via joint trajectory and power control. *IEEE Transactions on Wireless Communications, 18*(2), 1376–1389.
42. Kuhn, H. W. (1955). The hungarian method for the assignment and transportation problems. *Naval Research Logistics Quarterly, 2*, 83–97.

43. Yang, G., Yuan, D., Liang, Y., Zhang, R., & Leung, V. C. M. (2019). Optimal resource allocation in full-duplex ambient backscatter communication networks for wireless-powered IoT. *IEEE Internet of Things Journal, 6*(2), 2612–2625.
44. Wu, S. X., Ma, W., & So, A. M. (2013). Physical-layer multicasting by stochastic transmit beamforming and alamouti space-time coding. *IEEE Transactions on Signal Processing, 61*(17), 4230–4245.
45. Guo, J., Zhang, H., & Zhu, X. (2014). Theoretical analysis of RF-DC conversion efficiency for class-F rectifiers. *IEEE Transactions on Microwave Theory and Techniques, 62*(4), 977–985.
46. Mohammadi, M., Chalise, B. K., Suraweera, H. A., Ngo, H. Q., & Ding, Z. (2021). Design and analysis of full-duplex massive antenna array systems based on wireless power transfer. *IEEE Transactions on Communications, 69*(2), 1302–1316.

Chapter 6
Access Control for Energy Constrained Devices

6.1 Distributed Collision Avoidance Access Protocol: Performance Analysis

When multiplexing schemes are not used on the physical layer. The media access control (MAC) layer protocol needs to adopt some mechanisms to avoid collisions. The MAC mechanism is an operation that has been taken by nodes to gain access to the medium. For energy constrained devices with the capability of energy harvesting, MAC layer protocol plays a significant role in the design of data and energy integrated networks (DEIN). In fact, the MAC layer protocol consumes a large portion of energy due to message transmission/reception and channel sensing activities. This section introduces some widely used MAC protocols in DEINs, while describing some basic mechanisms in these protocols. Then, we introduce several energy-harvesting enhanced MAC protocols specifically designed for DEIN and review the basic methods for performance analysis of these protocols.

6.1.1 Some Basic Access Control Protocols

First, we introduce some classical MAC protocols, which are the basic of energy-harvesting enhanced MAC protocols.

6.1.1.1 ALOHA

It is the most basic MAC protocol, allowing a node to transmit a data frame whenever it pleases. Every successful reception must be acknowledged by the recipient to the transmitter. The transmitter will retransmit the same frame after a random waiting period (referred to as backoff) subject to a maximum number of retransmissions, if it is unable to receive the acknowledgment within a specific window of time. Slotted ALOHA (S-ALOHA) prevents collisions between new transmissions and ongoing transmissions, which may happen in ALOHA, by dividing the time into equal-sized slots. New attempts are only permitted at the start of a slot.

6.1.1.2 Carrier Sensing Multiple Access (CSMA)

Before making any transmissions, a node first senses the carriers. A perceived carrier prevents the node from transmitting. How to behave produces numerous CSMA versions, including CSMA with collision detection (CSMA/CD) and CSMA with collision avoidance (CSMA/CA), when no carrier is sensed. When using CSMA/CD, the node broadcasts right away while continuing to monitor the medium for collisions. Although this protocol is used for early Ethernet, wireless networks cannot use it since it is difficult to execute "listening-while-talking" operation. A node using CSMA/CA delays transmission for an arbitrary backoff period and only transmits when the medium is still perceived as being idle at the end of the backoff period. CSMA/CA is used by several well-known MAC protocols for wireless networks such as IEEE 802.11 and IEEE 802.15.4.

J. Hu, K. Yang, *Integrated Data and Energy Transfer in Wireless Networks*, Textbooks in Telecommunication Engineering,
https://doi.org/10.1007/978-3-031-60514-7_6

1. IEEE 802.11: It consists of two functions: Distributed Coordination Function (DCF) and Point Coordination Function (PCF). DCF follows CSMA/CA and uses Rotation Backoff Competition Window (CW) jointly. The main goal of CSMA/CA is to avoid conflicts and to minimize their occurrence. Its basic theory is that when a medium suddenly changes from busy to idle, it is most likely to conflict because many STAs are waiting for the medium to change from busy to idle. The solution to this high potential conflict is the random fallback mechanism. Random fallback does not fundamentally resolve the conflict but only reduces the possibility of conflict. Random fallback allows the node generates a pseudo-random integer between $[0, CW]$ before its each transmission attempt. This pseudo-random integer minus one every time slot until it is zero. Only when it decreases to zero can the node transmite a packet. The range of CW is $[CWmin, CWmax]$. The initial value of CW is $CWmin$, and each failed transfer will make the CW value larger than the next value (the previous value multiplied by 2 and then plus 1) until $CWmax$. Any successful data frame or management frame transmission resets the CW to $CWmin$.

   PCF is designed based on DCF, therefore, they are compatible and function with a mechanism of alternating. This alternating cycle is known as CFP repetition interval. The main idea of PCF is that AP acts as the central coordinating controller (PC) and polls the connected nodes (CF-Pollable STA) in turn according to their internal polling list to check whether they have data to be transmitted. During CFP time, nodes cannot actively compete for channels due to NAV mechanism. Therefore, unless the base station polls the node and requests its feedback data, the node cannot actively carry out transmission actions. In PCF, PCs poll nodes in ascending order in the order of polling tables. During a race-free period, stations cannot transmit data unless the base station requests a polling frame.
2. IEEE 802.15.4: It is designed for low data rate wireless personal area networks (PANs). 802.15.4 networks can operate in beacon or non-beacon modes. In beacon mode, the coordinator broadcasts beacons regularly to achieve device synchronization and other purposes. In the non-beacon mode, the coordinator transmits the beacon to the device when it requests it. 802.15.4 has the option of a superframe structure, which is defined by the coordinator. The superframe is divided into 16 equal-sized time slots, including active and inactive part. During the inactive period, the coordinator goes into a low voltage state, the beacon frame is located in the first time slot of the superframe. If the coordinator does not want to use the superframe structure, it turns off the beacon transmission. The beacons are used to synchronize connected devices, to distinguish the network, and to describe the superframe structure. Any device that wants to communicate during contested access must use the CSMA/CA mechanism, and all communication transmissions must be completed when the next beacon comes. In applications with low latency or requiring special data bandwidth, the PAN coordinator can allocate active time slots in the superframe to such applications. These time slots are called guaranteed time slots (GTSs). Multiple GTSs in the same superframe make up the contention-free time slots (CFPs), which are often located at the end of the active part of the superframe structure. A PAN coordinator can allocate up to seven such GTSs in a superframe, and a single GTS can occupy multiple time slots (but a single GTS cannot span two superframes). Although there can be more than seven GTSs slots in a superframe, the majority of slots in the superframe are reserved for competing access slots (CAPs) for devices that are connected or want to be connected. All information used for contention is transmitted before the arrival of the CFP. Similarly, the data in the GTSs slots must be transmitted before the next superframe arrives.

6.1.2 Some Basic Mechanisms for Access Control Protocols

There are many basic mechanisms used in different MAC protocols. One or more different mechanisms may be employed in one MAC protocol. We present in this section some of the mechanisms that are often employed in MAC protocols and that may be useful in designing new protocols.

1. Free access: It includes random access and probabilistic access. For the former, nodes can access the medium at any time without relying on any other method, such as ALOHA. For the latter, nodes have ω probability in accessing the medium. This mechanism is relatively simple, reliable, and applicable to all network topologies but with low medium utilization due to frequent conflicts.
2. Slotted access: The time slot is divided into multiple time slots and any new media access is allowed only at the beginning of each time slot, as is the case for slotted ALOHA. It blocks new attempts with active traffic which may cause interfere. To achieve this, a network-level SYN is needed to ensure that each node has the same time reference. Similar to TDMA, protection time must be set between consecutive time slots to prevent conflicts.
3. Backoff: It requires nodes to wait for a predetermined amount of time before accessing the medium, whereas CSMA/CA in IEEE 802.11 uses random fallback. In addition, media access can be prioritized by giving a shorter fallback time to the node with higher priority. If the fallback period is too long and there are no other competitors, bandwidth may be wasted.

4. Signaling: In order to interfere or stop a competitor's ongoing broadcast, the node intentionally sends a short signal. The receiver does not have to decode the received signal, making it independent of the quality of the channel, but lacking information. It can be used with any topology and uses various signaling lengths to prioritize the media access nodes.
5. Messaging: Similar to signaling, a node intentionally sends a message to a node carrying explicit information such as RTS/CTS. It is more informative than signaling notification since the received signal can be successfully decoded. This backloaded scheme allows data frames to carry information incidentally, to reduce protocol overhead.
6. Polling: A central node can poll potential senders, and only a polled node can transmit. This is suitable for a star topology network. To avoid idle nodes from being polled, the central node needs the information on transmission requests of the potential senders. This can be achieved by allowing each node to submit its request in advance via random access such as ALOHA.

6.1.3 Special MAC Protocols for Energy Constrained Devices

Forces on MAC protocols for energy constrained devices, we need mention that synchronous MAC protocols may not suitable. For example, a node with none energy may not be able to function as promised by synchronous schedule. Furthermore, frequent message exchange for synchronizing may exhaust the energy faster. We exemplify several specific MAC protocols for DEINs.

6.1.3.1 Probabilistic Polling

This protocol is proposed in [1]. The coordinator will broadcast a competition probability to all nodes, which can be chosen based on the current energy level of all nodes, and nodes below this probability will be automatically dropped from the pool of competing nodes and wait for the next round of voting. Once a node does not have enough energy to participate in the voting, it will not compete but will switch to the charging state to collect enough energy for future operations. This contention probability follows Additive Increase Multiplicative Decrease (AIMD) policy for the number of active nodes. If the coordinator does not receive any data packet from any node in response to polling, it would increase the polling probability of next cycle considering that there are not enough nodes in contention. Conversely, polling probability would be decreased if there are collisions within the system. This yields higher throughput, scalability, and fairness and it handles the collision situations by probability adjustment leading to the fair source allocation.

6.1.3.2 On-Demand Medium Access Control (ODMAC)

ODMAC [2] is a classical MAC protocol for DEINs that the coordinator sends periodic beacons toward the intended senders. The beacons inform the senders about readiness of receiver to receive their packets. As soon as the senders receive these beacons, the transmission would instantly be started. This protocol is based on the duty cycle adjustment and opportunistic forwarding techniques. Duty cycle can further be adjusted by two different mechanisms: beacon period adjustment and sensing period adjustment. A node with extra harvested energy can decrease either the beacon or sensing interval (time between consecutive beacons or sensing operations) and the deficient energy nodes can contrarily increase either of these parameters based on the type of application and its requirements. The node can further tune either of the intervals in case of extreme energy shortage. Each node maintains a list of relay nodes in opportunistic forwarding and sends the packet instantly whenever it receives a beacon by the list members avoiding the need of keep waiting for a particular receiver.

6.1.3.3 Energy Harvesting MAC Protocol (EH-MAC) and RF-MAC

EH-MAC is also a polling scheme protocol. Its main goal is to maximize throughput using contention probability concept. EH-MAC increases the network capacity when the energy harvested is augmented since the data to be transmitted increases. This is achieved by defining the duty cycle duration as a function of the number of packets transmitted by the charging cycles. RF-MAC is a protocol based on EH-MAC that uses RF energy transfer. Two important aspects related to the available energy are considered in RF-MAC, including charging threshold and prioritization. The former is related to the adaptive charging threshold, where each node calculates its maximum charging threshold based on the communication activity. RF-MAC establishes a minimum voltage as a threshold. Only the node with energy more than this threshold has priority access to the channel. The latter is related to the prioritization of energy over exchange of information. RF-MAC presents the priority in the data transfer process. In this case, the prioritization is linked to the duration of the backoff mechanism, which is defined according to the residual energy and the contention window.

6.1.4 Basic Analysis Method

Modeling based on the specific behavior of devices using different MAC protocols often enables quantitative analysis of their performance through some classic mathematical tools. In this section, we introduce some classic mathematical tools.

(1) Probability Theory and Statistical Geometry
Classical probability theory is a necessary tool for analyzing a MAC protocol. In a duty cycle based MAC protocol, nodes periodically take actions such as transmitting, receiving, and sleeping, which may have the probability of P_t, P_r, and P_s, respectively. We represent the node may consume E_t, E_r, and E_s during these states. Accordingly, we can derive the average energy consumption rate of the node as:

$$E_{node} = P_t E_t + P_r E_r + P_s E_s. \tag{6.1}$$

This is an indicator that may affect the energy performance of the device. Furthermore, when all devices are competing for the same channel, the channel has two states, e.g. occupy and idle. When the channel is occupied, it can also be divided into occupied by a successful transmission or a collision. We denote the probability of channel in idle state, successful transmission state, and collision state as P_f, P_s, and P_c, respectively. Accordingly, the throughout can be expressed as:

$$R = \frac{P_s L[PCK]}{P_f T_f + P_s T_s + P_c T_c}, \tag{6.2}$$

where $L[PCK]$ is the average length of packets, T_f, T_s, and T_c is the duration of the related channel states. However, the expressions of T_f, T_s, T_c, P_t, P_r, and P_s should be analyzed according to the specific protocol. In [3], the authors just use probability theory and statistical geometry to analyze the performance of a network with an energy harvesting enhanced Wi-Fi network.

(2) Markov Analysis
When analyzing the CSMA based MAC protocol, Bian's [4] analysis framework is widely adopted for modeling the behaviors of nodes. Consider there are N nodes competing the same channel. In saturation conditions, all nodes have full transmission buffer, in other words it always has packets for transmission. CSMA/CA protocol is adopt by all nodes. Each transmission attempt needs to wait for a random backoff time before transmitting. Let $b(t)$ and $s(t)$ be the backoff counter and backoff stage of the node during time t, respectively. A discrete timescale is adopted. Since the value of the backoff counter of each station depends also on its transmission history, $b(t)$ itself is non-Markovian. However, when we focus on a bidimensional process $\{s(t), b(t)\}$ is possible to be modeled as Markov chain. The key approximation is that, at each transmission attempt, each packet collides with constant and independent probability p. In saturation conditions, it is intuitive that this approximation will more accurate when the number of nodes get larger. Therefore, according to DCF protocol, the state transition diagram can be depicted as Fig. 6.1. The one-step transition probability can be accordingly derived as:

$$\begin{cases} P\{i, k|i, k+1\} = 1, k \in (0, W_i - 2)\, i \in (0, H) \\ P\{0, k|i, 0\} = \frac{(1-p)}{W_0}, k \in (0, W_0 - 1)\, i \in (0, H) \\ P\{i, k|i-1, 0\} = \frac{(p)}{W_i}, k \in (0, W_i - 1)\, i \in (1, H) \\ P\{H, k|H, 0\} = \frac{(p)}{W_H}, k \in (0, W_H - 1). \end{cases} \tag{6.3}$$

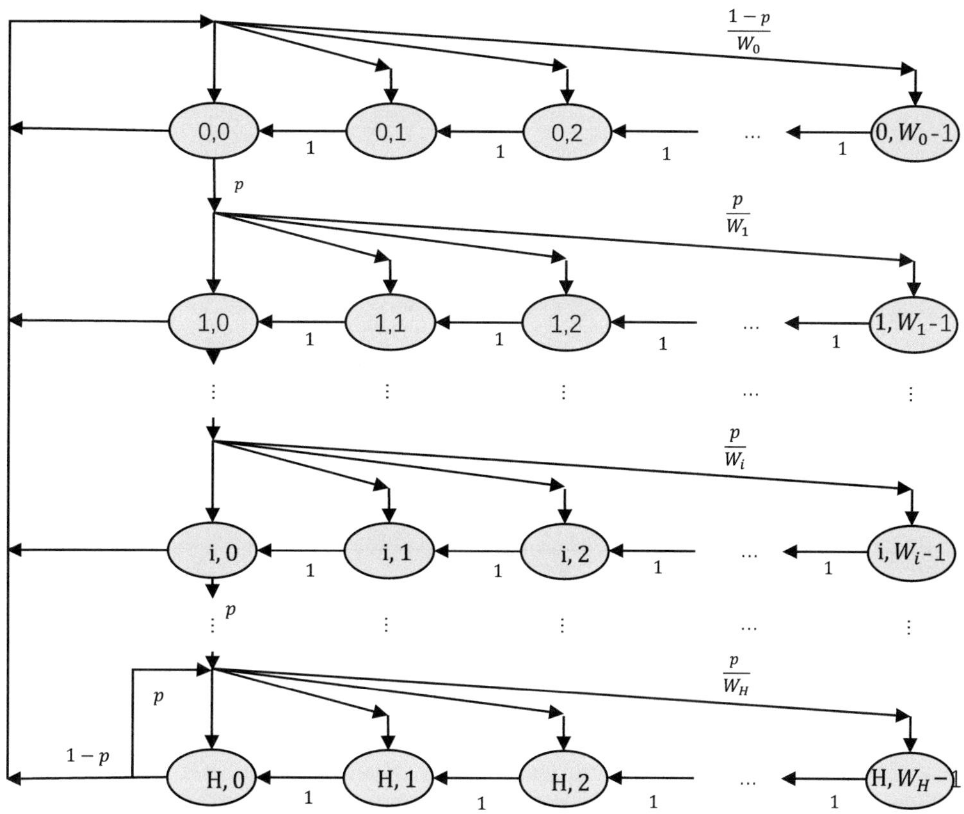

Fig. 6.1 State transition diagram of the Markov chain

Stationary probabilities $\pi_{i,k}$ can be obtained by $\lim_{n\to\infty} P^n$. We denote the probability of the node attempting to transmit data as τ, which can be expressed as:

$$\tau = \sum_{i=0}^{H} \pi_{i,0}, \tag{6.4}$$

since the node only transmits when its backoff counter has reduced to zero.

6.2 Access Protocol Design for Batteryless Communication Devices

Radio frequency (RF) signals are capable of simultaneously transferring data and energy from a hybrid access point (HAP) toward battery-powered and batteryless wireless devices. Battery-powered and batteryless wireless devices with the capability of RF energy harvesting need a distributed access control protocol with collision avoidance to achieve higher energy efficiency. We study the performance of a data and energy integrated network (DEIN) that adopts an enhanced carrier sensing multiple access with collision avoidance (CSMA/CA) protocol. Each device in this network can switch to RF energy harvesting mode or data reception mode according to HAP's instruction and freezes its backoff counter when energy storage is insufficient. By invoking a three-dimensional (3D) Markov chain, we model the operating behaviors of batteryless wireless devices and an HAP in a DEIN. Apart from backoff operations of devices, the 3D Markov chain also depicts their dynamic energy changes, including RF energy harvesting and energy consumption. Wireless devices consume energy harvested from the HAP's downlink transmissions for powering their data upload and random backoff. With the aid of the 3D Markov chain, the upload throughput of devices can be obtained in semi-closed-form. Moreover, a decoupling method is proposed to approximate throughput performance with low complexity. The accuracy of our theoretical model is validated by simulation results. By characterizing the impact of various parameters on throughput performance, a design guideline for a DEIN with a distributed batteryless access protocol is provided.

6.2.1 Introduction

6.2.1.1 Background and Motivation

Many Internet of Everything (IoE) devices are powered by embedded batteries with limited capacity. In order to maintain their regular operations, these batteries have to be frequently replaced, which substantially increases network maintenance cost. Harvesting energy from radio frequency (RF) signals [5, 6] is capable of prolonging the lifetime of wireless devices. Different from uncontrollable ambient energy harvesting [7–10], RF energy harvesting can be controlled by allowing base stations (BSs) and access points (APs) to actively emit RF signals, which is regarded as RF signal based wireless energy transfer (WET). Coordinating traditional wireless data transfer (WDT) together with wireless energy transfer yields a novel data and energy integrated network (DEIN) [11]. This is a promising technique to achieve energy sustainability in future 6G [12]. One of the typical applications of DEIN is to provide far-field recharging and communications services in wireless sensor networks [13].

In a DEIN [11], wireless devices can harvest RF energy from transmitters and thus achieve a longer lifetime. Many works focus on improving the efficiency of wireless data and energy transfer in DEIN from a physical layer aspect. For example, Clerckx et al. [14] designed optimal waveform to improve the efficiency of wireless energy transfer. Moreover, many works analyzed the network performance of DEIN [15, 16]. Lu et al. [15] analyzed the energy outage probability of a wireless device in a DEIN, where RF transmitters were distributed by following a Ginibre point process. Liu et al. [16] studied the impact of network deployment on the average power harvested by wireless devices in a multi-tier heterogeneous DEIN. However, the abovementioned works did not consider the impact of devices' access behaviors on the probabilities of energy transmitters transmitting RF energy, which further affect devices' RF energy harvesting performance. Moreover, access behaviors directly affect how much energy is consumed by wireless devices for data transmission, data reception, sleeping, and some other necessary operations. Therefore, access control protocols of battery-powered and batteryless wireless devices should be carefully considered for achieving energy sustainability in DEINs [17]. Moreover, we need accurate theoretical performance analysis in network planning stage of DEIN, before it can be practically deployed. Otherwise, engineers have to rely on time-consuming Monte-Carlo simulation for estimating network performance. Bianchi [18] first proposed a Markov chain based framework for analyzing the performance of a CSMA/CA protocol, which stimulated wide interest in relevant research. However, Bianchi's analytical modeling is only suitable for wireless devices with sufficient energy storage. In line with this motivation, our chapter aims to study the effect of access control behaviors on the performance of battery-powered and batteryless wireless devices in a DEIN system, while analyzing its performance in semi-closed-form.

6.2.1.2 Related Works

Centrally controlled access protocols have been widely adopted in DEINs [19, 20], in which central controllers may allocate orthogonal time or frequency resources to wireless devices based on their channel state information and their status reported. However, centrally controlled protocols require frequent information exchange between wireless devices and central controllers for time synchronization, channel state information feedback and resource allocation results. This may quickly consume the limited energy storage of wireless devices.

Apart from centrally controlled networks, such as cellular networks, many devices access the network for wireless data transfer in a distributed manner [21, 22]. Specifically, Liang et al. [21] proposed a duty cycle based access control protocol to achieve low latency for wireless devices by adjusting sleep windows. Zhi et al. [22] proposed a probabilistic polling based random access protocol for multi-hop RF energy harvesting aided wireless sensor networks. However, these random access protocols did not consider collision avoidance, which made wireless devices consume more energy for frequent retransmission. Their limited energy storage might be quickly drained.

The classic CSMA/CA protocol effectively reduces collision probabilities. Each device with a transmission request has to wait for a random period to avoid any collision. The length of this random waiting period is determined by backoff counter. The distributed and collision avoidance characteristics of CSMA/CA make it suited for battery-powered and batteryless devices. Based on Bianchi's [18] analysis framework, many works studied CSMA/CA based access control protocols for different networks with a Markov chain [23–26]. Specifically, Li et al. [23, 24] proposed a CSMA/CA based protocol for an air-ground integration network. They divided ground devices in the coverage into different clusters according to their communication duration. A 3D Markov chain was exploited to model backoff stages, the backoff counters, and the number of clusters. Li et al. [25] studied a block access control protocol in a blockchain-based wireless local-area network (B-WLAN). Moreover, Choi et al. [26] designed a contention window control scheme for the CSMA/CA protocol, while they analyzed its performance in terms of the collision probability, collision time, and backoff time. Unfortunately, these works only study the wireless devices which always have sufficient energy to support their operations, while totally ignoring either their practical energy consumption or their energy harvesting.

The CSMA/CA protocol was adopted for wireless devices enabled by RF energy harvesting capabilities [27, 28]. Naderi et al. [27] studied an enhanced CSMA/CA protocol named RF-MAC to optimize the energy delivery of wireless devices. Wireless devices could actively request energy from HAP. By responding to these requests, HAP recharged these wireless devices on demand. Based on RF-MAC, Tamilarasi et al. [28] studied a DEIN system that allocated the same rate to all wireless devices regardless of their different channel states. However, they did not consider theoretical analysis, all performances were derived by time-consuming Monte-Carlo simulation.

The Bianchi [18]'s Markov chain based analysis framework was also adopted for modeling the behaviors of wireless devices enabled by RF energy harvesting capabilities [3, 29–31]. Specifically, Arshad et al. [29] studied a wireless powered communication network assisted by a relay, where the relay transmitted both wireless data and energy to wireless devices. A 2D Markov chain was relied upon for modeling a distributed access control protocol of wireless devices. However, the energy storage of devices was inaccurately simplified as several discrete states, while the zero-energy state was completely ignored. In practice, a batteryless wireless device inevitably depletes its limited energy storage. In this situation, it cannot transmit any data packets or take any backoff operations, which further reduces its throughput performance. In this state, wireless devices cannot transmit or backoff, which further affects throughput performance. Khairy et al. [3] studied the AP's beacon frequency and the recharging period of wireless devices in Wi-Fi based on Bianchi's analytical results. However, battery-powered and batteryless wireless devices suffer both collision and energy insufficient problems, which further affect the data transmission performance of devices. Therefore, Bianchi's analytical modeling is only suitable for wireless devices with sufficient energy storage. In [30, 31], the throughput performance of a CSMA/CA protocol based network with batteryless wireless devices was analyzed with an incomplete energy consumption model. Only the energy consumption for data upload was considered, while the energy consumption for the random backoff process was totally ignored. Moreover, they impractically assume that a wireless device depleted all its energy storage for uploading a single data packet.

6.2.1.3 Novel Contributions

Against the drawbacks we mentioned above, we considered a DEIN system with an RF energy harvesting enhanced CSMA/CA protocol and further analyze the performance of batteryless devices with this protocol. To the best of our knowledge, no studies accurately characterize dynamic energy states and the wireless data transfer process of a batteryless wireless device with a distributed access control protocol. The main contributions are summarized as follows:

- We analyze the performance of wireless devices with an RF energy harvesting enhanced CSMA/CA protocol in a DEIN, where each device may switch to RF energy harvesting mode or data reception mode according to HAP's instruction and freeze its backoff counter when its energy storage is insufficient. All these devices' data uploading is powered by energy harvested from the downlink transmission of the HAP.
- A 3D Markov chain is exploited to model the operations of wireless devices. This 3D Markov chain accurately characterizes RF energy harvesting of wireless devices, and their energy consumption for data upload and for collision avoidance oriented backoff process, which is never considered in existing works.
- In order to reduce the complexity, the 3D Markov chain is decoupled in the data-domain and the energy domain which help us derive approximated access probabilities and the throughput performance of both the HAP and devices in semi-closed-form.

The rest of this chapter is organized as follows. After introducing the system model in Sect. 6.2.2, our 3D Markov chain based modeling and the theoretical analysis are described in Sect. 6.2.3. We then provide simulation results in Sect. 6.2.4. Unless specified, battery-powered and batteryless wireless devices are referred to as devices in the rest of this chapter.

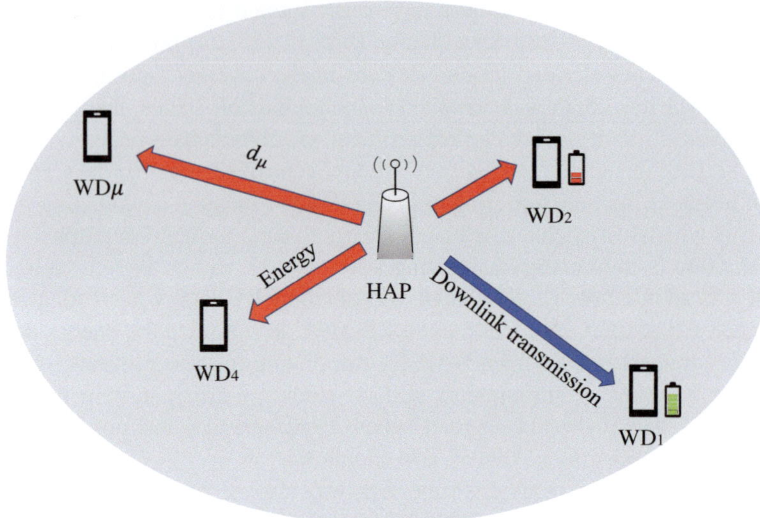

Fig. 6.2 System model

6.2.2 *Batteryless Distributed Access Control Protocol*

6.2.2.1 System Model

As shown in Fig. 6.2, our DEIN system consists of a single HAP with a stable energy source, and N batteryless wireless devices (WD), which are indexed by an integer $\mu \in \{0, 1, \ldots, N\}$. Wireless devices are randomly deployed around the HAP, and the distance between WD$_\mu$ and HAP is d_μ. Note that all devices are equipped with supercapacitors for energy storage. Wireless devices can only communicate with the HAP. Direct communications among themselves are not allowed. All devices and HAP are equipped with a single antenna. HAP always transmits with its maximum transmit power P_{AP}, and the WD$_\mu$'s transmit power depends on its distance to the HAP. The physical layer model and MAC layer model of the system are separately described in the following subsections.

The channel model is constituted by a distance-dependent path-loss and an additional random term ζ_μ accounting for small-scale fading. A similar channel model is used in [32]. The instantaneous channel power gain between the HAP and the WD$_\mu$ can be then expressed as

$$h_\mu = G_R G_T \left(\frac{c}{4\pi f_0 d_0}\right)^2 \left(\frac{d_0}{d_\mu}\right)^\beta \zeta_\mu, \tag{6.5}$$

where G_T and G_R are the transmit and receive antennas' gains, c is the speed of the light, and β is the path-loss exponent. Moreover, in Eq. (6.5), f_0 is the carrier frequency. Furthermore, d_0 represents the near-field range of the HAP. $\zeta_\mu \sim \text{Exp}(1)$ is an exponential random variable with unit mean accounting for small-scale Rayleigh fading. Due to the duration of the require to send (RTS) and clear to send (CTS) packets is very small, the HAP can only transfer sufficient energy toward devices during the data packet transmissions. We consider the device as a time switching RF energy harvesting receiver [11]. When the device is harvesting energy, it will not receive any packets. The duration of the data packet transmission is denoted as T_{PCK}. By considering a nonlinear energy harvesting model [33, 34], the energy harvested by WD$_\mu$ can be expressed as

$$E_\mu^H = \left[\frac{P_{\max}}{\exp(-AP_0 + B)}\left(\frac{1 + \exp(-AP_0 + B)}{1 + \exp(-AP_{AP}\overline{h}_\mu + B)} - 1\right)\right]\text{T}_{\text{PCK}}, \tag{6.6}$$

where $\overline{h}_\mu = \mathbf{E}[h_\mu]$ is the average channel power gain; $P_{\max}$ is the maximum harvestable power when RF energy harvesting circuit is saturated; P_0 is the harvester's sensitivity threshold; A and B are both constants related to the RF energy harvesting circuit's resistance, capacitance and diode's turn-on voltage.

Let us now consider the data packet uploading of WD_μ. Only when the signal-to-noise-ratio (SNR) is higher than a predefined threshold γ_{th}, the data packet transmitted from WD_μ can be correctly decoded by the HAP. We denote the energy consumption of WD_μ for its data packet transmission E_μ^{PCK}, which should satisfy the following outage constraint as:

$$P_r\left(\frac{E_\mu^{\mathrm{PCK}}h_\mu}{BN_0\mathrm{T_{PCK}}}\leq\gamma_{\mathrm{th}}\right)\leq\xi, \tag{6.7}$$

where B is bandwidth, N_0 is the power spectral density of the additive white Gaussian noise (AWGN), and ξ is the symbol error rate (SER) threshold which the outage probability should satisfy. Therefore, by solving Eq. (6.7), E_μ^{PCK} can be obtained as

$$E_\mu^{\mathrm{PCK}}=\frac{BN_0\mathrm{T_{PCK}}\gamma_{\mathrm{th}}}{F_{h_\mu}^{-1}(\xi)}, \tag{6.8}$$

where $F_{h_\mu}^{-1}(.)$ represents the inverse cumulative distribution function of the channel power coefficient h_μ. Note that Eq. (6.8) is used for deriving the energy consumption of data packet transmission. By considering path-loss, a device with data transmission requests chooses a proper transmit power to ensure that all data packets can be successfully transmitted in uplink. HAP always transmits with its maximum power P_{AP}, which also results in a zero packet loss rate in the downlink data packet transmission.

6.2.2.2 Protocol Description

All wireless devices and the HAP operate at the same frequency, only one device (or HAP) is allowed to transmit at the same time. When a device gains the highest priority to access the channel, it also has to have sufficient energy for powering its data transmission. The general operation of a device in this protocol is summarized as the following steps:

- *Step 1*: A device with data packets to be transmitted senses the channel.
- *Step 2*: When the channel is free, the device initializes its backoff stage as $i=0$. It also initializes the backoff counter as a random number within $[0, W_0-1]$, where W_0 represents the initial backoff window size.
- *Step 3*: The backoff counter reduces by one during every backoff interval, if the channel is still free. Otherwise, the device freezes its backoff counter. Moreover, the device continuously consumes energy during the backoff process. If the energy storage of the device is not sufficient for backoff, it freezes the backoff counter until sufficient energy is harvested.
- *Step 4*: After the backoff counter reduces to zero, it starts its RTS/CTS handshake for data packet transmissions if it has gained sufficient energy for data packet transmissions.
- *Step 5*: If the transmission suffers from a collision, the device then increases the backoff stage to $i=i+1$, while initializing the backoff counter as a random number within $[0, W_i-1]$, where we have $W_i=2W_{i-1}$.
- *Step 6*: When the device receives an RTS packet from the HAP, it switches to data packet reception mode if it is the destination of this RTS packet. Otherwise, it switches to the RF energy harvesting mode. It then switches back to the data packet reception mode, once the HAP finishes its data packet transmission.

Moreover, when the HAP transmits data packets toward a device, the other devices harvest energy carried by downlink RF signals. The energy harvested can be obtained by Eq. (6.6). A device can operate either in a data packet reception mode or in an RF energy harvesting mode in a time switching manner. The RTS packet can also be regarded as a mode-switching instruction. When WD_μ receives an RTS packet from HAP, it switches to the RF energy harvesting mode, if it is not the destination of this RTS packet. Otherwise, it switches to the data packet reception mode instead. Only when the received RF power satisfies the threshold of the device's energy harvester, can the device convert the RF energy into DC energy. We assume that devices can only harvest energy from the downlink transmission of the HAP, they cannot harvest energy from their peers.

We also exemplify the energy dynamics of the devices with our distributed batteryless access control protocol in Fig. 6.3, where we have a single HAP and two devices. Figure 6.3 is a simplified schematic diagram, that we do not consider a realistic model for energy harvesting and energy consumption. The duration of each stage is also not proportional in Fig. 6.3. During Stage 1, the channel is free, all devices are taking backoff operations and continuously consume energy for backoff and channel estimation. At the end of Stage 1, the HAP's backoff counter first reduces to zero. It then starts to communicate

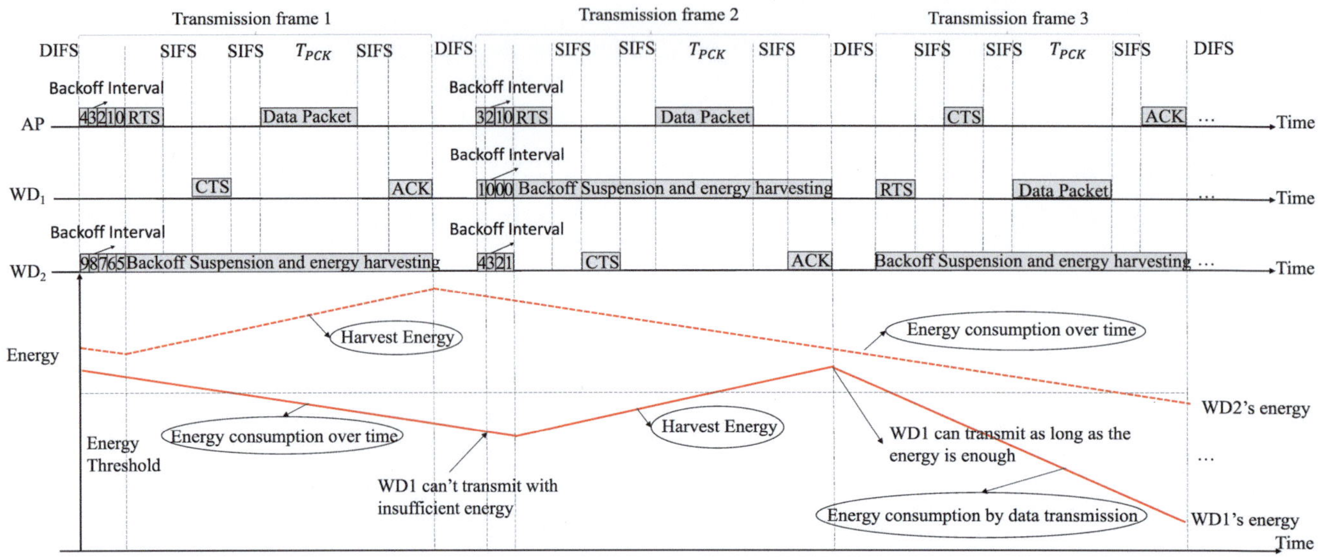

Fig. 6.3 Protocol description

with WD_2 after the RTS/CTS handshaking. During Stage 2, WD_1 switches to the RF energy harvesting mode and harvests energy from the downlink transmission of the HAP to replenish its energy storage, while WD_2's energy storage reduces for powering its own data packet reception. Note that WD_2 cannot harvest energy at this stage because it switches to the data packet reception mode rather than the RF energy harvesting mode. After a duration of T_{PCK}, all devices restart their backoff process. At the end of Stage 2, although WD_2's backoff counter reduces to zero, it cannot transmit any data packet because its energy storage is lower than the minimum requirement for data packet uploading. Then after several backoff intervals, the HAP also has a zero backoff counter, and it is allowed for downlink data packet transmissions. During Stage 4, WD_2 may harvest energy, while WD_1 consumes its energy storage for data packet reception. As shown in Fig. 6.2, at the end of Stage 4, WD_2 harvests sufficient energy to support its data packet uploading. Since its backoff counter is zero, it can immediately commence the data packet uploading process during Stage 5, which may consume the amount E_{PCK} of WD_2's energy storage. Observe from Fig. 6.1 that, WD_1 cannot harvest energy during Stage 5, when WD_2 uploads its data packet. Therefore, WD_1 only consumes energy for the backoff process.

6.2.3 Modeling and Throughput Analysis

All devices and the HAP are assumed to have full buffers. Therefore, they always have data packets pending to be transmitted. Moreover, the HAP has equal probabilities to transmit data packets to all devices. Under this assumption, we derive all the upper-bounded system performance of the distributed batteryless access control protocol.

6.2.3.1 3D Markov Modeling

In order to obtain the transmission probabilities of the devices and the HAP, a specific device's operational behaviors can be modeled as a discrete 3D Markov chain, as shown in Fig. 6.4. A tuple $[s(t), c(t), e(t)]$ with three elements defines the state of a device during the t-th time slot. Specifically, $s(t) = i \in [0, H]$ represents the backoff stage, $c(t) = j \in [0, W_i - 1]$ represents the backoff counter, while $e(t) = e \in [0, \overline{E}_{\mu}^{\max}]$ represents the energy state of WD_{μ}. Actually, the length of a time slot is not a fixed value and is related to WD_{μ}'s operations, which include successful transmission, transmission collision, and no transmission. We denote the length of a time slot in these three operations as T_s, T_c, and T_f, respectively. When WD_{μ} does not transmit any data packets (no transmission), it may take a backoff operation or take none operation if it does not have any energy storage. Moreover, we denote the average length of a time slot as $\mathbf{E}[time - slot]$. We introduce a discrete energy unit E_{δ}. A positive integer $\overline{E}_{\mu}^{\max} = Q/E_{\delta}$ is defined as the maximum discrete energy capacity, where Q is the actual capacity of the energy storage. The discrete energy threshold for data packet transmissions can be expressed

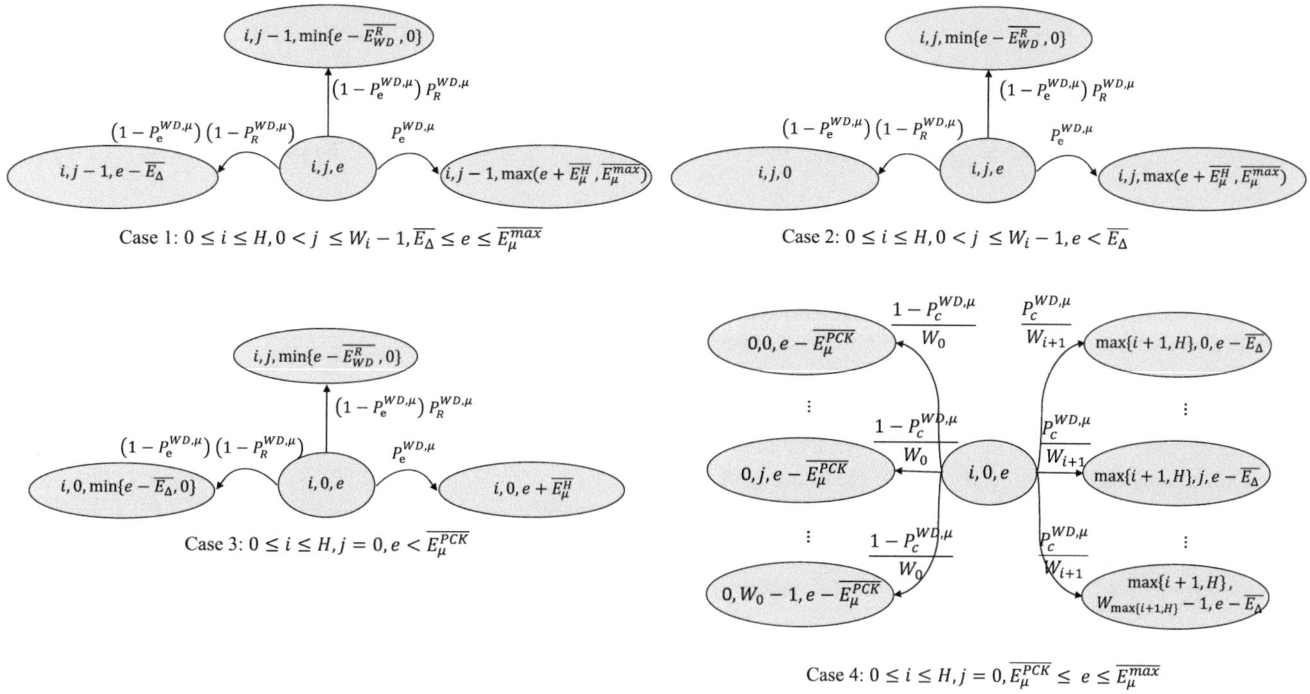

Fig. 6.4 3D Markov chain model of WD_μ

as $\overline{E_\mu}^{\mathrm{PCK}} = \lceil E_\mu^{\mathrm{PCK}}/E_\delta \rceil$, while the discrete energy harvested by WD_μ can be expressed as $\overline{E_\mu}^{H} = \lceil E_\mu^{H}/E_\delta \rceil$. We denote the average energy consumption of the backoff operation and channel detection operation of WD_μ in a transmission frame as E_Δ. Its discrete version can be expressed as $\overline{E}_\Delta = \lceil E_\Delta/E_\delta \rceil$. Moreover, when a device receives a packet from HAP, we denote that it will consume E_{WD}^{R} energy, which should be proportional to the length of $\mathrm{T_{PCK}}$, and its discrete version is expressed as $\overline{E}_{\mathrm{WD}}^{R} = \lceil E_{\mathrm{WD}}^{R}/E_\delta \rceil$. The probability of a device harvesting energy when it is not transmitting packets is denoted as $P_e^{\mathrm{WD},\mu}$. If a device has sufficient energy storage and its backoff counter reduces to zero, it may send an RTS packet to the HAP. However, concurrent transmission attempts of multiple devices may induce collisions. We denote the probability of transmission collisions when WD_μ transmits a packet as $P_c^{\mathrm{WD},\mu}$. When WD_μ is not transmitting, the probability of WD_μ receiving a packet from the HAP is denoted as $P_R^{\mathrm{WD},\mu}$. Note that $P_c^{\mathrm{WD},\mu}$, $P_e^{\mathrm{WD},\mu}$ and $P_R^{\mathrm{WD},\mu}$ can be calculated in Sect. 6.2.3.4. Moreover, all these probabilities of WD_μ are jointly determined by the behaviors of the other devices and the HAP. Therefore, they are independent of the device's own states.

When the backoff counter j of WD_μ does not reduce to zero, it cannot transmit any data packets, as shown in Cases 1–2 of Fig. 6.4. Specifically, in Case 1 of $0 \le i \le H$, $0 < j \le W_i - 1$, and $\overline{E}_\Delta \le e \le \overline{E_\mu}^{\max}$, WD_μ's backoff counter j reduces by one. It may either harvest $\overline{E_\mu}^{H}$ energy with a probability of $P_e^{\mathrm{WD},\mu}$ or consume energy for data reception or backoff operation. The HAP may choose WD_μ as the packet destination. Therefore, WD_μ may consume $\overline{E}_{\mathrm{WD}}^{R}$ energy with a probability of $(1 - P_e^{\mathrm{WD},\mu})P_R^{\mathrm{WD},\mu}$. If the HAP may not choose WD_μ as the packet destination, WD_μ may consume $\overline{E}_\Delta$ energy for backoff operation with a probability of $(1 - P_e^{\mathrm{WD},\mu})(1 - P_R^{\mathrm{WD},\mu})$. Moreover, the energy state e of WD_μ can only transit to zero, when WD_μ does not have sufficient energy to consume. In Case 2 of $e < \overline{E}_\Delta$, the WD_μ freezes its backoff counter because its energy is exhausted. The transition duration of each state is T_s, when WD_μ harvests energy or receives a packet. It is T_f, when WD_μ only takes a single backoff operation.

When the backoff counter j reduces to 0, WD_μ still cannot transmit any data packets with insufficient energy storage, as shown in Cases 3 of Fig. 6.4. Normally, data packet transmissions have a longer duration $\mathrm{T_{PCK}}$ than the backoff intervals. Therefore, the energy consumption $\overline{E_\mu}^{\mathrm{PCK}}$ is higher than $\overline{E}_\Delta$. Moreover, due to the path-loss of RF signals' propagation, the energy harvested by WD_μ should be much lower than $E_\mu^{\max}$. Without loss of generality, we assume $\overline{E_\mu}^{\mathrm{PCK}} + \overline{E_\mu}^{H} < \overline{E_\mu}^{\max}$. Therefore, in Case 3 of $0 \le i \le H$, $j = 0$, and $e < \overline{E_\mu}^{\mathrm{PCK}}$, WD_μ cannot transmit any data packets. We should consider the following three situations: 1) It may harvest $\overline{E_\mu}^{H}$ energy with a probability of $P_e^{\mathrm{WD},\mu}$; 2) It consumes $\overline{E}_\Delta$ energy with a probability of $(1 - P_e^{\text{WD},\mu})(1 - P_R^{\text{WD},\mu})$; 3) It may consume $\overline{E}_{\text{WD}}^R$ energy with a probability of $(1 - P_e^{\text{WD},\mu})P_R^{\text{WD},\mu}$. Moreover, the energy storage e of WD_μ can only reduce to 0 without harvesting any energy, when $e < \overline{E}_\Delta$. The transition duration of each state is T_s, when WD_μ harvests energy or receives a packet, while it is T_f when WD_μ only takes a single backoff operation.

When the backoff counter j reduces to 0 and WD_μ's energy storage is higher than E_μ^{PCK}, WD_μ transmits data packets in Cases 3 of Fig. 6.4. When no collision happens with a probability of $(1 - P_c^{\text{WD},\mu})$, WD_μ sets its backoff stage i to 0 and initializes its backoff counter j to a random number within $[0, W_0 - 1]$. WD_μ then consumes $\overline{E}_\mu^{\text{PCK}}$ energy for a data packet transmission. When the collision happens with a probability of $P_c^{\text{WD},\mu}$, WD_μ increases its backoff stage i by one, and chooses a random number within the range of $[0, W_{i+1} - 1]$ as its backoff counter j. It then consumes $\overline{E}_\Delta$ energy for the backoff process. When the backoff stage reaches the maximum H, it may not increase anymore. The transition duration of each state is T_s, when no collision happens, otherwise, it is T_c.

Observe from Fig. 6.4 that the transition probabilities between every pair of states in the 3D Markov chain are related to the collision probability $P_c^{\text{WD},\mu}$, reception probability $P_R^{\text{WD},\mu}$ and energy harvesting probability $P_e^{\text{WD},\mu}$. Therefore, the stationary probabilities of the 3D Markov chain are determined by $P_c^{\text{WD},\mu}$ $P_R^{\text{WD},\mu}$ and $P_e^{\text{WD},\mu}$. Moreover, these probabilities are also determined by the stationary probabilities and vice versa. In order to calculate $P_c^{\text{WD},\mu}$, $P_R^{\text{WD},\mu}$, $P_e^{\text{WD},\mu}$ and all the stationary probabilities, we first fix $P_c^{\text{WD},\mu}$, $P_R^{\text{WD},\mu}$ and $P_e^{\text{WD},\mu}$ to calculate the stationary probabilities and then use the calculated stationary probabilities to update $P_c^{\text{WD},\mu}$, $P_R^{\text{WD},\mu}$ and $P_e^{\text{WD},\mu}$. However, the 3D Markov chain has $\overline{E}_\mu^{\max} W_0(2^{H+1} - 1)$ states. The stationary probabilities can be denoted as an $\overline{E}_\mu^{\max} W_0(2^{H+1} - 1) \times 1$ vector Π. The state transition probabilities can be denoted as an $\overline{E}_\mu^{\max} W_0(2^{H+1} - 1) \times \overline{E}_\mu^{\max} W_0(2^{H+1} - 1)$ matrix P which can be derived by Fig. 6.4. Π can be obtained by solving the following equations:

$$\begin{cases} \Pi P = \Pi, \\ \sum \Pi = 1. \end{cases} \tag{6.9}$$

The complexity for solving Eq. (6.9) is $\mathcal{O}((\overline{E}_\mu^{\max} W(2^{H+1} - 1)^3))$. In order to reduce the complexity, the original 3D Markov chain is decoupled into an independent 2D counterpart for characterizing WD_μ's behavior in the data packet transmission and an independent 1D counterpart for reflecting WD_μ's dynamic energy storage.

6.2.3.2 Analysis in the Data Packet Transmission Domain

The state of WD_μ is represented by $[s(t) = i, c(t) = j]$ in the data packet transmission domain. Apart from the transmission collision probability $P_c^{\text{WD},\mu}$, we denote the probability of WD_μ having no energy at all as $b_n^{\text{WD},\mu}$, while denoting the probability of the WD_μ having sufficient energy for data packet transmissions as $b_e^{\text{WD},\mu}$. Furthermore, the probability of the WD_μ having sufficient energy for data packet receptions is denoted as $b_r^{\text{WD},\mu}$.

As depicted in Fig. 6.5, in the case of $0 \le i \le H$ and $j > 0$, the backoff counter continues to reduce by one, if WD_μ has some energy with a probability of $(1 - b_n^{\text{WD},\mu})$. WD_μ's state may transit to $[s(t), c(t) - 1]$ from $[s(t), c(t)]$. By contrast, WD_μ freezes its backoff counter if it does not have any energy with a probability of $b_n^{\text{WD},\mu}$. WD_μ may maintain its current state $[s(t), c(t)]$.

When the backoff counter reduces to zero, WD_μ is ready for data packet transmission. If it has sufficient energy and no collision happens, the data packet transmission is successful. WD_μ then sets its backoff stage to zero, while choosing a random number within the range of $[0, W_0 - 1]$ as its backoff counter. As depicted in Fig. 6.5, in the case of $0 \le i \le H$ and $j = 0$, the probability of a device successfully transmitting a data packet can be formulated as $b_e^{\text{WD},\mu}(1 - P_c^{\text{WD},\mu})$. WD_μ's state may transit to $[s(t) = 0, c(t) = j]$ from $[s(t), c(t) = 0]$, while j is a random number uniformly chosen from $[0, W_0 - 1]$. Since WD_μ uniformly chooses a number as its backoff counter, the state transition to every $[s(t) = 0, c(t) = j]$ for $\forall j \in [0, W_0 - 1]$ has the same probability of $\frac{b_e^{\text{WD},\mu}(1 - P_c^{\text{WD},\mu})}{W_0}$.

By contrast, if the data packet transmission fails, WD_μ increases its backoff stage by one and chooses a random number within the range of $[0, W_i - 1]$ as its backoff counter. As depicted in Fig. 6.5, in the case of $0 \le i < H$ and $j = 0$, the probability that a device has sufficient energy but some collisions happen can be formulated as $b_e^{\text{WD},\mu} P_c^{\text{WD},\mu}$. Similarly, since the backoff counter j is uniformly chosen, WD_μ's state may transit to $[s(t) = i + 1, c(t) = j]$ from $[s(t) = i, c(t) = 0]$

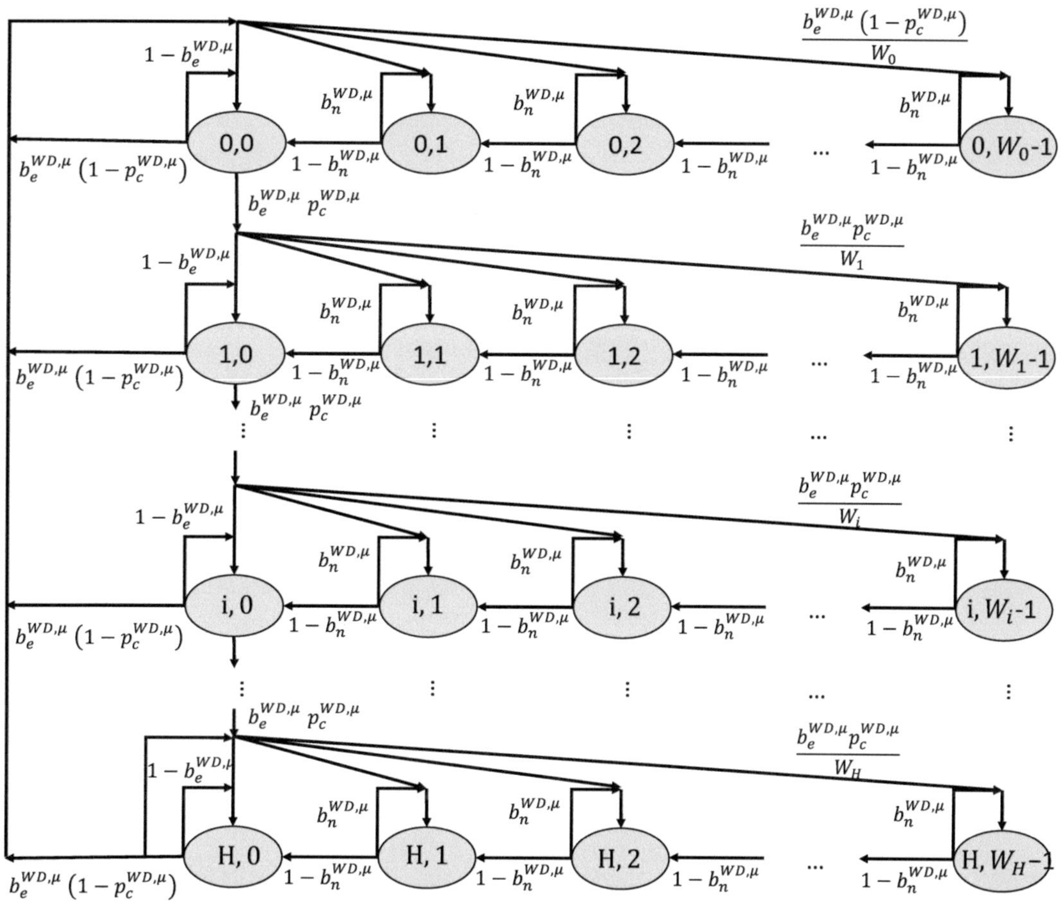

Fig. 6.5 2D Markov chain in the data transmission domain

with the same probability of $\frac{b_e^{\mathrm{WD},\mu} P_c^{\mathrm{WD},\mu}}{W_i}$. When the backoff stage reaches the maximum H, it may not increase anymore. In the case of $i = H$ and $j = 0$ as shown in Fig. 6.5, WD_μ's state may transit to $[s(t) = H, 0 \le c(t) = j \le W_H - 1]$ from $[s(t) = H, c(t) = 0]$ with the same probability of $\frac{b_e^{\mathrm{WD},\mu} P_c^{\mathrm{WD},\mu}}{W_H}$.

If WD_μ does not have sufficient energy, when the backoff counter reduces to zero, it maintains its backoff stage. As depicted in Fig. 6.5, in the case of $0 \le i \le H$ and $j = 0$, WD_μ may maintain its state $[s(t) = i, c(t) = 0]$ with a probability of $(1 - b_e^{\mathrm{WD},\mu})$.When the backoff counter reduces to zero, WD_μ is ready for data packet transmission. If it has sufficient energy and no collision happens, the data packet transmission is successful. WD_μ then sets its backoff stage to zero, while choosing a random number within the range of $[0, W_0 - 1]$ as its backoff counter. As depicted in Fig. 6.5, in the case of $0 \le i \le H$ and $j = 0$, the probability of a device successfully transmitting a data packet can be formulated as $b_e^{\mathrm{WD},\mu}(1 - P_c^{\mathrm{WD},\mu})$. WD_μ's state may transit to $[s(t) = 0, c(t) = j]$ from $[s(t), c(t) = 0]$, while j is a random number uniformly chosen from $[0, W_0 - 1]$. Since WD_μ uniformly chooses a number as its backoff counter, the state transition to every $[s(t) = 0, c(t) = j]$ for $\forall j \in [0, W_0 - 1]$ has the same probability of $\frac{b_e^{\mathrm{WD},\mu}(1-P_c^{\mathrm{WD},\mu})}{W_0}$.

By contrast, if the data packet transmission fails, WD_μ increases its backoff stage by one and chooses a random number within the range of $[0, W_i - 1]$ as its backoff counter. As depicted in Fig. 6.5, in the case of $0 \le i < H$ and $j = 0$, the probability that a device has sufficient energy but some collisions happen can be formulated as $b_e^{\mathrm{WD},\mu} P_c^{\mathrm{WD},\mu}$. Similarly, since the backoff counter j is uniformly chosen, WD_μ's state may transit to $[s(t) = i + 1, c(t) = j]$ from $[s(t) = i, c(t) = 0]$ with the same probability of $\frac{b_e^{\mathrm{WD},\mu} P_c^{\mathrm{WD},\mu}}{W_i}$. When the backoff stage reaches the maximum H, it may not increase anymore. In the case of $i = H$ and $j = 0$ as shown in Fig. 6.5, WD_μ's state may transit to $[s(t) = H, 0 \le c(t) = j \le W_H - 1]$ from $[s(t) = H, c(t) = 0]$ with the same probability of $\frac{b_e^{\mathrm{WD},\mu} P_c^{\mathrm{WD},\mu}}{W_H}$.

If WD$_\mu$ does not have sufficient energy, when the backoff counter reduces to zero, it maintains its backoff stage. As depicted in Fig. 6.5, in the case of $0 \le i \le H$ and $j = 0$, WD$_\mu$ may maintain its state $[s(t) = i, c(t) = 0]$ with a probability of $(1 - b_e^{\mathrm{WD},\mu})$. We denote the stationary probabilities of the 2D Markov chain in the data packet transmission domain as $\{\pi_{i,j}^D | \forall i = s(t), j = e(t)\}$. According to Fig. 6.5, these stationary probabilities obey the following equations:

$$\begin{cases} \pi_{i,0}^D = \pi_{i,0}^D(1 - b_e^{\mathrm{WD},\mu}) + \sum_{j=0}^{W_i-1} \pi_{i-1,0}^D \frac{b_e^{\mathrm{WD},\mu} P_c^{\mathrm{WD},\mu}}{W_i} (1 - b_n^{\mathrm{WD},\mu})^j, \\ \qquad 1 \le i \le H-1, \\ \pi_{H,0}^D = \pi_{H,0}^D(1 - b_e^{\mathrm{WD},\mu}) + \sum_{j=0}^{W_H-1} \pi_{H-1,0}^D \frac{b_e^{\mathrm{WD},\mu} P_c^{\mathrm{WD},\mu}}{W_H} (1 - b_n^{\mathrm{WD},\mu})^j \\ \quad + \sum_{j=0}^{W_H-1} \pi_{H,0}^D \frac{b_e^{\mathrm{WD},\mu} P_c^{\mathrm{WD},\mu}}{W_H} (1 - b_n^{\mathrm{WD},\mu})^j, \\ \qquad i = H. \end{cases} \tag{6.10}$$

Equation (6.10) can be further simplified as

$$\begin{cases} \pi_{i,0}^D = \sum_{j=0}^{W_i-1} \pi_{i-1,0}^D \frac{P_c^{\mathrm{WD},\mu}}{W_i} (1 - b_n^{\mathrm{WD},\mu})^j, & 1 \le i \le H-1, \\ \pi_{H,0}^D = \sum_{j=0}^{W_H-1} \pi_{H-1,0}^D \frac{P_c^{\mathrm{WD},\mu}}{W_H} (1 - b_n^{\mathrm{WD},\mu})^j & \\ \quad + \sum_{j=0}^{W_H-1} \pi_{H,0}^D \frac{P_c^{\mathrm{WD},\mu}}{W_H} (1 - b_n^{\mathrm{WD},\mu})^j, & i = H. \end{cases} \tag{6.11}$$

By recursively solving Eq. (6.11), we may derive

$$\begin{cases} \pi_{i,0}^D = \pi_{0,0}^D \prod_{m=1}^{i} \sum_{j=0}^{W_m-1} \frac{P_c^{\mathrm{WD},\mu}}{W_m} (1 - b_n^{\mathrm{WD},\mu})^j, & \\ \qquad 1 \le i \le H-1. & \\ \pi_{H,0}^D = \dfrac{\pi_{H-1,0}^D \sum_{j=0}^{W_H-1} \frac{P_c^{\mathrm{WD},\mu}}{W_H} (1-b_n^{\mathrm{WD},\mu})^j}{1 - \sum_{j=0}^{W_H-1} \frac{P_c^{\mathrm{WD},\mu}}{W_H} (1-b_n^{\mathrm{WD},\mu})^j}, & i = H. \end{cases} \tag{6.12}$$

Similarly, we may also derive the following equations according to Fig. 6.5:

$$\begin{cases} \pi_{0,j}^D = \sum_{m=0}^{W_0-j} \dfrac{\sum_{i=0}^{H} \pi_{i,0}^D b_e^{\mathrm{WD},\mu} (1-P_c^{\mathrm{WD},\mu})(1-b_n^{\mathrm{WD},\mu})^{m-1}}{W_0}, \\ \qquad 1 \le j \le W_0 - 1, \\ \pi_{i,j}^D = \pi_{0,0}^D \prod_{n=1}^{i} \sum_{m=0}^{W_i-j} \dfrac{b_e^{\mathrm{WD},\mu} P_c^{\mathrm{WD},\mu} (1-b_n^{\mathrm{WD},\mu})^{m-1}}{W_n}, \\ \qquad 1 \le i \le H, 1 \le j \le W_i. \end{cases} \tag{6.13}$$

Therefore, we can express all the stationary probabilities with $\pi_{0,0}^D$. By exploiting $\sum \pi_{i,j}^D = 1$, we may finally solve $\pi_{0,0}^D$ and derive all the stationary probabilities by leveraging Eqs. (6.12) and (6.13). We denote the probability of WD$_\mu$ attempting to transmit data as $\tau^{\mathrm{WD},\mu}$, which can be expressed as:

$$\tau^{\mathrm{WD},\mu} = b_e^{\mathrm{WD},\mu} \sum_{i=0}^{H} \pi_{i,0}^D, \tag{6.14}$$

since WD$_\mu$ only transmits when it has sufficient energy and its backoff counter has reduced to zero. Therefore, $\tau^{\mathrm{WD},\mu}$ can be calculated, when $P_c^{\mathrm{WD},\mu}$, $b_n^{\mathrm{WD},\mu}$, and $b_e^{\mathrm{WD},\mu}$ are given. Note that $b_n^{\mathrm{WD},\mu}$ and $b_e^{\mathrm{WD},\mu}$ can be obtained by solving the 1D Markov chain in the energy domain.

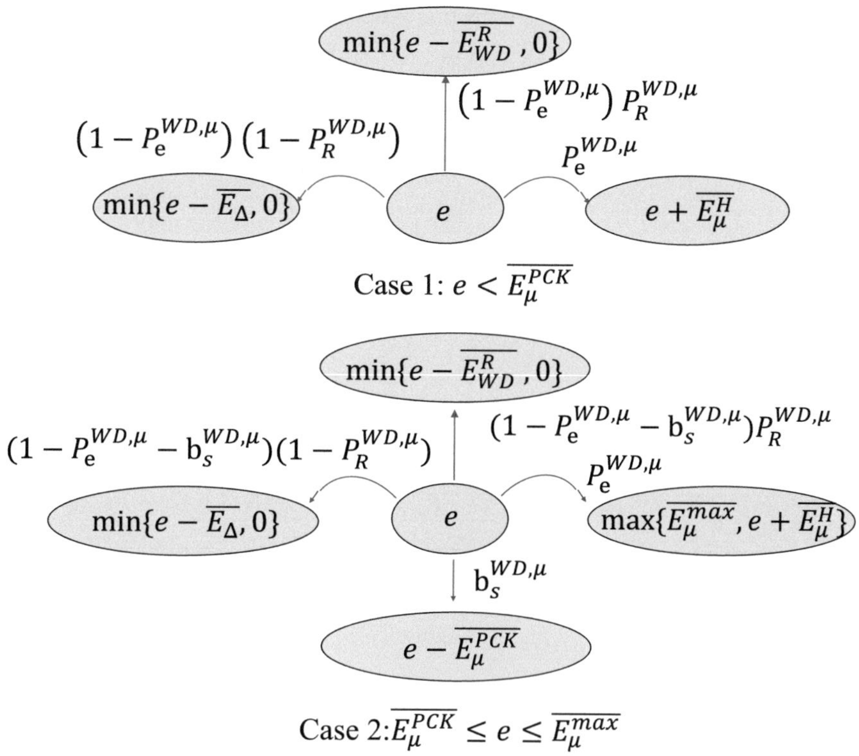

Fig. 6.6 1D Markov chain in the energy domain

6.2.3.3 Analysis in the Energy Domain

According to the analysis in the data transmission domain, when the Markov chain of Fig. 6.5 becomes stationary, we can formulate the probability $b_s^{\mathrm{WD},\mu}$ of WD_μ successfully transmitting a packet as:

$$b_s^{\mathrm{WD},\mu} = (1 - P_c^{\mathrm{WD},\mu}) \sum_{i=0}^{H} \pi_{i,0}^{D} = \frac{\tau^{\mathrm{WD},\mu}(1 - P_c^{\mathrm{WD},\mu})}{b_e^{\mathrm{WD},\mu}}. \tag{6.15}$$

The state of WD_μ is denoted as $[e(t) = e]$ in the energy domain. Furthermore, we still assume that $\overline{E_\mu}^{\mathrm{PCK}} > \overline{E}_\Delta$ and $\overline{E_\mu}^{\mathrm{PCK}} < E_\mu^{\max} - \overline{E_\mu}^H$. The state transitions of WD_μ's energy storage can be summarized into a Markov chain with 4 cases, as depicted in Fig. 6.6.

If WD_μ's energy storage is lower than $\overline{E_\mu}^{\mathrm{PCK}}$, as shown in Case 1 of Fig. 6.6, it does not have sufficient energy for data packet transmissions. Specifically, in Case 1 of $e < \overline{E_\mu}^{\mathrm{PCK}}$, WD_μ may harvest $\overline{E_\mu}^H$ energy with a probability of $P_e^{\mathrm{WD},\mu}$. WD_μ's state then transits to $[e + \overline{E_\mu}^H]$. Otherwise, the WD_μ may consume $\overline{E}_\Delta$ energy with a probability of $(1 - P_e^{\mathrm{WD},\mu})(1 - P_R^{\mathrm{WD},\mu})$, or it may consume $\overline{E}_{\mathrm{WD}}^R$ energy with a probability of $(1 - P_e^{\mathrm{WD},\mu})P_R^{\mathrm{WD},\mu}$. The energy storage of a device is lower-bounded by 0 as shown in Case 1 of Fig. 6.6.

If the energy storage is higher than $\overline{E_\mu}^{\mathrm{PCK}}$, as shown in Case 2 of Fig. 6.6, WD_μ attempts to transmit its data packets. Specifically, in Case 2 of $\overline{E_\mu}^{\mathrm{PCK}} \le e \le \overline{E_\mu}^{\max}$, WD_μ may harvest $\overline{E_\mu}^H$ energy and its state transits to $[e + \overline{E_\mu}^H]$ with a probability of $P_e^{\mathrm{WD},\mu}$. Furthermore, if WD_μ successfully transmits a data packet, its energy state transits to $[e - \overline{E_\mu}^{\mathrm{PCK}}]$ with a probability of $b_s^{\mathrm{WD},\mu}$. Otherwise, WD_μ may consume $\overline{E}_\Delta$ energy for the backoff process with a probability of $(1 - b_s^{\mathrm{WD},\mu} - P_e^{\mathrm{WD},\mu})(1 - P_R^{\mathrm{WD},\mu})$, or it may consume $\overline{E}_{\mathrm{WD}}^R$ energy for data reception with a probability of $(1 - b_s^{\mathrm{WD},\mu} - P_e^{\mathrm{WD},\mu})P_R^{\mathrm{WD},\mu}$. Similarly, WD_μ's energy storage is upper-bounded by $\overline{E_\mu}^{\max}$.

According to Fig. 6.6, we can further formulate the transition probabilities of all the energy states as

$$\begin{cases} P_r\{0|e\} = 1 - P_e^{\mathrm{WD},\mu}, 0 \le e < \overline{E}_\Delta, \\ P_r\{0|e\} = (1 - P_e^{\mathrm{WD},\mu}) P_R^{\mathrm{WD},\mu}, \overline{E}_\Delta \le e < \overline{E}_{\mathrm{WD}}^R, \\ P_r\{e - \overline{E}_{\mathrm{WD}}^R|e\} = (1 - P_e^{\mathrm{WD},\mu}) P_R^{\mathrm{WD},\mu}, \overline{E}_{\mathrm{WD}}^R \le e < \overline{E_\mu}^{\mathrm{PCK}}, \\ P_r\{e - \overline{E}_\Delta|e\} = (1 - P_e^{\mathrm{WD},\mu})(1 - P_R^{\mathrm{WD},\mu}), \overline{E}_\Delta \le e < \overline{E_\mu}^{\mathrm{PCK}}, \\ P_r\{e - \overline{E}_\Delta|e\} = (1 - P_e^{\mathrm{WD},\mu} - b_s^{\mathrm{WD},\mu})(1 - P_R^{\mathrm{WD},\mu}) \\ \qquad , \overline{E_\mu}^{\mathrm{PCK}} \le e \le E_\mu^{\max}, \\ P_r\{e - \overline{E}_{\mathrm{WD}}^R|e\} = (1 - P_e^{\mathrm{WD},\mu} - b_s^{\mathrm{WD},\mu}) P_R^{\mathrm{WD},\mu} \\ \qquad , \overline{E_\mu}^{\mathrm{PCK}} \le e \le E_\mu^{\max}, \\ P_r\{e - \overline{E_\mu}^{\mathrm{PCK}}|e\} = b_s^{\mathrm{WD},\mu}, \overline{E_\mu}^{\mathrm{PCK}} \le e \le E_\mu^{\max}, \\ P_r\{e + \overline{E_\mu^H}|e\} = P_e^{\mathrm{WD},\mu}, 0 \le e \le E_\mu^{\max} - \overline{E_\mu}^H, \\ P_r\{E_\mu^{\max}|e\} = P_e^{\mathrm{WD},\mu}, E_\mu^{\max} - \overline{E_\mu}^H \le e \le E_\mu^{\max}. \end{cases} \tag{6.16}$$

Stationary probabilities can be obtained by $\lim_{n\to\infty} \mathbf{P}^n$. Therefore, we can get the stationary probabilities of WD$_\mu$'s energy state by substituting the transition probabilities in Eq. (6.16) into a transition probability matrix $\mathbf{P}$, and solve the $\lim_{n\to\infty} \mathbf{P}^n$. Given the stationary probabilities, we may formulate $b_n^{\mathrm{WD},\mu}$, $b_e^{\mathrm{WD},\mu}$ and $b_r^{\mathrm{WD},\mu}$ as

$$\begin{cases} b_e^{\mathrm{WD},\mu} = \sum_{e=\overline{E_\mu^{\mathrm{PCK}}}}^{E_\mu^{\max}} \pi_e^E, \\ b_n^{\mathrm{WD},\mu} = \pi_0^E, \\ b_r^{\mathrm{WD},\mu} = \sum_{e=\overline{E_{WD}^R}}^{E_\mu^{\max}} \pi_e^E, \end{cases} \tag{6.17}$$

where π_e^E represents the stationary probability of WD$_\mu$ having an energy storage of e. Note that $b_n^{\mathrm{WD},\mu}$ represents the probability of WD$_\mu$ having no energy at all. Moreover, $b_e^{\mathrm{WD},\mu}$ represents the probability of WD$_\mu$'s energy storage being higher than $\overline{E_\mu}^{\mathrm{PCK}}$.

6.2.3.4 Transmission Probabilities and Throughput Analysis

The authors of [18] studied the transmission probability of a device which adopts a CSMA/CA protocol. According to its conclusion, the transmission probability is a function of its collision probability. In our system, the device adopts a similar protocol, but it may suffer from insufficient energy problem. Therefore, apart from collision probability, the device's transmission probability is also related to its energy status. However, as for the HAP, it is a device with a stable energy source. It does not suffer from any energy problems. The HAP's transmission probability is the same as [18], which is only related to the collision probability. It can be expressed as:

$$\tau^{AP} = \frac{2}{1 + W + P_c^{AP} W \sum_{i=0}^{H-1} (2P_c^{AP})^i}, \tag{6.18}$$

where P_c^{AP} is the collision probability of the HAP's transmission.

Given the transmission probabilities $\tau^{\mathrm{WD},\mu}$ of WDs and τ^{AP} of HAP, when the system is in a stationary state, we can further analyze the probability of WD$_\mu$ harvesting energy and the probabilities of WD$_\mu$'s and HAP's collisions.

The HAP can only transmit data packets successfully when there are no concurrent transmissions of all the WDs. Therefore, P_c^{AP} is expressed as:

$$P_c^{AP} = 1 - \prod_{\mu=1}^{N} (1 - \tau^{\mathrm{WD},\mu}). \tag{6.19}$$

Observe form Eqs. (6.18) and (6.19), the HAP's transmission probability τ^{AP} is related to all devices' transmission probabilities.

The device's transmission collision happens when there are some transmission attempts from other devices. When WD_μ is transmitting, the probability $P_c^{\text{WD},\mu}$ can be obtained as:

$$P_c^{\text{WD},\mu} = 1 - (1 - \tau^{AP}) \prod_{\mu'=1,\mu'\neq\mu}^{N} (1 - \tau^{\text{WD},\mu'}). \tag{6.20}$$

Moreover, WD_μ can harvest energy when the HAP is transmitting to another device. By assuming that the HAP has equal probabilities to transmit data packets to all devices when the WD_μ is not transmitting packets, the probability $P_e^{\text{WD},\mu}$ is expressed as:

$$\begin{aligned} P_e^{\text{WD},\mu} &= \frac{(N - 1/N)\tau^{AP} \prod_{\mu'=1}^{N}(1 - \tau^{\text{WD},\mu'})}{(1 - \tau^{\text{WD},\mu})} \\ &= \frac{N-1}{N}\tau^{AP} \prod_{\mu'=1,\mu'\neq\mu}^{N} (1 - \tau^{\text{WD},\mu'}). \end{aligned} \tag{6.21}$$

The WD_μ can receive a packet only when it has sufficient energy and the HAP transmit to WD_μ with no collisions. Since the HAP has equal probabilities to transmit data packets to all devices, when the WD_μ is not transmitting packets, its probability of receiving a packet from the HAP is expressed as:

$$P_R^{\text{WD},\mu} = \frac{b_r^{\text{WD},\mu}}{N}\tau^{AP} \prod_{\mu'=1,\mu'\neq\mu}^{N} (1 - \tau^{\text{WD},\mu'}). \tag{6.22}$$

Algorithm 6.1, which has the complexity of $\mathcal{O}((\overline{E}_\mu^{\max})^2 MN)$, is proposed for iteratively solving the transmission probabilities $\{\tau^{\text{WD},\mu}\}$ of all devices, that of the HAP, as well as the probability $P_e^{\text{WD},\mu}$ of a device harvesting energy, the probability $P_R^{\text{WD},\mu}$ of a device receiving packets and the transmission collision probabilities $P_c^{\text{WD},\mu}$ and P_c^{AP} of all devices and the HAP.

Algorithm 6.1 Iterative algorithm for calculating probabilities

Require: The maximum backoff stage H; The initial backoff window size W_0; The number of WDs N; Energy threshold $\overline{E_\mu^{\text{PCK}}}$; Energy consumption unit $\overline{E}_\Delta$; Energy harvesting amount $\overline{E_\mu^H}$;

Ensure: All WDs transmission probabilities $\tau^{\text{WD},\mu}$, the HAP's transmission probability τ^{AP}, All WDs energy harvesting probabilities $P_e^{\text{WD},\mu}$, All WDs transmission collisions probabilities $P_c^{\text{WD},\mu}$, All WDs receiving probabilities $P_R^{\text{WD},\mu}$ and the HAP's transmission collisions probability P_c^{AP};

Initialize each $\tau^{\text{WD},\mu}$ within the range of (0,1), τ^{AP} with in the range of (0,1), initial iteration number n=0;
while $\mu < N$ **do**
  Initialize $b_e^{\text{WD},\mu}$, $b_n^{\text{WD},\mu}$ and $b_r^{\text{WD},\mu}$;
  Obtain $P_c^{\text{WD},\mu}$, $P_R^{\text{WD},\mu}$ and $P_e^{\text{WD},\mu}$ according to (6.20), (6.21) and (6.22);
  while $n < M$ **do**
    Obtain $\tau^{\text{WD},\mu}$ and $b_s^{\text{WD},\mu}$ according to (6.14) and (6.15);
    Obtain $b_e^{\text{WD},\mu}$, $b_n^{\text{WD},\mu}$ and $b_r^{\text{WD},\mu}$ according to (6.17);
    $n = n + 1$;
  end while
  Obtain P_c^{AP} and τ^{AP} according to (6.19) and (6.18);
  Update $\mu = \mu + 1$;
end while
return

By assuming that all devices and the HAP always have packets to send, the system throughput R is formulated as

$$R = \frac{\mathbf{E}[PCK]}{\mathbf{E}[time - slot]}, \tag{6.23}$$

where $\mathbf{E}[PCK]$ is the average payload length transmitted in a slot time, while $\mathbf{E}[time - slot]$ represents the average duration of each state of the 3D Markov chain.

As exemplified in Fig. 6.2, when a transmission frame successfully delivers data packets from a source to a destination, it may have a duration expressed as

$$\begin{aligned} \mathrm{T}_s = \mathrm{T}_{\mathrm{RTS}} + \mathrm{T}_{\mathrm{SIFS}} + \mathrm{T}_\delta + \mathrm{T}_{\mathrm{CTS}} + \mathrm{T}_{\mathrm{SIFS}} + \mathrm{T}_\delta + \mathrm{T}_{\mathrm{PCK}} \\ + \mathrm{T}_{\mathrm{SIFS}} + \mathrm{T}_\delta + \mathrm{T}_{\mathrm{ACK}} + \mathrm{T}_{\mathrm{DIFS}} + \mathrm{T}_\delta, \end{aligned} \tag{6.24}$$

where T_δ represents a constant duration of a data packet transmission in the air. However, when a transmission collision happens, a transmission frame only has a duration expressed as

$$\mathrm{T}_c = \mathrm{T}_{\mathrm{RTS}} + \mathrm{T}_{\mathrm{DIFS}} + \mathrm{T}_\delta. \tag{6.25}$$

Let P_f denote the probability of the channel being free, which indicates that neither devices nor the HAP is trying to transmit. Therefore, we can formulate P_f as

$$P_f = (1 - \tau^{AP}) \prod_{\mu'=1}^{N} (1 - \tau^{\mathrm{WD},\mu'}). \tag{6.26}$$

Let P_s^{AP} and P_s^{WD} represent the probability of the channel being occupied by a successful transmission from the HAP and a device, respectively. When the channel is occupied by a successful transmission from the HAP, no device is allowed to access the channel and the destination of the downlink packet should have sufficient energy. Therefore, the probability P_s^{AP} of the HAP occupying the channel is expressed as

$$P_s^{AP} = \sum_{\mu=1}^{N} \frac{b_r^{\mathrm{WD},\mu}}{N} \tau^{AP} \prod_{\mu'=1}^{N} (1 - \tau^{\mathrm{WD},\mu'}), \tag{6.27}$$

Moreover, when a device occupies the channel, all the other devices and the HAP are not allowed to access the channel. The probability P_s^{WD} of the channel being occupied by a device is expressed as

$$P_s^{\mathrm{WD}} = \sum_{\mu=1}^{N} \tau^{\mathrm{WD},\mu} (1 - \tau^{AP}) \prod_{\mu'=1,\mu'\neq\mu}^{N} (1 - \tau^{\mathrm{WD},\mu'}). \tag{6.28}$$

As a result, the total probability of the channel being occupied is obtained as

$$\begin{aligned} P_s = & \sum_{\mu=1}^{N} \frac{b_r^{\mathrm{WD},\mu}}{N} \tau^{AP} \prod_{\mu'=1}^{N} (1 - \tau^{\mathrm{WD},\mu'}) \\ & + \sum_{\mu=1}^{N} \tau^{\mathrm{WD},\mu} (1 - \tau^{AP}) \prod_{\mu'=1,\mu'\neq\mu}^{N} (1 - \tau^{\mathrm{WD},\mu'}). \end{aligned} \tag{6.29}$$

Eventually, the average length of a slot time can be formulated as

$$\mathbf{E}[time - slot] = P_f \mathrm{T}_f + P_s \mathrm{T}_s + (1 - P_f - P_s)\mathrm{T}_c, \tag{6.30}$$

where the T_f is the duration of an empty slot time.

Furthermore, we assume that each data packet has a fixed length of L[PCK]. The average payload can be formulated as

$$\mathbf{E}[PCK] = \underbrace{P_s^{AP} L[\text{PCK}]}_{downlink} + \underbrace{P_s^{\text{WD}} L[\text{PCK}]}_{uplink}. \tag{6.31}$$

Based on Eqs. (6.23) and (6.30), both the downlink and the uplink throughputs are expressed as

$$\begin{cases} R_{downlink} = \frac{P_s^{AP} L[\text{PCK}]}{P_f \text{T}_f + P_s \text{T}_s + (1-P_f-P_s)\text{T}_c}, \\ R_{uplink} = \frac{P_s^{\text{WD}} L[\text{PCK}]}{P_f \text{T}_f + P_s \text{T}_s + (1-P_f-P_s)\text{T}_c}. \end{cases} \tag{6.32}$$

In our system, only the HAP is powered by a stable energy source, while the devices are recharged by the HAP's downlink transmissions. Therefore, the total power consumed by the whole network is from the HAP. With all the probabilities derived above, the average power consumed by the HAP can be analyzed. When the channel is occupied by the HAP, it transmits packets in the downlink with a power of P_{AP}. When the channel is occupied by a device, the HAP receives packets in the uplink. Its energy consumption is E_{AP}^R. Therefore, the average power consumed by the HAP can be expressed as:

$$P_{Totol} = \frac{P_s^{AP} P_{AP} \text{T}_{\text{PCK}} + P_s^{\text{WD}} E_{AP}^R + (1 - P_s) E_\Delta}{P_f \text{T}_f + P_s \text{T}_s + (1 - P_f - P_s)\text{T}_c}. \tag{6.33}$$

6.2.4 Simulation Results

Without specific statements, all parameters in our simulation are provided in Table 6.1. The MAC layer parameters are from the IEEE 802.11 standard [35]. The transmit power of the HAP is set to 1 W, while the antennas' gain is set to 6 dBi. The maximum capacity of the device's supercapacitor is 1 J [36].

First of all, our theoretical analysis on $P_c^{\text{WD},\mu}$, P_c^{AP} and $P_e^{\text{WD},\mu}$ by using Algorithm 6.1 are validated by Monte-Carlo simulation in Fig. 6.7. Both the transmission collision probability of a device and that of the HAP increase as we increase the number of devices. This is because more devices may compete to access the channel for their own transmissions. By contrast, the energy harvesting probability decreases as the number of devices increases. This is because the HAP may gain fewer opportunities to facilitate its downlink transmissions, which reduces the energy harvesting opportunities for the devices.

Observe from Fig. 6.8 that $P_e^{\text{WD},\mu}$, $P_c^{\text{WD},\mu}$, τ^{AP} and $\tau^{\text{WD},\mu}$ all reduces as we increase the initial window size W_0. Specifically, both the transmission collision probabilities τ^{AP} and $\tau^{\text{WD},\mu}$ decrease because an increasing W_0 prolongs the backoff process of the devices and the HAP, which reduces their opportunities of data packet transmissions. Observe from Fig. 6.8 that reducing τ^{AP} may let devices gain more opportunities to access the channel, which then increases their transmission probabilities $\tau^{\text{WD},\mu}$. However, further reducing τ^{AP} may also reduce the energy harvesting probability $P_e^{\text{WD},\mu}$ of devices. Since the amount of energy harvested is reduced, the transmission probability $\tau^{\text{WD},\mu}$ of devices may further reduce. Moreover, with an increasing initial window size W_0, concurrent transmissions can be substantially avoided. Therefore, the transmission collision probabilities $P_c^{\text{WD},\mu}$ and P_c^{AP} both reduces. Note that Algorithm 6.1 helps us to obtain all devices' transmission probabilities $\{\tau^{\text{WD},\mu}\}$ and transmission collision probabilities $\{P_c^{\text{WD},\mu}\}$ and P_c^{AP}. These

Table 6.1 Parameter setting

Parameter	Value	Parameter	Value
Duration of RTS packet T_{RTS}	288 μs	Bandwidth B	20 MHz
Duration of CTS packet T_{CTS}	240 μs	Operation frequency f_0	2.4 GHz
Duration of ACK packet T_{ACK}	240 μs	Maximum backoff stage H	6
Signal propagation delay T_δ	1 μs	Transmit power of the HAP	1 W
Duration of a time slot T_f	50 μs	Antenna gains G_R G_T	6 dBi
Duration of SIFS T_{SIFS}	28 μs	Length of PCK L[PCK]	8 Kb
Duration of DIFS T_{DIFS}	128 μs	Transmission bit rate	1 Mbps
Initial backoff window size W_0	16	Supercapacitor's maximum capacity Q	1 J (5 V 80 mF)

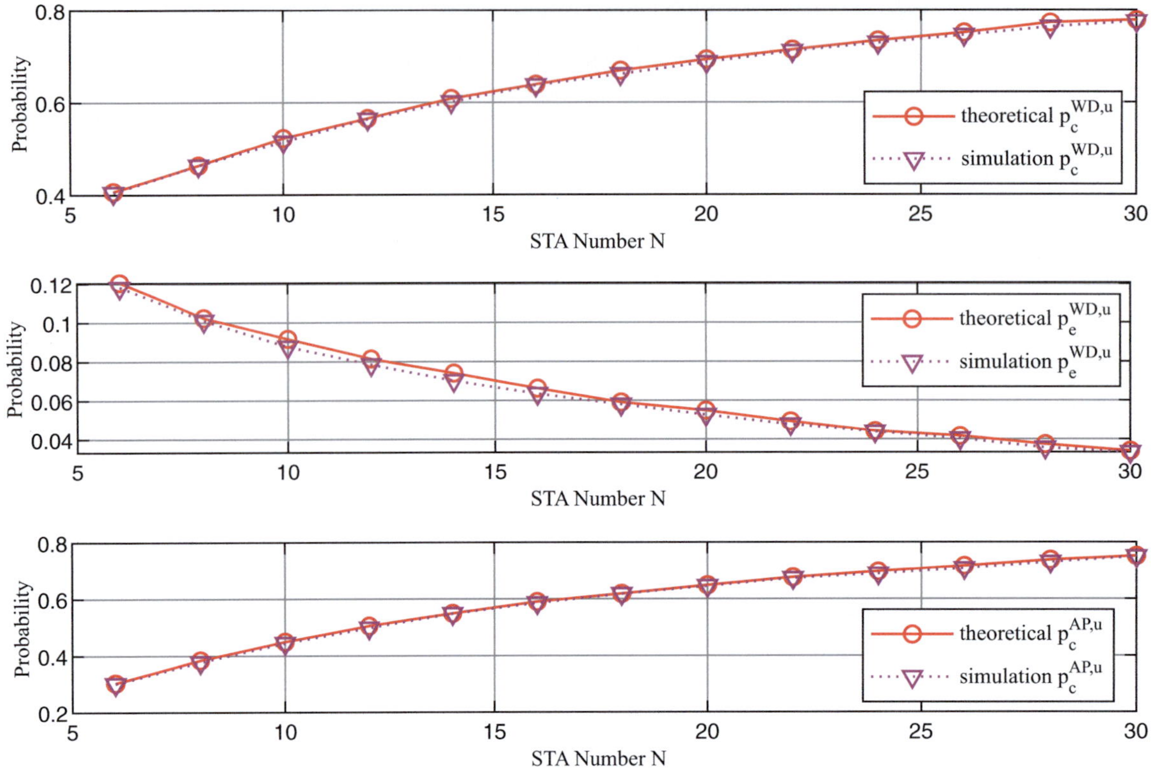

Fig. 6.7 Probabilities versus the number of wireless devices, where H = 4, W = 8, $E_{\mu}^{\max} = 1000$ and $\overline{E}_{\Delta} = 1$

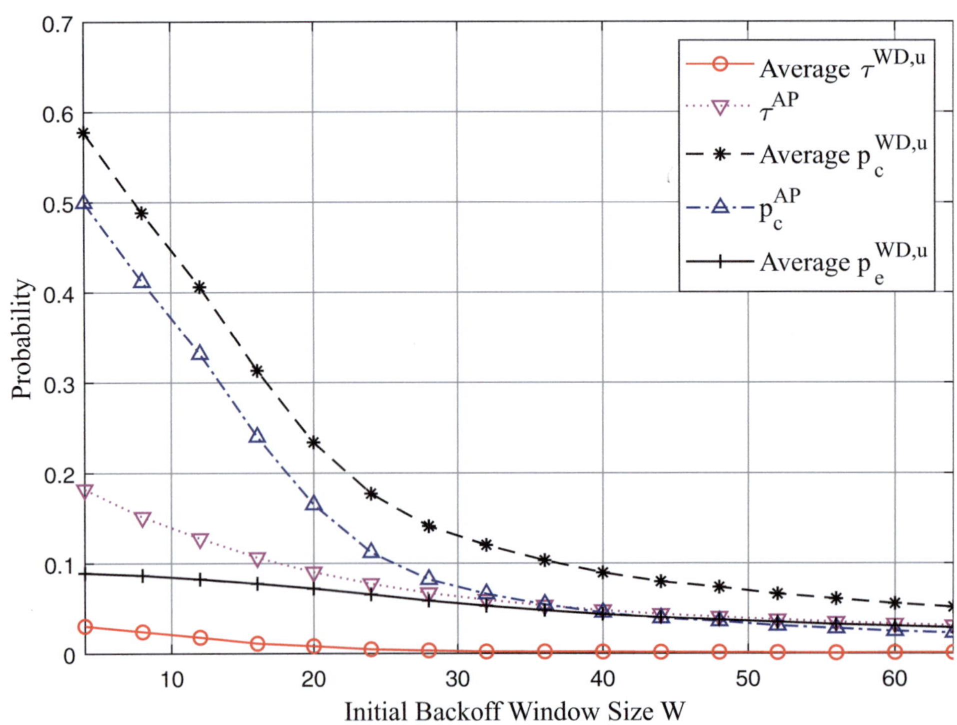

Fig. 6.8 Probabilities versus initial backoff window size W_0, where H = 4, W = 8, N = 20 and $\overline{E}_{\Delta} = 1$

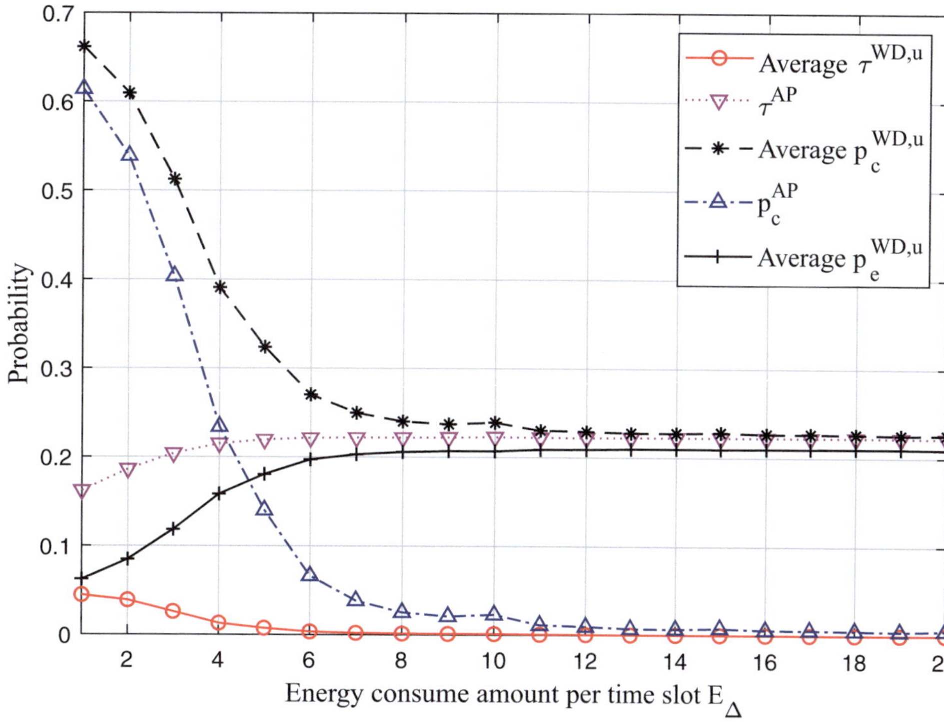

Fig. 6.9 Probabilities versus $\overline{E}_\Delta$, where H = 4, W = 8, N = 20

probabilities have similar trends. For clarity, we plot the average value of all these probabilities among all devices in Fig. 6.8. Similar approach is adopted in the rest of this section.

We study the impact of the energy consumption $\overline{E}_\Delta$ for backoff operation on both the transmission probabilities and collision probabilities of the devices and the HAP in Fig. 6.9. Observe from Fig. 6.9 that as we increase $\overline{E}_\Delta$, the device's transmission probability $\tau^{\mathrm{WD},\mu}$, the HAP's collision probability P_c^{AP} and the device's collision probability $P_c^{\mathrm{WD},\mu}$ all reduce and converge to a certain value. By contrast, the HAP's transmission probability τ^{AP} and the device's energy harvesting probability $P_e^{\mathrm{WD},\mu}$ both increase and converge to a certain value, as we increase $\overline{E}_\Delta$. This is because, with an increasing $\overline{E}_\Delta$, a device may quickly run out of its energy storage during the backoff and the packet exchange process. Therefore, it may not have sufficient energy to transmit any data packets, which results in the reduction of the transmission probability $\tau^{\mathrm{WD},\mu}$. As $\tau^{\mathrm{WD},\mu}$ decreases, the HAP gains more opportunity to transmit its own data packets in the downlink, since it is not constrained by the energy storage. Therefore, its transmission probability τ^{AP} increases, while its collision probability reduces. Moreover, frequent downlink transmissions of the HAP indicate that the devices have an increasing chance of harvesting energy, which results in an increasing energy harvesting probability $P_e^{\mathrm{WD},\mu}$. However, when $\overline{E}_\Delta$ becomes higher than the energy E_μ^H harvested by a device, it may not have any remaining energy storage to power the data transmissions, since all the energy harvested is consumed for the backoff and packet exchange process. As a result, we only have the HAP to transmit data packets in the system. Therefore, all these probabilities finally converge to a constant value.

Given all these transmission probabilities and collision probabilities, we may evaluate the throughput performance of the system, according to Eq. (6.30). Therefore, both the downlink and the uplink throughput are portrayed in Figs. 6.10 and 6.11. Interestingly, the total uplink throughput does not always increase as we increase the number N of the devices. This is because, with more devices, the channel is more efficiently used for data packet transmissions, which results in increasing uplink throughput. However, when the number of devices continuously increases, transmission collisions may frequently happen, which then reduces the uplink throughput. Moreover, observe from Fig. 6.10 that when we have fewer devices, a higher $\overline{E}_\Delta$ may result in a lower uplink throughput. This is because a higher energy consumption for packet exchange and backoff process may reduce the energy consumption for data transmissions. By contrast, when we have a higher number of devices, an increasing $\overline{E}_\Delta$ reduces the collision probability $P_c^{\mathrm{WD},\mu}$, which may increase the total uplink throughput. Observe from Fig. 6.11 that, as we increase the total number of devices, the downlink throughput of the HAP reduces all the time. This is because when more devices compete for data packet transmissions, the collision probability P_c^{AP} of the HAP increases, which thus reduces the downlink throughput. However, when $\overline{E}_\Delta$ increases, the devices do not have sufficient energy to support their own data packet transmissions. Therefore, the HAP may gain more opportunities to increase its downlink throughput.

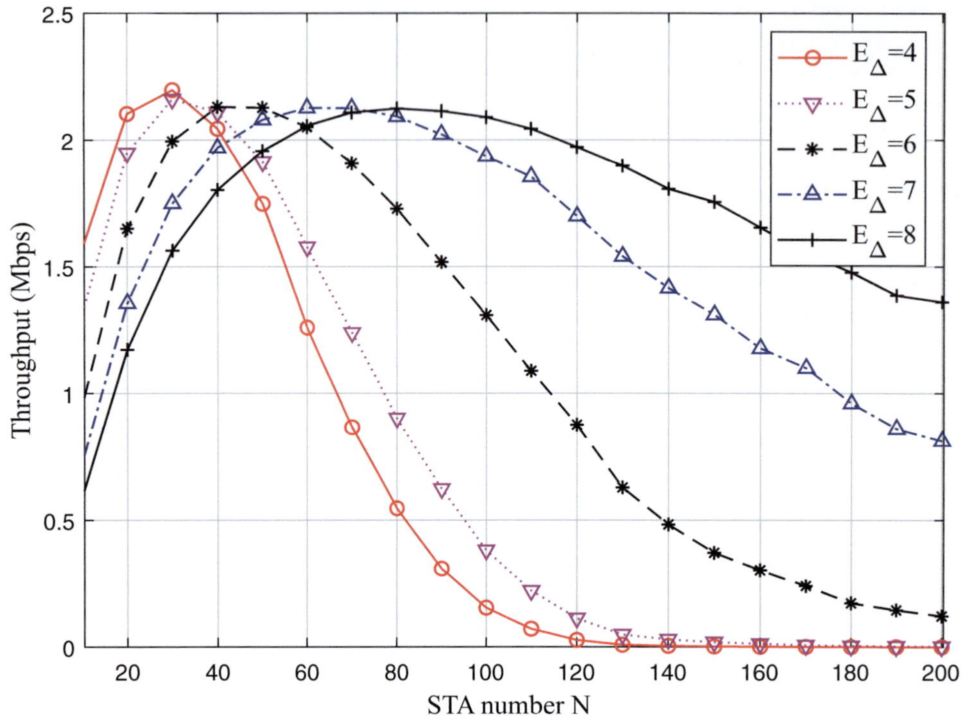

Fig. 6.10 Uplink throughput versus number of wireless devices N

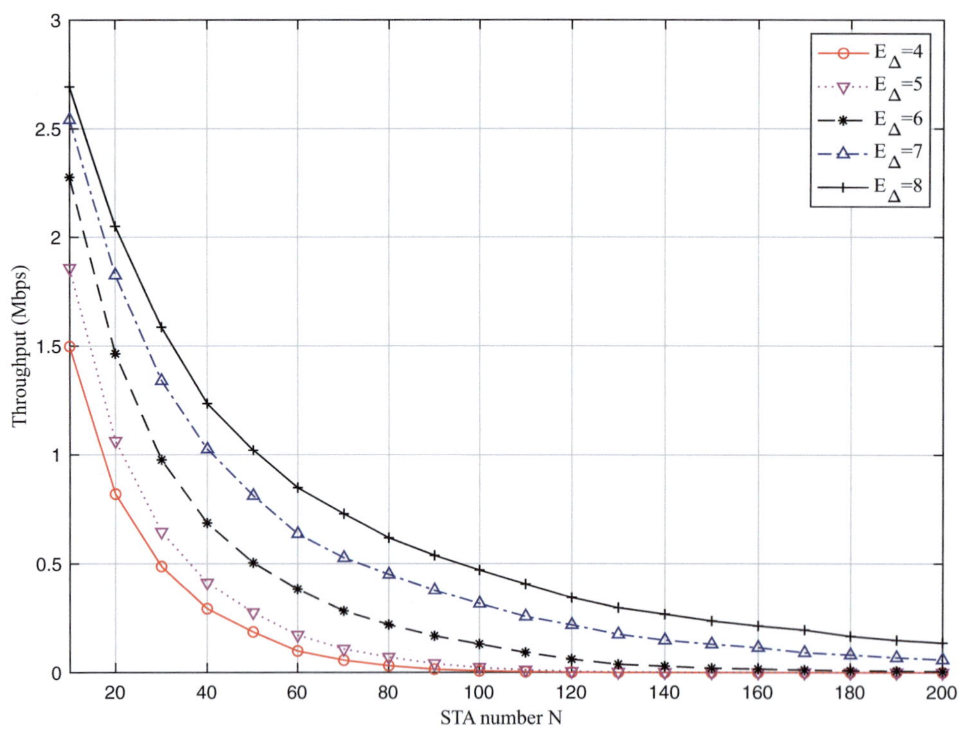

Fig. 6.11 Downlink throughput versus number of wireless devices N

In Fig. 6.12, we investigate the total throughput with different duration T_{PCK}. As we increase T_{PCK}, the total throughput firstly increases. This is because once the downlink transmission of the HAP commences, devices may harvest more energy for powering their data packet transmissions. However, as we continuously increase T_{PCK}, all devices may occupy the channel for a longer time, which thus increases the transmission collision probabilities. Therefore, the total throughput

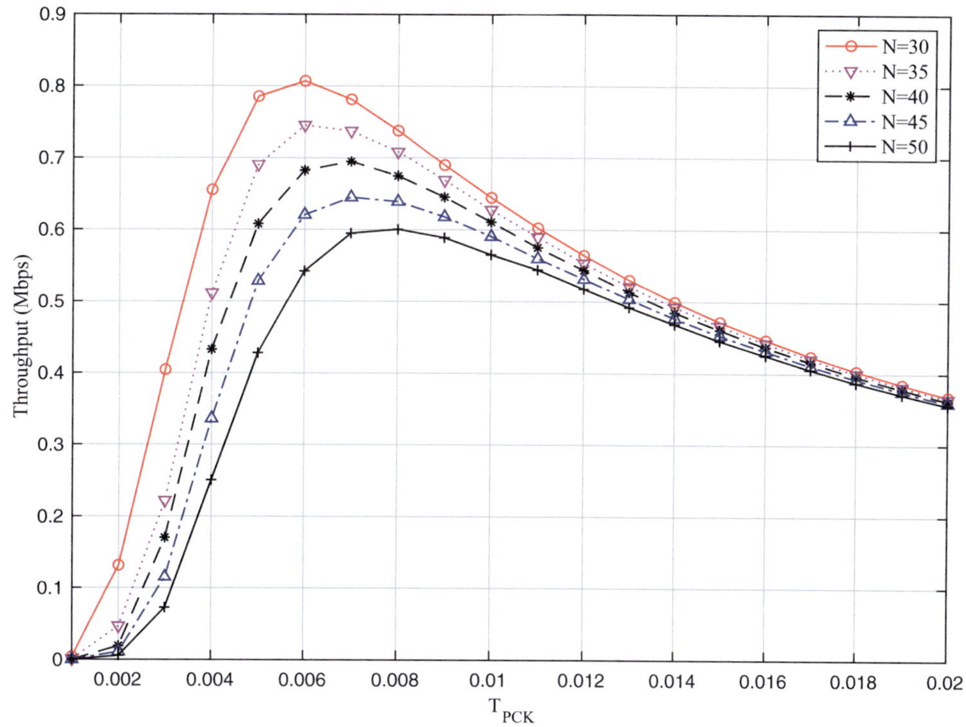

Fig. 6.12 Total throughput versus duration of a successful transmission T_{PCK}, where H = 4, W = 8 and $\overline{E}_{\Delta} = 4$

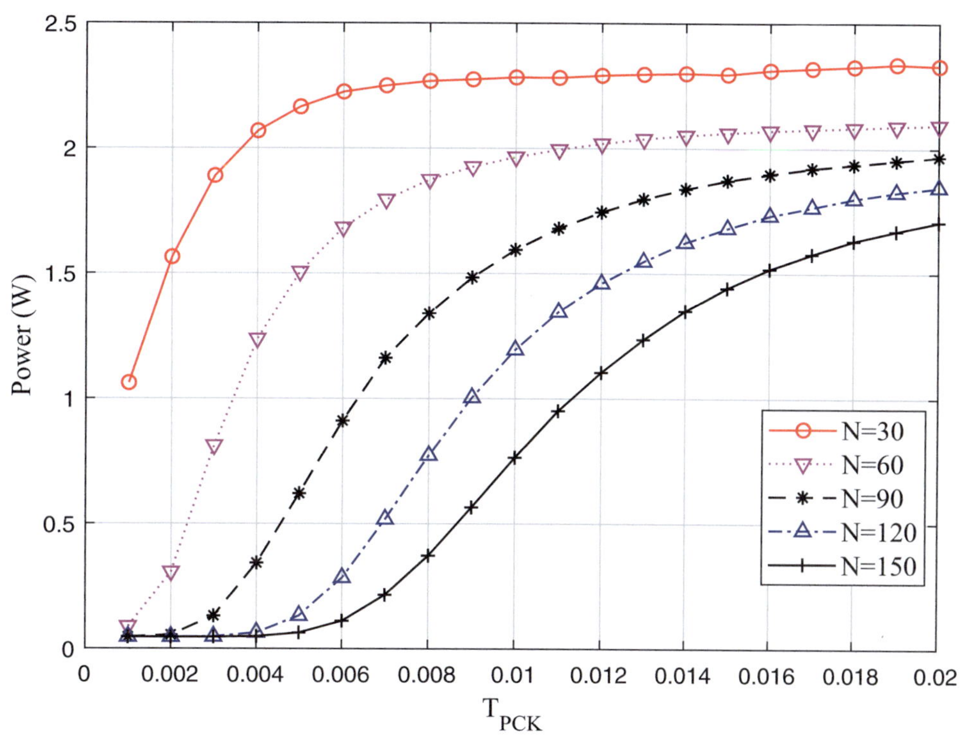

Fig. 6.13 Energy consumed by HAP versus duration of a successful transmission T_{PCK}, where H = 4, W = 8, $\overline{E}_{\Delta} = 4$ and $P_{AP} = 1$ W

reduces. Furthermore, a higher number of devices may increase their collision probabilities, which then decrease P_s^{WD} and P_s^{AP} in Eq. (6.30). Therefore, the total throughput reduces.

In Fig. 6.13, we study the impact of duration T_{PCK} on the power consumption of the HAP. By increasing T_{PCK}, the power consumption increases. This is because the HAP may consume more power to transmit downlink packets as T_{PCK}

increases. With the increasing T_{PCK}, devices further harvest more energy, and devices may get more opportunities to occupy the channel and P_s^{WD} may much larger than P_s^{AP} when T_{PCK} is sufficiently long. In this situation, the impact of T_{PCK} on the energy consumption is trivial, since the T_{PCK} is only related to the P_s^{AP} in Eq. (6.32). Moreover, by increasing the number of devices, the power consumption of the HAP decreases. This is because with the increasing number of devices, all devices suffer from more collisions, and the HAP thus takes more backoff operations than packet receptions. Therefore, the system consumes less energy.

6.3 Exercises

1. Compute the maximum throughput T for a pure ALOHA and a slotted ALOHA random access system, along with the load L that achieves the maximum in each case.
2. Consider a pure ALOHA system with a transmission rate of $R = 10$ Mbps. Compute the load L and throughput T for the system assuming 1000 bit packets and a Poisson arrival rate of $\lambda = 103$ packets per second. Also compute the effective data rate (rate of bits successfully received). What other value of load L results in the exact same throughput?
3. Consider a three-user uplink channel with channel power gains $g_1 = 1$, $g_2 = 3$, and $g_3 = 5$ from user k to the receiver, $k = 1, 2, 3$. Assume that all three users require a 10-dB SINR. The receiver noise is $n = 1$. Determine whether a feasible power vector exists for this system such that all users meet the required SINR constraints and, if so, find the optimal power vector $\mathbf{P}^*$ such that the desired SINRs are achieved with minimum transmit power. Assume $\rho = 0.1$.
4. Show that the K-user AWGN MAC capacity region is not affected if the k-th user's channel power gain g_k is scaled by α as long as the k-th user's transmit power P is also scaled by $1/\alpha$.
5. Consider a multiple access channel being shared by two users. The total system bandwidth is $B = 100$ kHz. The transmit power of user 1 is $P_1 = 3$ mW, while the transmit power of user 2 is $P_2 = 1$ mW. The receiver noise density is .001 μW/Hz. You can neglect any path-loss, fading, or shadowing effects.

   (a) Suppose user 1 requires a data rate of 300 kbps to see videos. What is the maximum rate that can be assigned to user 2 under time division? Under superposition coding with successive interference cancelation?
   (b) Compute the rate pair (R_1, R_2) where the frequency-division rate region intersects the region achieved by code division with successive interference cancelation (assuming $G = 1$).
   (c) Compute the rate pair (R_1, R_2) such that $R_1 = R_2$ (i.e., where the two users get the same rate) for time division and for spread-spectrum code division with and without successive interference cancelation for a spreading gain $G = 10$. Note: To obtain this region for $G > 1$ you must use the same reasoning on the MAC as was used to obtain the BC capacity region with $G > 1$.

6. Show that the sum-rate capacity of the AWGN MAC is achieved by having all users transmit at full power.
7. Derive the optimal power adaptation for a two-user fading MAC that achieves the sum-rate point.

References

1. Zhi, A. E., Tan, H. P., & Seah, W. (2011). Design and performance analysis of mac schemes for wireless sensor networks powered by ambient energy harvesting. *Ad Hoc Networks, 9*(3), 300–323.
2. Fafoutis, X., & Dragoni, N. (2011). ODMAC: An on-demand MAC protocol for energy harvesting-wireless sensor networks. In *Acm Symposium on Performance Evaluation of Wireless Ad Hoc*.
3. Khairy, S., Han, M., Cai, L. X., & Cheng, Y. (2019). Sustainable wireless iot networks with RF energy charging over wi-fi (cowifi). *IEEE Internet of Things Journal, 6*(6), 10205–10218. 2019.
4. Bianchi, G. (2000). Performance analysis of the IEEE 802.11 distributed coordination function. *IEEE Journal on Selected Areas in Communications, 18*(3), 535–547. ISSN: 0733–8716.
5. Lu, X., Wang, P., Niyato, D., Kim, D. I., & Han, Z. (2015). Wireless networks with RF energy harvesting: A contemporary survey. *IEEE Communications Surveys Tutorials, 17*(2), 757–789. ISSN: 1553-877X.
6. Padhy, A., Joshi, S., Bitragunta, S., Chamola, V., & Sikdar, B. (2021). A survey of energy and spectrum harvesting technologies and protocols for next generation wireless networks. *IEEE Access, 9*, 1737–1769.
7. Ma, D., Lan, G., Hassan, M., Hu, W., & Das, S. K. (2020). Sensing, computing, and communications for energy harvesting IoTs: A survey. *IEEE Communications Surveys & Tutorials, 22*(2), 1222–1250.
8. Akan, O. B., Cetinkaya, O., Koca, C., & Ozger, M. (2018). Internet of hybrid energy harvesting things. *IEEE Internet of Things Journal, 5*(2), 736–746.

9. Ruan, T., Chew, Z. J., & Zhu, M. (2017). Energy-aware approaches for energy harvesting powered wireless sensor nodes. *IEEE Sensors Journal, 17*(7), 2165–2173.
10. Ku, M. L., Li, W., Chen, Y., & Liu, K. J. R. (2016). Advances in energy harvesting communications: Past, present, and future challenges. *IEEE Communications Surveys Tutorials, 18*(2), 1384–1412. ISSN: 1553-877X.
11. Hu, J., Kun, Y., Guangjun, W., & Lajos, H. (2018). Integrated data and energy communication network: A comprehensive survey. *IEEE Communications Surveys Tutorials, 20*(4), 3169–3219.
12. Hu, J., Wang, Q., & Yang, K. (2020). Energy self-sustainability in full-spectrum 6G. *IEEE Wireless Communications, 28*, 104–111.
13. Yang, K., Yu, Q., Leng, S., Fan, B., & Wu, F. (2016). Data and energy integrated communication networks for wireless big data. *IEEE Access, 4*, 713–723. ISSN: 2169–3536.
14. Huang, Y., & Clerckx, B. (2018). Waveform design for wireless power transfer with limited feedback. *IEEE Transactions on Wireless Communications, 17*(1), 415–429.
15. Lu, X., Flint, I., Niyato, D., Privault, N., & Wang, P. (2016). Self-sustainable communications with RF energy harvesting: Ginibre point process modeling and analysis. *IEEE Journal on Selected Areas in Communications, 34*(5), 1518–1535.
16. Liu, C.-H., & Hsu, C.-S. (2019). Fundamentals of simultaneous wireless information and power transmission in heterogeneous networks: A cell-load perspective. *IEEE Journal on Selected Areas in Communications, 37*(1), 100–115.
17. Guruacharya, S., & Hossain, E. (2018). Self-sustainability of energy harvesting systems: Concept, analysis, and design. *IEEE Transactions on Green Communications and Networking, 2*(1), 175–192.
18. Dai, B., Liu, Y.-F., & Yu, W. (2018). Optimized base-station cache allocation for cloud radio access network with multicast backhaul. *IEEE Journal on Selected Areas in Communications, 36*(8), 1737–1750.
19. Chen, Z., Cai, L. X., Cheng, Y., & Shan, H. (2017). Sustainable cooperative communication in wireless powered networks with energy harvesting relay. *IEEE Transactions on Wireless Communications, 16*(12), 8175–8189.
20. Gao, J., Zhuang, W., Li, M., Shen, X., & Li, X. (2021). Mac for machine-type communications in industrial IoT–part I: Protocol design and analysis. *IEEE Internet of Things Journal, 8*(12), 9945–9957.
21. Liang, L., Liu, X., Wang, Y., Feng, W., & Yang, G. (2014). SW-MAC: A low-latency MAC protocol with adaptive sleeping for wireless sensor networks. *Wireless Personal Communications, 77*(2), 1191–1211. ISSN: 0929–6212.
22. Eu, Z. A., & Tan, H.-P. (2012). Probabilistic polling for multi-hop energy harvesting wireless sensor networks. In *2012 IEEE International Conference on Communications (ICC)* (pp. 271–275).
23. Li, B., Guo, X., Zhang, R., Du, X., & Guizani, M. (2020). Performance analysis and optimization for the MAC protocol in UAV-based IoT network. *IEEE Transactions on Vehicular Technology, 69*(8), 8925–8937.
24. Guo, X., Li, B., & Liu, K. (2020). Performance analysis for the CMSA/CA protocol in UAV-based IoT network. In *2020 IEEE 91st Vehicular Technology Conference (VTC2020-Spring)* (pp. 1–6).
25. Li, Y., Cao, B., Liang, L., Mao, D., & Zhang, L. (2021). Block access control in wireless blockchain network: Design, modeling and analysis. *IEEE Transactions on Vehicular Technology, 70*(9), 9258–9272.
26. Choi, J., Byeon, S., Choi, S., & Lee, K. B. (2017). Activity probability-based performance analysis and contention control for IEEE 802.11 WLANs. *IEEE Transactions on Mobile Computing, 16*(7), 1802–1814.
27. Naderi, M. Y., Nintanavongsa, P., & Chowdhury, K. R. (2014). RF-MAC: A medium access control protocol for re-chargeable sensor networks powered by wireless energy harvesting. *IEEE Transactions on Wireless Communications, 13*(7), 3926–3937.
28. Tamilarasi, P., & Lavenya, B. (2015). Energy and throughput enhancement in wireless powered communication networks using RF-MAC and CSMA. In *2015 International Conference on Innovations in Information, Embedded and Communication Systems (ICIIECS)* (pp. 1–4).
29. Iqbal, A., Kim, Y., & Lee, T.-J. (2018). Access mechanism in wireless powered communication networks with harvesting access point. *IEEE Access, 6*, 37556–37567.
30. Moon, K., Kim, K. M., Kim, Y., & Lee, T.-J. (2021). Device-selective energy request in RF energy-harvesting networks. *IEEE Communications Letters, 25*(5), 1716–1719.
31. Zhao, Y., Hu, J., Diao, Y., Yu, Q., & Yang, K. (2018). Modelling and performance analysis of wireless LAN enabled by RF energy transfer. *IEEE Transactions on Communications, 66*(11), 5756–5772. ISSN: 0090–6778.
32. Lyu, J., & Zhang, R. (2020). Spatial throughput characterization for intelligent reflecting surface aided multiuser system. *IEEE Wireless Communications Letters, 9*(6), 834–838.
33. Wang, S., Xia, M., Huang, K., & Wu, Y.-C. (2017). Wirelessly powered two-way communication with nonlinear energy harvesting model: Rate regions under fixed and mobile relay. *IEEE Transactions on Wireless Communications, 16*(12), 8190–8204.
34. Boshkovska, E., Ng, D. W. K., Zlatanov, N., & Schober, R. (2015). Practical non-linear energy harvesting model and resource allocation for SWIPT systems. *IEEE Communications Letters, 19*(12), 2082–2085.
35. IEEE Standard for Information technology–Telecommunications and information exchange between systems Local and metropolitan area networks–Specific requirements Part 11: Wireless LAN Medium Access Control (MAC) and Physical Layer (PHY) Specifications. *IEEE Std 802.11-2012 (Revision of IEEE Std 802.11-2007)* (pp. 1–2793).
36. Zheng, Y., Zhang, Y., Wang, Y., Hu, J., & Yang, K. (2020). Create your own data and energy integrated communication network: A brief tutorial and a prototype system. *China Communications, 17*(9), 193–209. ISSN: 1673–5447.

Chapter 7
Data and Energy Integrated Wireless Networking

With the wireless communication being more various in the future, it's becoming challenging to prolong the lifetime of many battery powered devices, since frequently replacing their batteries is a cumbersome job. An hybrid access point (H-AP) is capable of simultaneously operating wireless information transfer (WIT) and wireless energy transfer (WET) by exploiting the radio frequency (RF) signals. By jointly considering both the mobility of the user and the popularity of the sites, we focus on the design of the H-AP's deployment scheme. Specifically, a mobility model of the grid based city streets is exploited for characterizing the users' movements. Based on this mobility model, the impact of the deployment site's popularity on the WIT and WET efficiencies is firstly analyzed. Then, an H-AP deployment scheme for striking a balance between the WIT and the WET efficiencies is proposed, which is regarded as the B-deployment scheme. The simulation results demonstrate that the B-deployment scheme is more flexible for satisfying diverse requirement of the WIT and WET efficiencies.

With the development of wireless communication technologies, wearable and pocket communication devices, such as E-glasses, E-watches, smart belts, etc., have penetrated into our daily lives. All these communication devices require to exchange vital information with central servers, such as the users' physical condition, the location information, etc.

Delay tolerant networks (DTNs) [1] have been proposed for realizing communication in intermittent networks. In DTNs, a direct communication link between any pair of nodes is not always available, due to extreme terrestrial environments or movements of the communication nodes. In order to improve the performance of the information delivery, Akhtar et al. [2] have proposed to exploit the users' social properties for improving the information delivery performance of the DTNs, which yields a new concept of mobile social network (MSN). Nikolaos et.al. [3] proposed the architecture and social properties of MSN, while also threw some key research challenges. There are two communication types in MSNs, namely device to device (D2D) communication [4] and device to access point (D2A) communication, as portrayed in Fig. 7.1. In the D2D communication, a pair of different users may establish a direct communication link between their devices without the aid of any centralized infrastructure, if they enter each other's transmission range. For example, Jie et al. [5] have proposed a content dissemination scheme by exploiting the D2D communications among the users. By contrast, in the D2A communication, a user may communicate with a centralized AP, if it enters the AP's transmission range. For instance, Zhao et al. [6] have studied a joint AP deployment and routing scheme for efficiently delivering information to the users in MSNs. Later on, Ying et al. [7] have proposed another AP deployment scheme by exploiting the social popularity of specific locations. Furthermore, Fan et al. [8, 9] have developed a joint scheme for both the AP's deployment and the storage allocation in order to efficiently deliver popular content toward the users. In contrast to the conventional high-power cellular communication, the APs in the MSN often operate at a low transmit power. As a result, the transmission range of the APs is limited. Hence, users have to enter the transmission range of the APs so as to download important information to their wearable devices. Therefore, users' mobility largely determines the communication performance of the MSN.

Furthermore, due to their limited size, wearable devices are normally equipped with batteries having limited capacity. The batteries may quickly drain for the sake of powering complex signal processing and sensing tasks. Yang et.al. [10] proposed an approach to offload services for resource-constrained mobile devices, while the resource management can also be operated on the cloud to reduce the energy consumption [11]. However, this is not an ideal solution as some tasks are simply not suitable to run on servers and task offloading also incurs extra energy consumption. Therefore, it is essential to incorporate the function of wireless energy transfer (WET) to the conventional communication oriented APs, which yields the hybrid access points (H-APs). Many radio frequency (RF) signals based WET techniques have been studied in different OSI layers [12], such as the energy beamforming [13] and channel coding [14] in the physical layer, the network architecture [15] and the MAC protocol design [16, 17] as well as the resource allocation [18] in the network and MAC layer. Moreover, coordinating both the WIT and WET in the same RF spectral band has attracted much attention from both the academia and

J. Hu, K. Yang, *Integrated Data and Energy Transfer in Wireless Networks*, Textbooks in Telecommunication Engineering,
https://doi.org/10.1007/978-3-031-60514-7_7

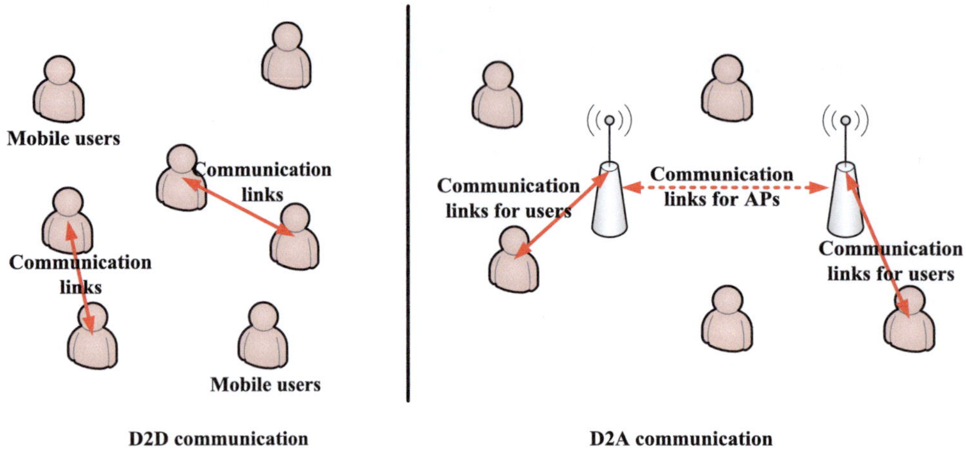

Fig. 7.1 Communication types in MSN

industry, which triggers substantial works in simultaneously wireless information and power transfer (SWIPT) and wireless powered communication (WPC). All these related works [19–21] aim for striking a balance between the WET and the WIT performance by exploiting the hot communication techniques, such as the massive MIMO technique [21], full duplex technique [22], and channel estimation technique [20].

However, many of the energy concerns focus in the future cellular networks [23], while few existing works study the WET in the scenario of MSNs, where both the WIT and WET performance are dominated by the users' mobility patterns. Fan et al. [24] have firstly proposed to exploit the so-called enhanced throwboxes for realizing both the WIT and WET in the MSN. However, their model is extremely simplified without considering a realistic mobility model, the wireless channel attenuation, and the distinctive features of the WIT and the WET. In order to fill this gap, our novel contribution can be summarized as follows

- A user mobility model in the city streets is exploited for characterizing the users' mobility patterns, in which users move randomly along the city streets and they may randomly opt to turn or to go straight, when they reach a crossroad.
- A two-dimensional Markov chain is exploited for analyzing the seminal metrics of the users' mobility patterns, including the users' total sojourn duration within the WIT range of the H-AP and that within the WET range.
- Both the WIT and WET efficiencies of the MSN studied are derived in close-form. Then, three different H-AP deployment schemes are proposed, namely the I-deployment scheme for maximizing the WIT efficiency, the E-deployment for maximizing the WET efficiency, and the B-deployment for striking a balance between the WIT and WET efficiencies. The algorithms for obtaining both the optimal and sub-optimal solutions are developed.

7.1 System Model

7.1.1 Network Model

We focus our attention on the typical D2A communication in a mobile social network (MSN), where users are equipped with wearable devices and moving around street blocks. Users may download important information and charge their batteries via RF signals emitted by the H-APs. Due to the limited transmission range of the H-APs, users' mobility and the popularity of the hot-spots (e.g. shopping mall, parks, and sport centers) may jointly determine the performance of both the WIT and WET. As a result, an optimal deployment scheme of the H-AP should consider both the users' mobility and the popularity of the hot-spots. We assume M users in the MSN studied, which are denoted by a set of $\{u_1, \ldots, u_m, \ldots, u_M\}$. We also assume that N sites in the MSN studied as the candidates for the deployment of the H-APs, which are denoted by a set of $\{s_1, \ldots, s_n, \ldots, s_N\}$. Furthermore, we have K H-APs in total pending to be deployed. A vector $\mathbf{y} = \{y_1, \ldots, y_n, \ldots, y_N\}$ is exploited for characterizing a specific deployment scheme of the H-APs, where $y_n = 1$ indicates that a H-AP is deployed at the site s_n, while $y_n = 0$ indicates that no H-AP is deployed at the site s_n. Obviously, if we have more H-APs than the number of sites, say $K \geq N$, each site can have a single H-AP to be deployed. By considering the deploying cost, we normally have fewer H-APs than the number of sites, say $K < N$. In order to maximize the benefit from these K H-APs, an optimal deployment scheme is required.

The H-APs are equipped with a pair of radio fronts operating in different spectral bands for facilitating the WIT and the WET, respectively. Specifically, the WET operates in a low frequency band, which may substantially reduce the path-loss induced channel attenuation [25]. By contrast, the WIT operates in a high frequency band in order to obtain a high bandwidth for increasing the throughput. The energy harvesting circuit and the information reception circuit at a wearable device have diverse requirement on the minimum power of the received RF signal. For example, the minimum power of the received RF signal at a wearable device can be -80 dBm for successful information recovery. By contrast, the minimum power requirement on the received RF signal is at least -40 dBm for activating the energy harvesting circuit [26]. As a result, given the same transmit power, the H-AP has distinct WIT range and the WET range. If a H-AP is deployed at the site s_n, its WIT range is denoted as r_n^D, while its WET range is denoted as r_n^E. Since the energy harvesting requires higher received signal power than the information decoding, we normally have $r_n^E < r_n^D$. Furthermore, homogeneous H-APs are conceived in the MSN studied. As a result, all the H-APs pending to be deployed have an identical WIT range, denoted as $r_n^D = r^D$, and they also have an identical WET range,[1] denoted as $r_n^E = r^E$. Moreover, in the MSN studied, when a user enters the WIT range of the H-AP, it can only download its requested information from the H-AP. If the user enters the WET range of the H-AP, it can simultaneously download the information and harvest energy from the H-AP, since we have $r^E < r^D$. In this MSN, we aim for deploying the K H-APs at the optimal sites in order to maximize the performance of both the WIT and WET.

7.1.2 Mobility Model

We assume that all the users move around a grid based city street blocks, by obeying a mobility model more generic than the conventional Manhattan mobility model [27]. In our mobility model, the users move along the streets. Once they arrive at a crossroad, they may choose any available direction to move forward. Since users from four different directions may gather at a crossroad, deploying a H-AP at a crossroad is capable of efficiently serving more users. As a result, all the crossroads are regarded as the potential sites for the deployment of the H-APs.

Our city street blocks are illustrated in Fig. 7.2, where the green lines indicate the streets in the city, while the red and blue circles represent the WIT ranges and the WET ranges of the H-APs, respectively. The entire area is equally divided into several square regions having the cross roads as their centers, as portrayed in Fig. 7.2. We assume that there are $X_I \times X_J = N$ crossroads in the city street blocks, where X_I represents the number of crossroads in the vertical dimension, while X_J represents the number of crossroads in the horizontal dimension. All the crossroads can be denoted by the set $\{s_{i,j} | 1 \leq i \leq X_I, 1 \leq j \leq X_J\}$, which are also potential sites for the deployment of the H-APs, and users' movements are bounded in the area studied. Generally, a crossroad may have arbitrary number of streets connected to it. However, in our model, we consider that a crossroad is only connected to its nearest peers in our grid based street blocks. Moreover, a crossroad also has a so-called crowded range for characterizing its popularity. Moving within the crowded range may substantially reduce the speed. A higher crowded range represents that more people gather at this crossroad. The crowded range of the crossroad $s_{i,j}$ is represented by an orange circle having the radius of $r_{i,j}^C$. The crowded range $r_{i,j}^C$ is an average value according to the statistics of the crossroad $s_{i,j}$. When a user u_m moves outside of the crowded ranges of the crossroads, its speed is assumed to be $v_{m,0}$. If u_m enters the crowded range of the crossroad $s_{i,j}$, it may reduce its speed to $v_{m,i,j}$, which is determined by the user u_m itself and by the popularity of the crossroad $s_{i,j}$. Without loss of generality, every street connecting a pair of neighboring crossroads has an identical length of l. Since the short-range communication technique, such as Wi-Fi, Bluetooth, etc., is invoked for enabling the D2A communication in the MSN, the WIT range r^D is relatively low. As a result, we may reasonably assume that the street length satisfy the condition of $l > 2\max(r^D, r^E, r_{i,j}^C)$.

[1] The wireless channel power gain is mainly affected by the path-loss as well as the multipath fading. We assume that the multipath fading has the same statistical properties at each crossroad. Hence, given the information outage probability of the received WIT SNR lower than an identical threshold, the WIT ranges of the H-APs are identical. So are the WET ranges of the H-APs. Therefore, we assume that the WIT and the WET ranges of all the H-APs are identical.

7.2 User Mobility Analysis

In this section, we model the users' mobility pattern by a two-dimensional discrete Markov chain, which may help us in analyzing a user's sojourn duration within the WET range of a H-AP and that within its WIT range, respectively.

7.2.1 Two-Dimensional Markov Chain Modeling

As portrayed in Fig. 7.2, the entire area is equally divided into a set of square regions having cross roads as their centers. As a result, a single street is halved into two segments belonging to a pair of neighboring square regions. The side length of each square region is l, which is equal to the length of a street. The square region having the crossroad $s_{i,j}$ as its center is denoted by $w_{i,j}$. All the direction choosing probabilities of the user u_m can be denoted by a set of $\mathbb{P}^m = \{p^m_{(i_1,j_1),(i_2,j_2)} | s_{i_1,j_1}$ and s_{i_2,j_2} are neighboring crossroads$\}$. Specifically, $p^m_{(i_1,j_1),(i_2,j_2)}$ represents the probability of the user u_m moving from the crossroad s_{i_1,j_1} to its neighbor s_{i_2,j_2}. By defining the user u_m's stay within the square region $w_{i,j}$ as one of its mobility states, we may model the user u_m's mobility pattern by a two-dimensional Markov chain, which is illustrated in Fig. 7.3. Consequently, $p^m_{(i_1,j_1),(i_2,j_2)}$ also represents the probability of u_m transiting from the state w_{i_1,j_1} to the state w_{i_2,j_2}.

The stationary probabilities of all the states in the Markov chain of u_m can be denoted by a $1 \times X_I X_J$ vector $\boldsymbol{\phi}^m$, whose element $\phi^m_{(i-1)X_J+j}$ represents the corresponding stationary probability of the state $w_{i,j}$. Similarly, we define a $X_I X_J \times X_I X_J$ matrix of the state transition probabilities $\mathbf{P}^{m'} = \{p^{m'}_{i,j} | 1 \leq i \leq X_I X_J, 1 \leq j \leq X_I X_J\}$ by re-arranging the entries in the original set $\mathbb{P}^m$ of the state transition probabilities. The entry $p^m_{(i_1,j_1),(i_2,j_2)}$ of $\mathbb{P}^m$ is then mapped on the entry $p^{m'}_{(i_1-1)X_J+j_1,(i_2-1)X_J+j_2}$ of $\mathbf{P}^{m'}$. Given that our Markov chain is ergodic and irreducible, the stationary distribution $\boldsymbol{\phi}^m$ can be readily calculated by

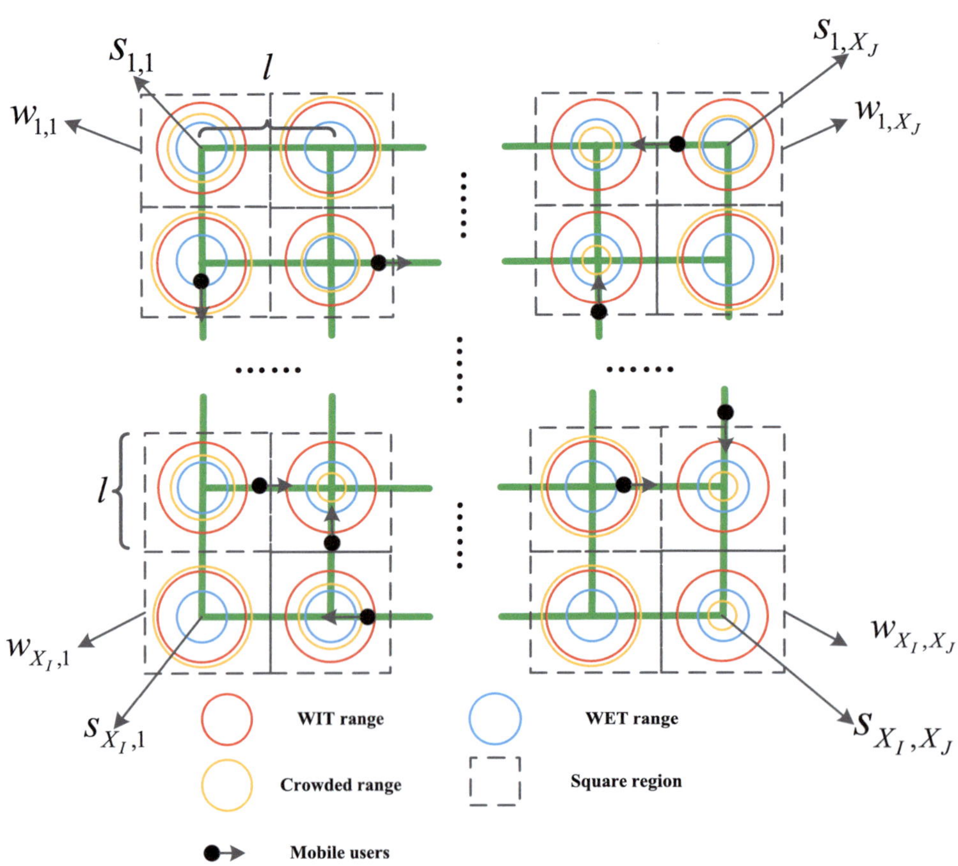

Fig. 7.2 Mobility scenario of city streets

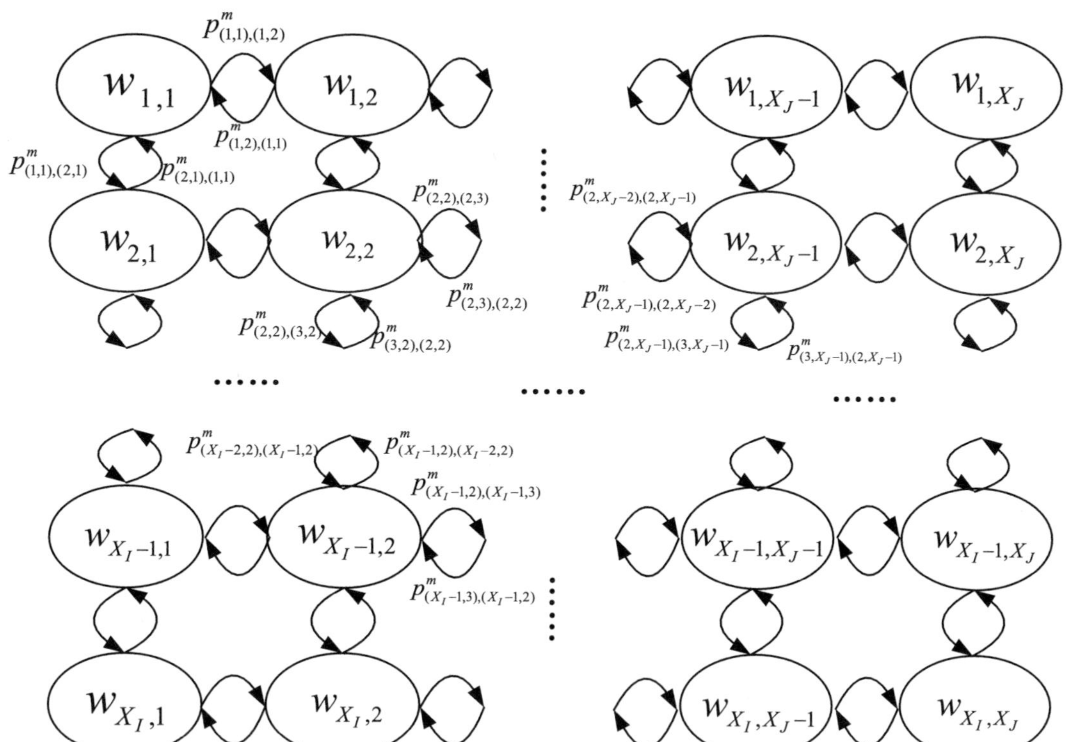

Fig. 7.3 Markov chain model of u_m

$$\begin{cases} \boldsymbol{\phi}^m \mathbf{P}^{m'} = \boldsymbol{\phi}^m \\ \sum \boldsymbol{\phi}^m = 1. \end{cases} \quad (7.1)$$

In the rest of the chapter, we denote the stationary probability of the state $w_{i,j}$ by $\phi^m_{i,j}$, which is equal to the element of $\phi^m_{(i-1)X_J+j}$ in the vector $\boldsymbol{\phi}^m$.

7.2.2 Total Sojourn Duration of the Square Regions

Since the users only change their moving directions when they arrive at any of the crossroads, the travel distance of the user u_m within any of the square regions $w_{i,j}$ is l. For a specific square region $w_{i,j}$, since we have $r^C_{i,j} < l/2$, the crowded range of the crossroad $s_{i,j}$ is also included in this square region. As a result, when the user u_m visits the square region $w_{i,j}$, its sojourn duration $D^S_{m,i,j}$ within this region is derived as

$$D^S_{m,i,j} = 2\left(\frac{r^C_{i,j}}{v_{m,i,j}} + \frac{l/2 - r^C_{i,j}}{v_{m,0}}\right). \quad (7.2)$$

The sojourn distribution of the user u_m sojourning within a specific square region can be defined as the set $\{\pi^S_{m,1,1}, \ldots, \pi^S_{m,i,j}, \ldots, \pi^S_{m,X_I,X_J}\}$. Note that the probability $\pi^S_{m,i,j}$ is different from the stationary probability $\phi^m_{i,j}$ of the state $w_{i,j}$ in the Markov chain of Fig. 7.3, since the derivation of $\phi^m_{i,j}$ only considers the state transition probabilities but the derivation of $\pi^S_{m,i,j}$ further considers the time duration of u_m staying in a specific state. Based on the steady probability $\phi^m_{i,j}$ of the user u_m staying within the square regions $w_{i,j}$ and its corresponding sojourn durations $D^S_{m,i,j}$, the probability of the user u_m sojourning within the square region $w_{i,j}$ can be derived as

$$\pi^S_{m,i,j} = \frac{\phi^m_{i,j} D^S_{m,i,j}}{\sum\limits_{i',j'} \phi^m_{i',j'} D^S_{m,i',j'}}. \tag{7.3}$$

Given a specific observation duration T, the total sojourn duration $\tau^S_{m,i,j}$ of the user u_m in the square region $w_{i,j}$ can be expressed as $\tau^S_{m,i,j} = \pi^S_{m,i,j} T$.

7.2.3 Total Sojourn Duration Within the WIT and WET Ranges

We firstly aim for obtaining the sojourn duration of the user u_m within the WIT range of the H-AP, if it is deployed at the crossroad $s_{i,j}$. The travel distance of the user u_m within the WIT range should always be $2r^D$. If the WIT range r^D is shorter than the crowded range $r^C_{i,j}$, the user u_m always move within the WIT range at a speed of $v_{m,i,j}$. If the WIT range r^D is longer than the crowded range $r^C_{i,j}$, the user u_m may move for a distance of $2r^C_{i,j}$ at a speed of $v_{m,i,j}$ and it may move for another distance of $(2r^D - 2r^C_{i,j})$ at a speed of $v_{m,0}$. As a result, the sojourn duration of the user u_m within the WIT range can be expressed as:

$$D^D_{m,i,j} = \begin{cases} \dfrac{2r^D}{v_{m,i,j}}, & r^D < r^C_{i,j} \\[2ex] \dfrac{2r^C_{i,j}}{v_{m,i,j}} + \dfrac{2\left(r^D - r^C_{i,j}\right)}{v_{m,0}}, & r^D \geq r^C_{i,j} \end{cases}. \tag{7.4}$$

By sticking to the same methodology, the sojourn duration $D^E_{m,i,j}$ of the user u_m within the WET range is formulated as:

$$D^E_{m,i,j} = \begin{cases} \dfrac{2r^E}{v_{m,i,j}}, & r^E < r^C_{i,j} \\[2ex] \dfrac{2r^C_{i,j}}{v_{m,i,j}} + \dfrac{2\left(r^E - r^C_{i,j}\right)}{v_{m,0}}, & r^E \geq r^C_{i,j} \end{cases}. \tag{7.5}$$

Furthermore, the total sojourn duration $\tau^D_{m,i,j}$ of the user u_m within the WIT range of the H-AP deployed at the crossroad $s_{i,j}$ during the entire observation duration T is derived as:

$$\tau^D_{m,i,j} = \tau^S_{m,i,j} \frac{D^D_{m,i,j}}{D^S_{m,i,j}}. \tag{7.6}$$

Similarly, the total sojourn duration $\tau^E_{m,i,j}$ of the user u_m within the WET range of the H-AP deployed at the crossroad $s_{i,j}$ during the entire observation duration T can be derived by substituting $D^E_{m,i,j}$ for $D^D_{m,i,j}$ into (7.6). According to the above analysis, the following remark is obtained:

Remark 7.1 For a specific crossroad s_{i_0,j_0}, when the crowded range of other crossroads are fixed, the total sojourn duration of the user u_m in the WIT range of s_{i_0,j_0} is a convex function with respect to $r^C_{i_0,j_0}$. The total sojourn duration $\tau^D_{m,i,j}$ can be maximized when the crowded range $r^C_{i_0,j_0}$ is equal to the WIT range r^D. By contrast, the total sojourn duration of the user u_m within the WIT range of any other crossroad is a monotonously decreasing function with respect to the crowded range $r^C_{i_0,j_0}$ of the crossroad s_{i_0,j_0}. Same conclusion can be made on the total sojourn duration of the user u_m within the WET range of the H-AP.

7.3 Problem Formulation

By considering the users' mobility pattern and the popularity of the crossroads, we aim for obtaining an optimal H-AP deployment scheme for maximizing either the overall WIT or the overall WET performance of the MSN.

7.3.1 *WIT Efficiency*

The successful WIT efficiency is defined as the ratio between the actually delivered information amount and the maximum possible delivered information amount of all the users in the MSN studied. We assume that a user may download the information from the H-AP with a constant rate μ, which is readily realized by invoking power control, the WIT efficiency of all the M users during the observation duration T is formulated as

$$\eta = \frac{\sum\limits_{m}\sum\limits_{i,j} \mu \tau^{D}_{m,i,j} y_{i,j}}{\sum\limits_{m}\sum\limits_{i,j} \mu \tau^{D}_{m,i,j}} = \frac{\sum\limits_{m}\sum\limits_{i,j} \tau^{D}_{m,i,j} y_{i,j}}{MT}, \tag{7.7}$$

where $y_{i,j}$ denotes the deployment state of the H-AP at the crossroad $s_{i,j}$. Furthermore, $y_{i,j} = 1$ indicates that an H-AP is deployed at $s_{i,j}$, while $y_{i,j} = 0$ indicates that no H-AP is deployed at $s_{i,j}$. All the $y_{i,j}$ form an H-AP deployment matrix $\mathbf{Y}$, having a size of $X_I \times X_J$. The numerator of (7.7) indicates the actually delivered information amount of all the users, given a specific deployment scheme of the H-APs, while the denominator of (7.7) indicates the theoretically achievable information amount of all the users, if at least an H-AP is deployed at each crossroad. According to the above analysis, we have the following remark:

Remark 7.2 If no H-AP is deployed at the crossroad s_{i_0,j_0}, the WIT efficiency η is a monotonously decreasing function with respect to $r^{C}_{i_0,j_0}$. By contrast, if an H-AP is deployed at the corssroad s_{i_0,j_0}, the WIT efficiency η is a convex function with respect to $r^{C}_{i_0,j_0}$. Furthermore, the WIT efficiency η can be maximized, when we have $r^{C}_{i_0,j_0} = r^{D}$.

As a result, in order to improve the WIT efficiency, we should make the crossroads less crowded if no H-AP is deployed, while we should let $r^{C}_{i,j} = r^{D}$, if an H-AP is deployed at the crossroad $s_{i,j}$.

7.3.2 *WET Efficiency*

We define the WET efficiency as the total harvested energy amount by all the users in the network. Due to the path-loss incurred channel attenuation, the power of the received RF signals at the users is determined by the signals' propagation distance. For example, when the user u_m enters the WET range of the H-AP deployed at the crossroad $s_{i,j}$, the distance from u_m to the H-AP firstly reduces. Since the user keeps moving, the distance between u_m and the H-AP at $s_{i,j}$ will first decrease from r^E to 0 and increase to r^E again. According to [28], the path-loss between the H-AP and u_m, when they are ς_m meters away from each other, can be expressed as

$$h_m = G_t G_r \left(\frac{c}{4\pi f \varsigma_0}\right)^2 \left(\frac{\varsigma_0}{\varsigma_m}\right)^{\beta}, \quad \varsigma_m > \varsigma_0, \tag{7.8}$$

where G_t and G_r are the transmit and receive antenna gains, respectively, while c is the speed of the light, f is the carrier frequency, and β is the path-loss exponent. Furthermore, ς_0 and ς_m are the reference distance and the distance between u_m and the H-AP, respectively. Denoting the path-loss at the distance of ς_0 as h_0, the path-loss model of (7.8) can be reformulated as

$$h_m = h_0 \left(\frac{\varsigma_0}{\varsigma_m}\right)^{\beta}, \quad \varsigma_m > \varsigma_0. \tag{7.9}$$

In the near-field of the transmitters, say $\varsigma_m \leq \varsigma_0$, we assume that the path-loss is $h_m = h_0$. For the short-range communication, ς_0 is normally a single meter [28], which is even lower than the minimum among the WIT range r_D, the WET range r^E and the crowded range $r^C_{i,j}$. As a result, the energy harvested by the user u_m, when it enters the transmission range of the H-AP deployed at the crossroad $s_{i,j}$, can be formulated as

$$\varepsilon_{m,i,j} = \begin{cases} 2\delta\left(\int_{\varsigma_0}^{r^E} \frac{Ph_m}{v_{m,i,j}} d(\varsigma_m) + \frac{Ph_0\varsigma_0}{v_{m,i,j}}\right), & r^E \leq r^C_{i,j} \\ 2\delta\left(\int_{\varsigma_0}^{r^C_{i,j}} \frac{Ph_m}{v_{m,i,j}} d(\varsigma_m) + \int_{r^C_{i,j}}^{r^E} \frac{Ph_m}{v_{m,0}} d(\varsigma_m)\right. & \\ \left. + \frac{Ph_0\varsigma_0}{v_{m,i,j}}\right), & r^E > r^C_{i,j} \end{cases}, \tag{7.10}$$

where P indicates the transmitting power of the H-AP, while δ indicates the efficiency of rectifying the received RF signals to the direct current (DC) and we make $\delta = 80\%$ without loss of generality. Finally, the closed-form equation of the energy harvested by the user u_m can be derived from (7.10) as

$$\varepsilon_{m,i,j} = \begin{cases} 2\delta \frac{Ph_0\varsigma_0}{(1-\beta)v_{m,i,j}} \left(\left(\frac{r^E}{\varsigma_0}\right)^{1-\beta} - \beta\right), & r^E \leq r^C_{i,j} \\ 2\delta \frac{Ph_0\varsigma_0}{(1-\beta)} \left(\left(\frac{1}{v_{m,i,j}} - \frac{1}{v_{m,0}}\right)\left(\frac{r^C_{i,j}}{\varsigma_0}\right)^{1-\beta}\right. & \\ \left. + \frac{1}{v_{m,0}}\left(\frac{r^E}{\varsigma_0}\right)^{1-\beta} - \frac{\beta}{v_{m,i,j}}\right), & r^E > r^C_{i,j} \end{cases}. \tag{7.11}$$

As shown in (7.11), the amount $\varepsilon_{m,i,j}$ of energy harvested by the user u_m at the crossroad $s_{i,j}$ increases, as the crowded range $r^C_{i,j}$ of $s_{i,j}$ increases. If we keep increasing $r^C_{i,j}$ beyond $r^E_{i,j}$, the energy $\varepsilon_{m,i,j}$ converges to a constant value.

Given the total observation duration T, the sojourn duration of the user u_m within the WET range of the H-AP deployed at the crossroad $s_{i,j}$ is derived as $\tau^E_{m,i,j}$. As a result, the number of times that u_m moves within the WET range of the H-AP deployed at the crossroad $s_{i,j}$ during the entire observation duration T can be calculated as $\frac{\tau^E_{m,i,j}}{D^E_{m,i,j}}$. Therefore, the total energy harvested by the user u_m during the total observation duration T can be formulated as

$$\varepsilon_m = \sum_{i,j} \frac{\tau^E_{m,i,j}}{D^E_{m,i,j}} \varepsilon_{m,i,j} y_{i,j}. \tag{7.12}$$

Since the energy harvested by the users have to be stored in their batteries, energy harvesting may not operate, once their batteries are fully charged. Without loss of generality, we assume that all the users' battery capacity are Q. Hence, the actual energy harvested by the user u_m can be expressed as $\min(\varepsilon_m, Q)$. Furthermore, the WET efficiency can be obtained as

$$\varepsilon = \sum_m \min\left(\varepsilon_m, Q\right). \tag{7.13}$$

According to the above analysis, we have the remark as follows:

Remark 7.3 When the crowded range of the crossroad s_{i_0,j_0} satisfies the condition of $r^C_{i_0,j_0} > r^E$, the WET efficiency does not increase as the crowded range $r^C_{i_0,j_0}$ increases, regardless of whether an H-AP is deployed at the crossroad s_{i_0,j_0} or not.

Remark 7.3 tells us that in order to improve the WET efficiency, we should make the crowded range $r^C_{i_0,j_0}$ of the crossroad s_{i_0,j_0} smaller, when it satisfies the condition $r^C_{i_0,j_0} > r^E$.

7.3.3 *H-AP Deployment Schemes*

We will study three deployment schemes of the H-APs for serving different purposes, namely the balanced deployment scheme (B-deployment), the energy harvesting oriented deployment scheme (E-deployment) and the information downloading oriented deployment scheme (I-deployment).

In order to maximize the WIT efficiency η by ensuring that the total energy harvested by the users is higher than a threshold, we propose the B-deployment scheme, which can be formulated as the following optimisation problem:

$$(\text{P7.1.1}): \quad \max_{\mathbf{Y}} \ \eta, \tag{7.14}$$

$$\text{s.t.}: \quad \sum_{i,j} y_{i,j} = K, \tag{7.14a}$$

$$y_{i,j} = 0 \ or \ 1, \tag{7.14b}$$

$$\varepsilon \geq \alpha \varepsilon_{\max}, \tag{7.14c}$$

where the parameter $\alpha \in [0, 1]$ is introduced for adjusting the designer's preference on the WIT performance and the WET performance. A higher value of α indicates that the designer is inclined to obtain a higher WET performance, while a lower value of α indicates that the design is inclined to the WIT performance. The parameter $\varepsilon_{\max}$ in (7.14c) is the maximum energy that can be harvested by the users, which can be obtained by solving the following optimisation problem:

$$(\text{P7.1.2}): \quad \max_{\mathbf{Y}} \ \varepsilon, \tag{7.15}$$

$$\text{s.t.} \quad \sum_{i,j} y_{i,j} = K, \tag{7.15a}$$

$$y_{i,j} = 0 \ or \ 1. \tag{7.15b}$$

The optimisation problem (7.15) is regarded as the E-deployment scheme. Note that if we have $\alpha = 1$ in (7.14c), the optimisation problem (7.14) is degenerated to (7.15). If we have $\alpha = 0$ in (7.14c), the B-deployment scheme is degenerated to the I-deployment scheme, which can be formulated as

$$(\text{P7.1.3}): \quad \max_{\mathbf{Y}} \ \eta, \tag{7.16}$$

$$\text{s.t.} \quad \sum_{i,j} y_{i,j} = K, \tag{7.16a}$$

$$y_{i,j} = 0 \ or \ 1. \tag{7.16b}$$

7.4 Solutions of the H-AP Deployment

7.4.1 *Exhaustive Searching Method*

Since the deployment solution $\mathbf{Y}$ is an integer matrix, the above three problems cannot be solved in a polynomial time. The exhaustive search is capable of traversing all the possible deployment solutions and it thus chooses the optimal one. This is the most inefficient method since its complexity is as high as $\mathbf{C}^K_{X_I X_J}$. However, when the solution space is small, exhaustive search can be adopted for finding the optimal solution with the tolerant complexity. Let us define $g(\mathbf{Y})$ as the

Algorithm 7.1 exhaustive searching method

```
Define all the $\mathbf{C}_{X_I X_J}^K$ possible H-AP deployment set as $\Psi = \{\mathbf{Y}_1, \mathbf{Y}_2, \ldots, \mathbf{Y}_{\mathbf{C}_{X_I X_J}^K}\}$, in which each element in $\Psi$ is an optional deployment.
  Define $\mathbf{Y}^* = \mathbf{Y}_1$, $g(\mathbf{Y}^*) = 0$. Initialize deployment subscript $p = 1$;
while $p \le \mathbf{C}_{X_I X_J}^K$ do
    Calculate $g(\mathbf{Y}_p)$;
    If $g(\mathbf{Y}_p) \ge g(\mathbf{Y}^*)$ and $\mathbf{Y}_p$ satisfies all the constraints, let $\mathbf{Y}^* = \mathbf{Y}_p$ and $g(\mathbf{Y}^*) = g(\mathbf{Y}_p)$;
    $p = p + 1$;
end while
return The solution of H-AP deployment $\mathbf{Y}^*$.
```

Algorithm 7.2 Directional searching method

```
Calculate all the value of $k_{i,j}$ in (7.17). Initialize deployment subscript $p = 1$, and deployment matrix $\mathbf{Y}$;
while $p \le K$ do
    Find the max $k_{i,j}$;
    Let $y_{i,j} = 1$;
    Let $k_{i,j} = 0$;
end while
return The solution of H-AP deployment $\mathbf{Y}$.
```

generic objective function with respect to the H-AP's deployment indicator matrix $\mathbf{Y}$. Hence, we have $g(\mathbf{Y}) = \eta$ for both the I-deployment problem and the B-Deployment scheme. We also have $g(\mathbf{Y}) = \varepsilon$ for the E-deployment problem. The pseudocode of the exhaustive search is provided in Algorithm 7.1.

Since the exhaustive search is an aimless searching method, we have to traverse all the possible deployment solution before finding the optimal one. Since the objective function of the I-deployment problem (7.16) is linear, a directional searching method can be adopted for reducing the complexity. By letting $k_{i,j} = \sum_m \tau_{m,i,j}^D / MT$, the WIT efficiency η expressed in (7.7) can be reformulated as

$$\eta = \sum_{i,j} k_{i,j} y_{i,j}. \tag{7.17}$$

In order to maximize the WIT efficiency η, we should deploy an H-AP at the crossroad $s_{i,j}$ associated with a higher coefficient $k_{i,j}$. The pseudocode of the directional searching is provided in Algorithm 7.2.

By adopting the directional searching, the complexity can be reduced to $\mathcal{O}(K)$. By contrast, since the objective functions in the E-deployment problem is nonlinear, the directional searching cannot be adopted for finding the optimal deployment solution. Furthermore, since there is a nonlinear constraint on the energy harvesting performance, the directional searching cannot be adopted for finding the optimal solution of the B-deployment problem.

7.4.2 *H-AP Deploying Probability*

Observe from (7.14)–(7.16) that the optimisation problems of the B-deployment scheme, the E-deployment scheme and the I-deployment scheme all belong to the integer programming, since the deployment indicator $y_{i,j}$ at the crossroad $s_{i,j}$ is either zero or one, which cannot be solved in polynomial time. As a result, we transform these three integer programming problem into the real number based optimisation problem by relaxing the original integer constraint on $y_{i,j}$ to the constraint that $y_{i,j}$ is a real number within the region [0, 1]. This relaxation is reasonable since a real valued $y_{i,j}$ can be regarded as the probability of an H-AP being deployed at the crossroad $s_{i,j}$.

We will elaborate on how the constraint relaxation operates in solving the optimisation problem of the B-deployment as an example, since both the E-deployment scheme and the I-deployment scheme are special cases of the B-deployment scheme. We introduce a new variable $R_m = \min(\varepsilon_m, Q)$ for representing the actual energy harvested by the user u_m. The constraint (7.14c) can be then rewritten as

$$\begin{cases} R_m \le \varepsilon_m \\ R_m \le Q, \\ \sum_m R_m \ge \alpha \varepsilon_{\max}. \end{cases}$$

Accordingly, the original optimisation problem of (7.14) can be reformulated as

$$\text{(P7.1.4)}: \quad \max_{\mathbf{Y}, R_m} \sum_{i,j} k_{i,j} y_{i,j}, \tag{7.18}$$
$$\text{s.t.} \quad \sum_{i,j} y_{i,j} = K,$$
$$0 \leq y_{i,j} \leq 1,$$
$$R_m \leq \varepsilon_m,$$
$$R_m \leq Q,$$
$$\sum_m R_m \geq \alpha \varepsilon_{\max},$$

where $k_{i,j}$ is defined in Sect. 7.4.1. Since both the objective function and the constraints are linear with respect to $y_{i,j}$ and R_m, the optimisation problem (P7.1.4) can be solved by adopting the simplex method [29].

Similarly, the reformulated optimisation problem of the E-deployment scheme can be expressed as

$$\text{(P7.1.5)}: \quad \max_{\mathbf{Y}, R_m} \sum_{m} R_m, \tag{7.19}$$
$$\text{s.t.} \quad \sum_{i,j} y_{i,j} = K,$$
$$0 \leq y_{i,j} \leq 1,$$
$$R_m \leq \varepsilon_m,$$
$$R_m \leq Q,$$

and the reformulated optimisation problem of the I-deployment scheme can be expressed as

$$\text{(P7.1.6)}: \quad \max_{\mathbf{Y}} \sum_{i,j} k_{i,j} y_{i,j}, \tag{7.20}$$
$$\text{s.t.} \quad \sum_{i,j} y_{i,j} = K,$$
$$0 \leq y_{i,j} \leq 1.$$

Both these two optimisation problems can also be solved by adopting the simplex method.

7.4.3 *Branch-and-Bound Approach*

Since the exhaustive searching cannot solve the optimisation problem of the H-AP's deployment in a polynomial time and the directional searching is only capable of solving the optimisation problem of the I-deployment, we need a generic low-complexity algorithm for finding the optimal solution of all the three deployment scheme. As a result, the branch-and-bound approach is adopted for solving these optimisation problems by sacrificing the optimality to some extent. The key steps of the branch-and-bound approach are summarized as follows, where the B-deployment optimisation problem is chosen as an example:

STEP 1: Denote the upper bound and the lower bound of the H-AP's deployment indicator $y_{i,j}$ at the crossroad $s_{i,j}$ as $\Delta^U_{i,j}$ and $\Delta^D_{i,j}$, respectively. Let us initialize $\Delta^U_{i,j} = 1$ and $\Delta^D_{i,j} = 0$ for each $y_{i,j}$. The set of all $\Delta^U_{i,j}$ is denoted as $\boldsymbol{\Delta}^U$, while the set of all $\Delta^D_{i,j}$ is denoted as $\boldsymbol{\Delta}^D$.

STEP 2: We then transform original integer programming problem into the following real valued optimisation problem as

Algorithm 7.3 Branch-and-bound algorithm to obtain the solution of H-AP deployment

Define the upper and lower constraint bound of the optimisation variable as $\boldsymbol{\Delta}^U$ and $\boldsymbol{\Delta}^D$, respectively. Initialize the value of each element in these two constraint sets, as stated at STEP 1;
Solve the transformed real optimisation problem (7.21) and obtain the H-AP deployment $\mathbf{Y}^*$, as stated at STEP 2;
while At least one element $y_{i,j}$ in $\mathbf{Y}^*$ is not an integer **do**
  Separate constraints $\boldsymbol{\Delta}^U$ and $\boldsymbol{\Delta}^D$ into two divided parts, as stated at STEP 3;
  Solve two sub-problem (7.22) and (7.23), and obtain the corresponding H-AP deployment $\mathbf{Y}_1^*$ and $\mathbf{Y}_2^*$;
  Compare two objective value with $\mathbf{Y}_1^*$ and $\mathbf{Y}_2^*$. If $g(\mathbf{Y}_1^*) > g(\mathbf{Y}_2^*)$, let $\boldsymbol{\Delta}^U = \boldsymbol{\Delta}^{U1}$, $\boldsymbol{\Delta}^D = \boldsymbol{\Delta}^{D1}$, and $\mathbf{Y}^* = \mathbf{Y}_1^*$. Else let $\boldsymbol{\Delta}^U = \boldsymbol{\Delta}^{U2}$, $\boldsymbol{\Delta}^D = \boldsymbol{\Delta}^{D2}$ and $\mathbf{Y}^* = \mathbf{Y}_2^*$, as stated at STEP 4;
end while
return The solution of H-AP deployment $\mathbf{Y}^*$.

$$
\begin{aligned}
(\text{P7.1.7}):\quad & \max_{\mathbf{Y}, R_m} \; g(\mathbf{Y}), \\
\text{s.t.}\quad & \Delta_{i,j}^{D} \le y_{i,j} \le \Delta_{i,j}^{U}, \forall s_{i,j}, \\
& \sum_{i,j} y_{i,j} = K, \; (7.14\text{c}),
\end{aligned}
\tag{7.21}
$$

where $g(\mathbf{Y}) = \sum_{i,j} k_{i,j} y_{i,j}$. We can use the same method for solving the real programming problem in Sect. 7.4.2, while the optimal solution of all the $y_{i,j}$ of (7.21) is denoted as the set $\mathbf{Y}^*$.

STEP 3: If the resultant in $\mathbf{Y}^*$ is an integer matrix, $\mathbf{Y}^*$ is the final solution. Otherwise, let $\boldsymbol{\Delta}^{U1} = \boldsymbol{\Delta}^{U2} = \boldsymbol{\Delta}^U$ and $\boldsymbol{\Delta}^{D1} = \boldsymbol{\Delta}^{D2} = \boldsymbol{\Delta}^D$. Then, arbitrarily choose a non-integer element $y_{i,j}$ in $\mathbf{Y}^*$. Let $\Delta_{i,j}^{U1} = \lfloor y_{i,j} \rfloor$ and $\Delta_{i,j}^{D2} = \lceil y_{i,j} \rceil$, where $\lfloor x \rfloor$ and $\lceil x \rceil$ indicates the largest integer smaller than x and the smallest integer larger than x, respectively.

STEP 4: Divide the optimisation problem (7.21) into the following two sub-problems

$$
\begin{aligned}
(\text{P7.1.8}):\quad & \max_{\mathbf{Y}, R_m} \; g(\mathbf{Y}), \\
\text{s.t.}\quad & \Delta_{i,j}^{D1} \le y_{i,j} \le \Delta_{i,j}^{U1}, \forall s_{i,j}, \\
& \sum_{i,j} y_{i,j} = K, \; (7.14\text{c}),
\end{aligned}
\tag{7.22}
$$

and

$$
\begin{aligned}
(\text{P7.1.9}):\quad & \max_{\mathbf{Y}, R_m} \; g(\mathbf{Y}), \\
\text{s.t.}\quad & \Delta_{i,j}^{D2} \le y_{i,j} \le \Delta_{i,j}^{U2}, \forall s_{i,j}, \\
& \sum_{i,j} y_{i,j} = K, \; (7.14\text{c}).
\end{aligned}
\tag{7.23}
$$

Both of these two sub-problems can be solved by exploiting the simplex method of Sect. 7.4.2. The optimal solutions of these two sub-problems are denoted as $\mathbf{Y}_1^*$ and $\mathbf{Y}_2^*$, respectively. If we have $g(\mathbf{Y}_1^*) > g(\mathbf{Y}_2^*)$, we have $\boldsymbol{\Delta}^U = \boldsymbol{\Delta}^{U1}$, $\boldsymbol{\Delta}^D = \boldsymbol{\Delta}^{D1}$, and $\mathbf{Y}^* = \mathbf{Y}_1^*$. Otherwise, we have $\boldsymbol{\Delta}^U = \boldsymbol{\Delta}^{U2}$, $\boldsymbol{\Delta}^D = \boldsymbol{\Delta}^{D2}$ and $\mathbf{Y}^* = \mathbf{Y}_2^*$.

The pseudocode of our branch-and-bound approach is provided in Algorithm 7.3. The complexity of the branch-and-bound approach is $\mathcal{O}(2K)$, which is much lower than that of the exhaustive searching method.

7.5 Performance Evaluation

In this section, numerical results are provided for evaluating the performance of the three deployment schemes of the H-AP. The users in the MSN studied move within the city street blocks having 5×5 crossroads in our simulation and each crossroad

Table 7.1 Parameter settings

Number of users M	100
Total simulation time T	10 hours
Number of H-APs K	8
Street length l	200 m
WIT range r^D	50 m
WET range r^E	10 m
Crowd range $r^C_{i,j}$	5 m–60 m
Reference distance λ_0	1 m
Path-loss at λ_0	0.003
Battery capacity Q	1J
Importance parameter α	0.97
Energy transmitting power P	1W

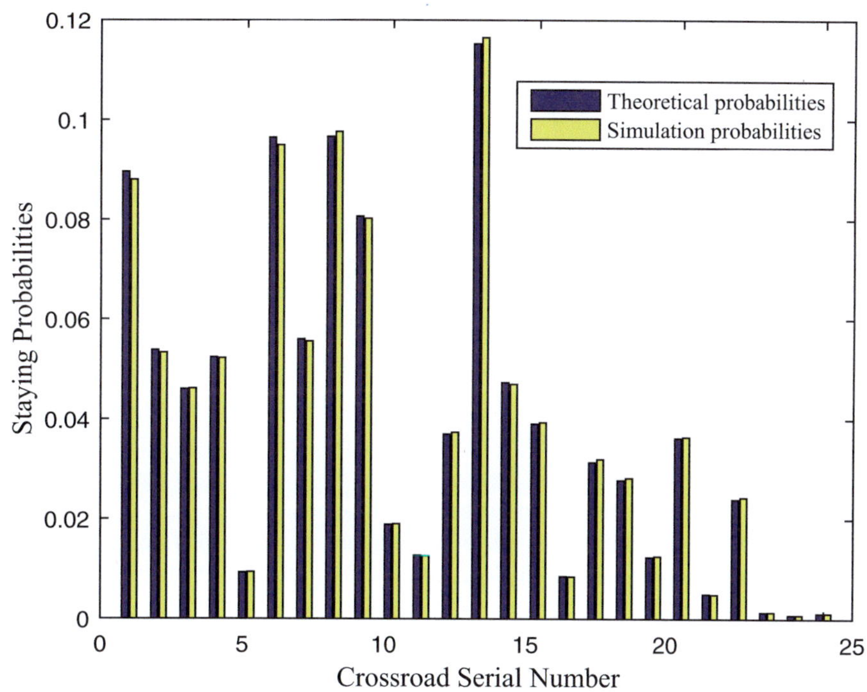

Fig. 7.4 Stationary probabilities of a user staying in different crossroads

has a different crowded range. The moving speed of the users in different spaces are arbitrary. However, the speed of a user moving within a crowded range is lower than the speed moving outside of a crowded range. Furthermore, the probability of a user changing its moving direction at a crossroad is also arbitrarily chosen. The parameter settings are summarized in Table 7.1, if there is no specific statement.

7.5.1 *Accuracy of Markov Analysis*

We evaluate a single user's stationary probability of staying at different crossroads by both the Markov chain based analysis and by the Monte-Carlo simulation. Observe from Fig. 7.4 that our analytical results perfectly match the simulation results, which demonstrates the accuracy of our Markov chain aided analysis.

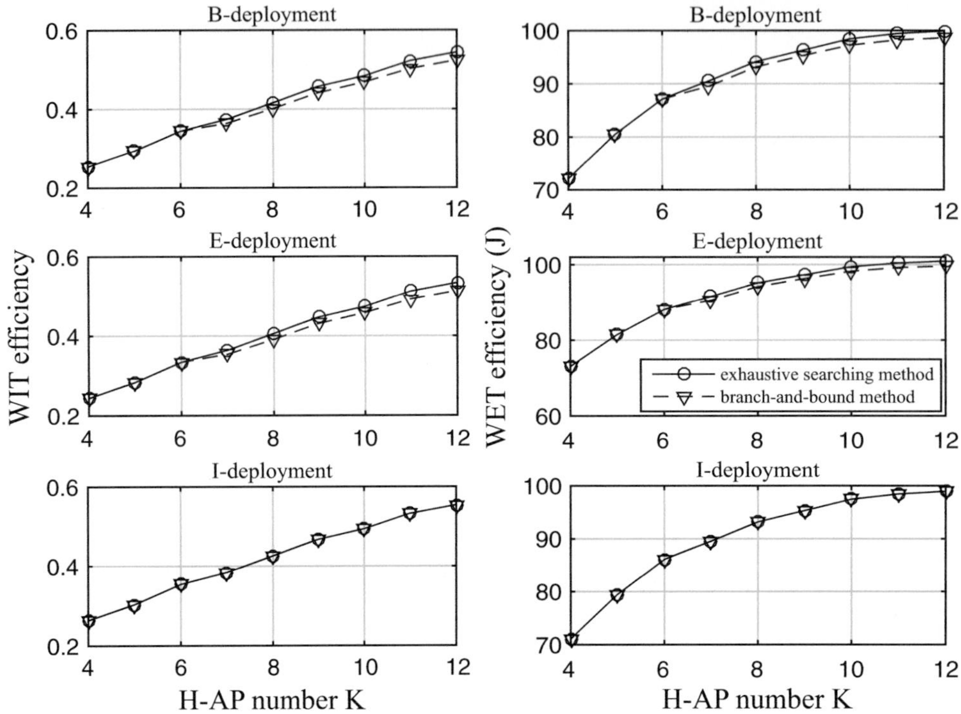

Fig. 7.5 Solutions between two methods

7.5.2 *Exhaustive Searching v.s. Branch-and-Bound*

We firstly compare the exhaustive searching method to the branch-and-bound counterpart in terms of the WIT and the WET efficiency. Specifically, the exhaustive searching is capable of finding the optimal deployment solution, while the branch-and-bound based method may only find the sub-optimal deployment solution. Observe from Fig. 7.5 that the solutions obtained by relying on the branch-and-bound approach are always worse than those of exhaustive searching method, since the branch-and-bound approach sacrifices the optimality for a lower complexity. Note that in the I-deployment scheme, the solutions of these two methods perform the same. This is because the first iteration of the branch-and-bound approach is capable of producing an integer deployment solution $\mathbf{Y}$. Moreover, observe from Fig. 7.5 that the performance gaps between the solutions of the branch-and-bound method and those of the exhaustive searching method are negligible. Hence, when the solution space is very large, the branch-and-bound approach can be adopted for reducing the computational complexity without remarkable performance degradation.

7.5.3 *Popularity of Crossroads*

We randomly choose a crossroad and change its crowded range for evaluating its impact on the users' sojourn duration and the probability of an H-AP being deployed at this crossroad, while the crowded ranges of other crossroads are fixed. Observe from Fig. 7.6 that, users' sojourn duration within the WIT range and the WET range of the H-AP deployed at the chosen crossroad both firstly increase and then reduce as its crowded range increases. Furthermore, when the crowded range is the same as the WIT range r^D (or the same as the WET range r^E), the sojourn duration within the WIT range (or the WET range) is maximized. By contrast, the user's sojourn duration within the WIT range and that within the WET range of the H-AP deployed at another crossroad both reduces as we increase the crowded range of the chosen crossroad. These observations are in line with Remark 7.1.

Furthermore, we plot the probabilities of an H-AP being deployed at the chosen crossroad against its crowded range in Fig. 7.7. Observe from Fig. 7.7 that the probability of the H-AP being deployed is a non-decreasing function with respect to the crowded range of the chosen crossroad, regardless of any specific deployment scheme. Moreover, the H-AP's deploying

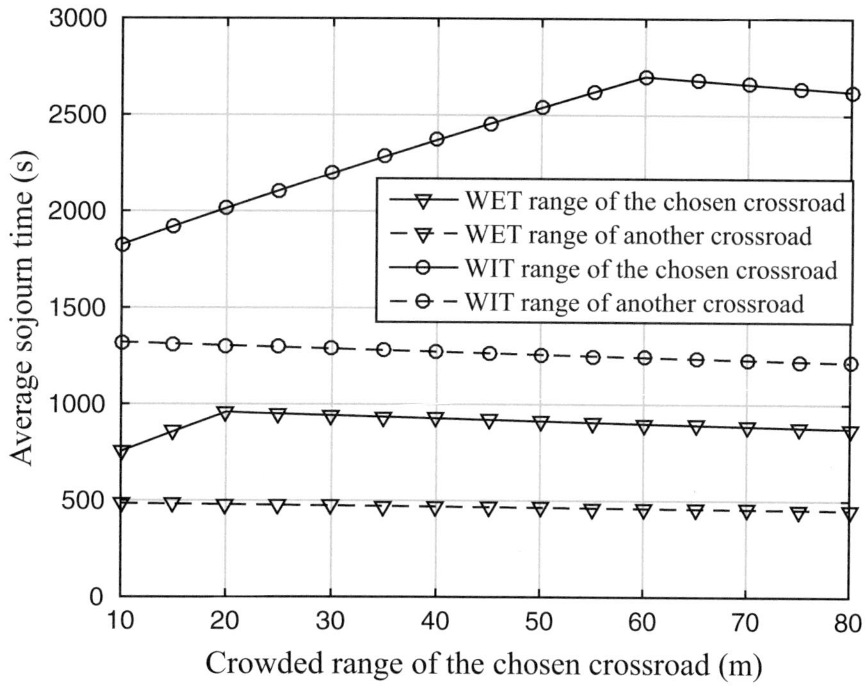

Fig. 7.6 Sojourn time of users at crossroads versus crowded ranges

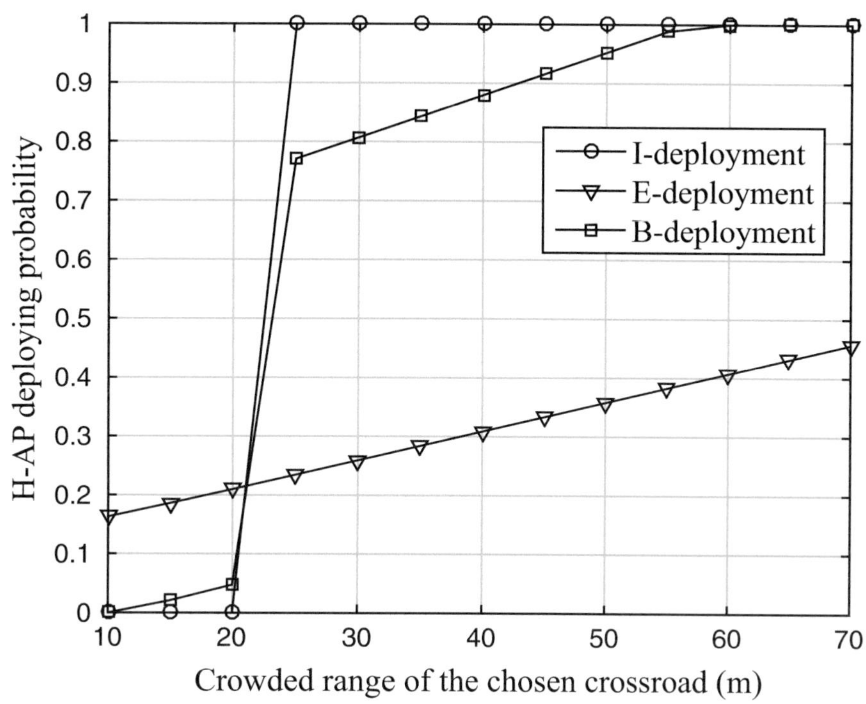

Fig. 7.7 H-AP deploying probabilities versus crowded ranges

probability in the I-deployment scheme is either 0 or 1. Furthermore, the deploying probability of the H-AP in the B-deployment scheme is a compromise between the I-deployment scheme and the E-deployment scheme, since it aims for balancing both the information downloading requirement and the energy harvesting requirement of the users.

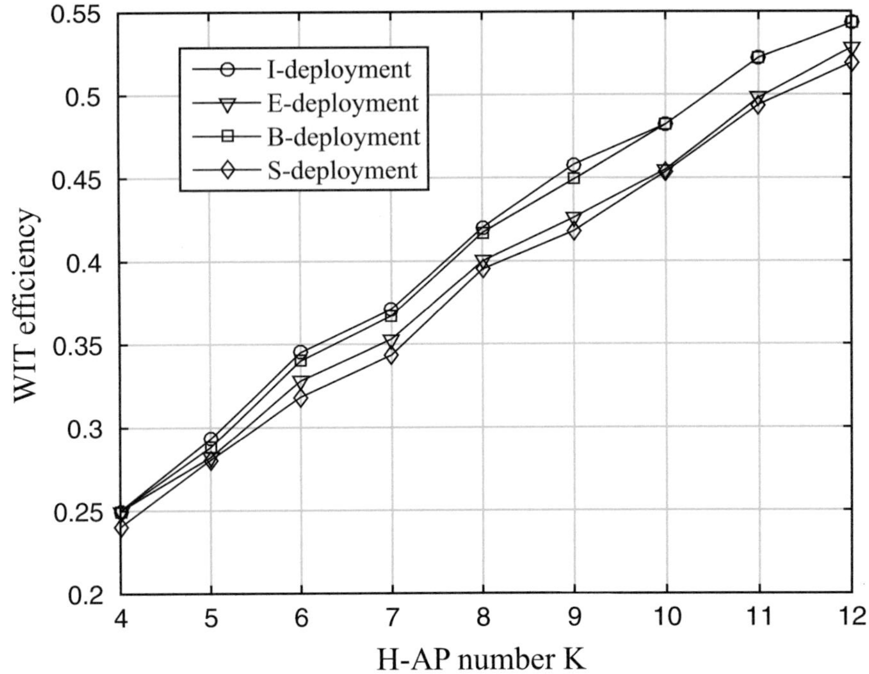

Fig. 7.8 WIT efficiency versus number of H-APs

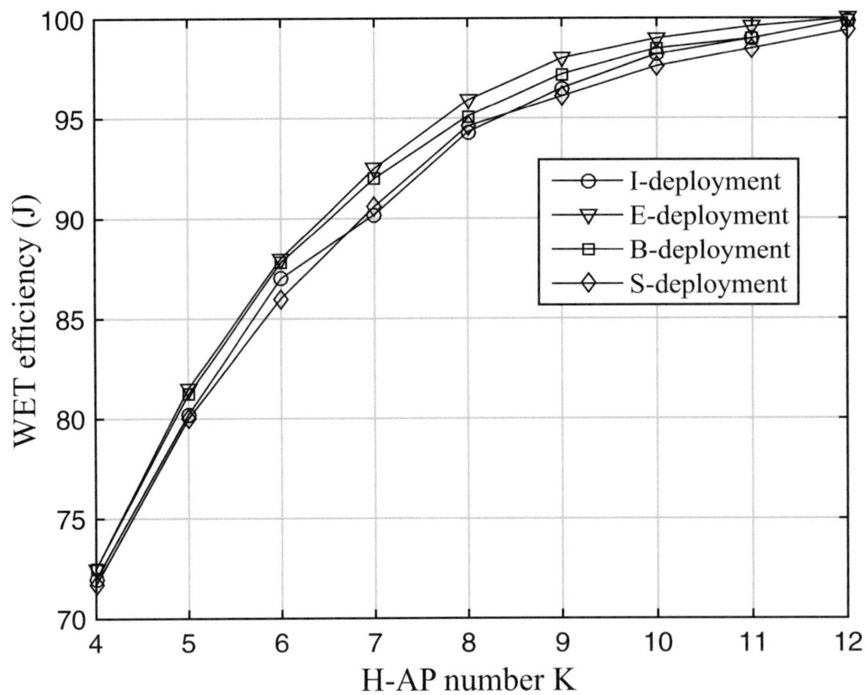

Fig. 7.9 WET efficiency versus number of H-APs

7.5.4 Information Downloading and Energy Harvesting

In Figs. 7.8 and 7.9, we plot both the WIT and the WET efficiencies against the number of the H-APs pending to be deployed. We choose the S-deployment scheme originally proposed by Ying et al [7]. In this scheme, the H-APs are deployed by only considering the visiting frequency of users at each crossroad. We can readily obtain that our proposed H-AP deployment

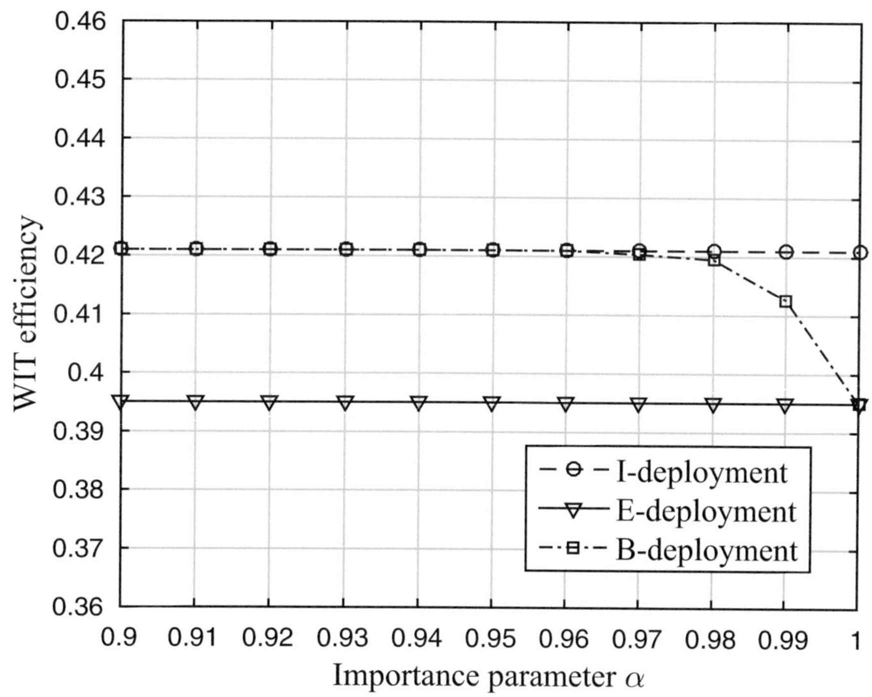

Fig. 7.10 WIT efficiency versus importance parameter α

schemes are better than the S-deployment in terms of both the WET and WIT efficiencies. Observe from Fig. 7.8 that the I-deployment scheme outperforms its counterparts in terms of the WIT efficiency, while the E-deployment scheme performs best in terms of the WET efficiency, as illustrated in Fig. 7.9. Furthermore, as shown in both Figs. 7.8 and 7.9, the B-deployment scheme achieves a compromise performance between the I-deployment scheme and the E-deployment one. Moreover, both the WIT and WET efficiencies increase, as we increase the number of the H-APs.

In Figs. 7.10 and 7.11, we investigate the impact of the preference parameter α on the performance of the B-deployment scheme. Observe from Fig. 7.10 that the I-deployment performs best in terms of the information downloading performance. When α is lower than 0.95, which indicates a relatively relaxed energy constraint, the B-deployment scheme has the same solution as the I-deployment scheme. By contrast, as α continuously grows, the information downloading performance of the B-deployment reduces until reaches that of the E-deployment. In Fig. 7.11, the opposite trend is observed. As the preference parameter α continues to increase beyond 0.96, the energy harvesting performance of the B-deployment scheme increases until it reaches that of the E-deployment.

7.6 Exercises

1. Under what circumstances will B-deployment scheme achieve optimal WET and WIT performance simultaneously? How does the importance parameter α affect the performance of WET and WIT? Define c_1 and c_2 as the performance weights for WIT and WET, respectively. When there is a relationship between alpha and c_1 and c_2, the overall performance of the two will be optimal?
2. What is the decode-and-forward (DF) and amplify-and-forward (AF) protocol? Moreover, what are the strengths and weaknesses of each respective protocol? In Sect. 7.2, the comparison of AoI performance is explored using these two relay protocols assuming there is sufficient energy. Could a comparison of the two protocols be made under conditions of limited energy or those that are energy-saving?
3. What is the secure communication and how does the behavior of UAV in SWIPT affect the performance indicators of secure communication? Covert communication is a popular trend that can achieve information blocking for eavesdroppers, you can look up information about it.
4. Assume that the UAV is equipped with two antennas: one for communication and the other for energy transmission. Would it be possible to explore the trajectory of a UAV equipped with a single antenna?

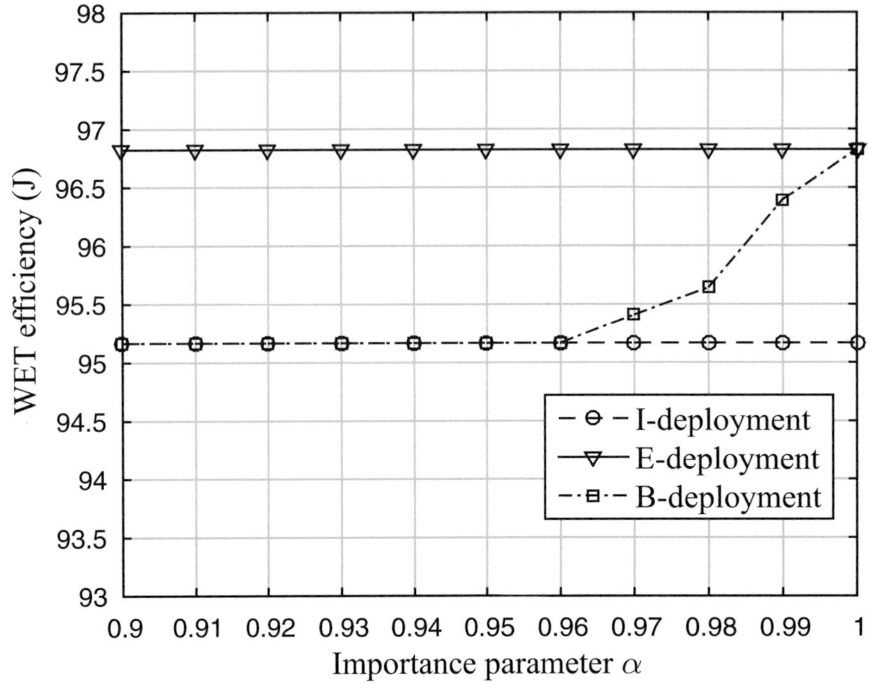

Fig. 7.11 WET efficiency versus importance parameter α

5. The author designed an adaptive method to set the initial iteration point by observing the structures of P7.3.16 and P7.3.17. Can you consider using the exhaustive method to find a suitable initial point and compare it with the above method?
6. Considering that the energy consumption model of the UAV consists of three parts, i.e., the computing energy, the wireless powering energy and the hovering energy. However, the power consumption of UAV also comprises energy usage during flight, which can greatly affect the UAV's hovering duration. How can you consider this in the energy consumption model of drones?

References

1. Wei, K., Liang, X., & Xu, K. (2014). A survey of social-aware routing protocols in delay tolerant networks: Applications, taxonomy and design-related issues. *IEEE Communications Surveys Tutorials, 16*(1), 556–578.
2. Akhtar, R., Leng, S., Memon, I., Ali, M., & Zhang, L. (2015). Architecture of hybrid mobile social networks for efficient content delivery. *Wireless Personal Communications, 80*(1), 85–96.
3. Vastardis, N. & Yang, K. (Third 2013) Mobile social networks: Architectures, social properties, and key research challenges. *IEEE Communications Surveys Tutorials, 15*(3), 1355–1371. ISSN: 1553-877X.
4. Jiang, L., Tian, H., Xing, Z., Wang, K., Zhang, K., Maharjan, S., Gjessing, S., & Zhang, Y. (2016). Social-aware energy harvesting device-to-device communications in 5g networks. *IEEE Wireless Communications, 23*(4), 20–27. ISSN: 1536–1284.
5. Hu, J., Yang, L. L., & Hanzo, L. (2015). Distributed multistage cooperative-social-multicast-aided content dissemination in random mobile networks. *IEEE Transactions on Vehicular Technology, 64*(7), 3075–3089. ISSN: 0018–9545.
6. Zhao, W., Chen, Y., Ammar, M., Comer, M. D., Levine, B. N., & Zegura, E. (2006). Capacity enhancement using throwboxes in DTNs. In *Proceedings of the IEEE MASS* (pp. 31–40).
7. Ying, Z., Zhang, C., & Wang, Y. (2014). Social based throwbox placement in large-scale throwbox-assisted delay tolerant networks. In *IEEE ICC* (pp. 2472–2477).
8. Fan, B., Leng, S., Shao, C., Zhang, Y., & Yang, K. (2015). Joint optimization of throwbox deployment and storage allocation in mobile social networks. In *2015 IEEE International Conference on Communications (ICC)* (pp. 1213–1218).
9. Fan, B., Leng, S., Yang, K., & Zhang, Y. (2015). Optimal storage allocation on throwboxes in mobile social networks. *Computer Networks, 91*, 90–100.
10. Yang, K., Ou, S., & Chen, H. H. (2008). On effective offloading services for resource-constrained mobile devices running heavier mobile internet applications. *IEEE Communications Magazine, 46*(1), 56–63. ISSN: 0163–6804.
11. Yu, R., Zhang, Y., Gjessing, S., Xia, W., & Yang, K. (2013). Toward cloud-based vehicular networks with efficient resource management. *IEEE Network, 27*(5), 48–55. ISSN: 0890–8044.

12. Yang, K., Yu, Q., Leng, S., Fan, B., & Wu, F. (2016). Data and energy integrated communication networks for wireless big data. *IEEE Access, 4*, 713–723. ISSN: 2169–3536.
13. Shi, Q., Liang, L., Xu, W., & Rui, Z. (2014). Joint transmit beamforming and receive power splitting for MISO SWIPT systems. *IEEE Transactions on Wireless Communications, 13*(6), 3269–3280.
14. Fouladgar, A. M., Simeone, O., & Erkip, E. (2014). Constrained codes for joint energy and information transfer. *IEEE Transactions on Communications, 62*(6), 2121–2131. ISSN: 1558–0857.
15. Huang, K. & Lau, V. K. N. (2014). Enabling wireless power transfer in cellular networks: Architecture modeling and deployment. *IEEE Transactions on Wireless Communications, 13*(2), 902–912.
16. Nintanavongsa, P., Naderi, M. Y., & Chowdhury, K. R. (2013). Medium access control protocol design for sensors powered by wireless energy transfer. In *IEEE INFOCOM-Mini Conference* (pp. 150–154)
17. Zhao, Y., Hu, J., Leng, S., & Yang, K. (2016). Transmission probability analysis of energy harvesting enabled 802.11 protocol. In *2016 2nd IEEE International Conference on Computer and Communications (ICCC)* (pp. 2036–2041)
18. Yu, Q., Zhao, Y., Zhang, L., Yang, K., Leng, S., & Wu, F. (2016). Fair energy-efficient resource allocation based on queue balancing in data and energy integrated communication networks. In *2016 IEEE Conference on Computer Communications Workshops (INFOCOM WKSHPS)* (pp. 812–819)
19. Ju, H. & Zhang, R. (2014). Throughput maximization in wireless powered communication networks. *IEEE Transactions on Wireless Communications, 13*(1), 418–428. ISSN: 1536–1276.
20. Liu, C. F., Maso, M., Lakshminarayana S., Lee, C. H., & Quek T. Q. S. (2015). Simultaneous wireless information and power transfer under different CSI acquisition schemes. *IEEE Transactions on Wireless Communications, 14*(4), 1911–1926.
21. Liu, L., Zhang, R., & Chua, K.-C. (2014). Multi-antenna wireless powered communication with energy beamforming. *IEEE Transactions on Communications, 62*(12), 4349–4361. ISSN: 0090–6778.
22. Ju, H. & Zhang, R. (2014). Optimal resource allocation in full-duplex wireless-powered communication network. *IEEE Transactions on Communications, 62*(10), 3528–3540. ISSN: 0090–6778.
23. Ge, X., Yang, J., Gharavi, H., & Sun, Y. (2017). Energy efficiency challenges of 5G small cell networks. *IEEE Communications Magazine, 55*(5), 184–191. ISSN: 0163–6804.
24. Fan, B., Leng, S., Yang, K., & Yu, Q. (2016). Socially-aware e-box deployment schemes for joint data forwarding and energy harvesting. In *2016 IEEE International Conference on Communications (ICC)* (pp. 1–6).
25. Grover, P. & Sahai, A. (2010). Shannon meets tesla: Wireless information and power transfer. In *2010 IEEE International Symposium on Information Theory* (pp. 2363–2367).
26. Liu, H., Li, X., Vaddi, R., Ma, K., Datta, S., & Narayanan, V. (2014). Tunnel FET RF rectifier design for energy harvesting applications. *IEEE Journal on Emerging and Selected Topics in Circuits and Systems, 4*(4), 400–411. ISSN: 2156–3357.
27. Markoulidakis, J. G., Lyberopoulos, G. L., Tsirkas, D. F., & Sykas, E. D. (1997). Mobility modeling in third-generation mobile telecommunications systems. *IEEE Personal Communications, 4*(4), 41–56. ISSN: 1070–9916.
28. Hu, J., Zhang, Y., Yu, Q., & Yang, K. (2016). Towards practical implementation of data and energy integrated networks. *ZTE Communications, 14*(3), 45–54.
29. Sullivan, F. & Dongarra, J. (2000). Guest editors' introduction: The top 10 algorithms. *Computing in Science and Engineering, 2*, 22–23. ISSN: 1521–9615.

Chapter 8
Prototype and Platform Design

In indoor environments, various battery-powered Internet of Things (IoT) devices, such as remote controllers and electronic tags on high-level shelves, require efficient energy management. However, manually monitoring remaining energy levels and battery replacement is both adequate and costly. This chapter introduces an energy management system for indoor IoT, which includes a mobile energy station (ES) for enabling on-demand wireless energy transfer (WET) in radio frequency (RF), energy receivers (ERs) and a cloud server. By implementing a two-stage positioning system and embedding energy receivers into traditional IoT devices, we robustly manage their energy storage. The experimental results demonstrate that the energy receiver can harvest a minimum power of 58 mW.

8.1 Background

The popularity of Internet of Things (IoT) technology continues to grow, and many wireless IoT devices require frequent charging due to their limited battery life [1]. In some indoor scenarios, it may be impractical to deploy wired–powered devices due to the associated costs of wiring. The employment of batteries becomes a solution. However, it is inadequate for battery-powered devices to replace batteries manually. Wireless energy transfer (WET) offers an ideal solution to alleviate the energy supply pressure of battery-operated devices [2–4]. There are three main types of WET: inductive coupling [5], resonant energy transfer [6], and radio frequency (RF) energy transfer. Compared to inductive coupling and magnetic resonance, RF energy transfer can fulfill the charging demands of most low-power devices, offering better fault tolerance and longer range energy coverage [7, 8].

However, long-distance transmission using RF-based WET often suffers from poor efficiency. Fortunately, there are several promising technologies that can help improve harvesting efficiency. These technologies include (i) enhancement of hardware modules such as RF-DC converters [9] and antennas [10, 11], (ii) energy beamforming and waveform design [8], and (iii) the assistance of mobile energy station (ES) [12, 13]. An RF-based WET system typically requires precise knowledge of the target's location. In ii), an accurate position will determine the direction of the beam. In (iii), an accurate position can shorten the distance between the target, which results in a lower path-loss.

In our previous work [14], we explored the use of a WET system assisted by an unmanned aerial vehicle (UAV) where the receivers' positions were determined using the global positioning system (GPS). However, this approach is not applicable to indoor environments for two reasons. Firstly, GPS signals have difficulty penetrating buildings, rendering them unreliable indoors. Secondly, indoor WET requires precise positioning accuracy at the decimeter level and three-dimensional coordinates [15]. Therefore, the absence of an energy management system with specifically designed positioning methods has motivated our work.

In this work, we have developed and implemented an energy management system for indoor IoT devices using RF-based technology. We introduce a two-stage positioning method that accurately determines the receiver's location. The proposed positioning approach takes into account both accuracy and practicality, making it cost-effective and easy to deploy. Moreover, we have also designed a mobile ES and energy receiver (ER) for this system. The energy receiver can be conveniently integrated between the energy container and the circuit of the device, allowing traditional devices to harvest energy efficiently.

The current research on RF-based WET focuses on improving the performance of its hardware components, such as the antenna or RF-DC converter. In a study by Liu et al. [10], a novel hybrid antenna was designed and verified for synchronous radio frequency energy harvesting (RF-EH). This antenna was specifically designed for WET systems that

J. Hu, K. Yang, *Integrated Data and Energy Transfer in Wireless Networks*, Textbooks in Telecommunication Engineering,
https://doi.org/10.1007/978-3-031-60514-7_8

target multiple frequency bands commonly used in wireless communication. Another study by Shen et al. [11] introduced a novel architecture for far-field WET based on a distributed antenna system. The system was experimentally validated and showed increased output DC energy by taking advantage of space and frequency diversity to compensate for path-loss. In addition, in a study by Tjukovs et al. [9], the authors explored the potential of utilizing LoRa technology and RF-DC converters with varying parameters on the receiver. This was done to create autonomous sensor nodes for far-field WET within a range of $1-3.5$ m.

Improving the conversion efficiency of these sub-modules undoubtedly contributes to enhancing the overall efficiency of the system. However, it is worth noting that there is currently a scarcity of comprehensive WET prototypes specifically designed with consideration for the target's location. We explored several RF-based WET prototypes.

In a study conducted by the authors in [9], the energy harvesting efficiency of a system using different antennas (LP0410 and VERT900) and RF-DC modules (HSMS285C and P2110B) was tested at varying distances ranging from 1 to 3.2 m. However, it was observed that the results were heavily influenced by the distance.

In a different study by the authors in [16], the expansion and application of ultra-wideband (UWB) communication for information exchange in the wireless charging process of electric vehicles was explored. The authors employed UWB technology to locate the target, utilizing Time-Difference-of-Arrival (TDOA) technology for precise linear distance measurement with minimal errors. However, it is worth noting that UWB technology is characterized by its high cost and relatively high energy consumption compared to Bluetooth-based technology [17]. This poses a challenge because IoT devices, unlike vehicles, typically require low energy consumption and cost-effectiveness.

An interesting data and energy integrated approach was proposed by the authors in [18]. This approach, integrated into the existing Wi-Fi network, reduced charging latency by 31% and improved sensor lifespan by 15% through simulation. However, it should be noted that this approach overlooks another important issue in practice, which is the requirement of advanced knowledge of the device's position for beamforming.

In a study by Buyle et al. [19], a dual-band solution was employed, where one frequency band provided a base charge to multiple nodes through a single omnidirectional antenna, and another band allowed for more power-hungry operations through a directional antenna. A hybrid RF-acoustic ranging module served as the payload. The results of the study showed that when using an omnidirectional antenna with a power of 0.5 W, it took 40 s to update the module's position, while with a directional antenna with a power of 1.5 W, it took only 6 s. Although an omnidirectional antenna provided the necessary energy for updating the position message of the target, it had significantly lower energy efficiency compared to a directional antenna.

In a study by Koonen et al. [20], a novel infrared-enhanced device localization concept was proposed, which achieved a completely passive function at the user device without draining its battery. This system offered high positioning precision and speed, but it is still limited to the research stage and incurs high deployment costs in IoT scenarios.

In another study, the authors in [21] utilized a camera to identify the position of a mobile phone on a table and controlled servo motors to direct the beam toward the target. However, this WET system had a limited range, confined to the unobstructed tabletop area.

It is important to note that existing WET systems have yet to achieve autonomous charging capabilities and relatively reliable conversion efficiency [19]. Therefore, considering the limitations of the aforementioned research studies, we have designed a cost-effective RF-based energy management system specifically tailored for indoor IoT applications.

In this work, we propose the design of an RF-based energy management system for indoor IoT devices, along with a two-stage positioning method. This method aims to reduce the distance between the ES and the ER and enhance alignment between the directional antennas. The main contributions of this study are outlined as follows:

- We develop an RF-based energy management system, consisting of a cloud server, a mobile ES, and some ERs for indoor use. The ER can easily be integrated into existing IoT devices. We also validate the effectiveness of the prototype and evaluate its performance, considering factors such as charging distance, angle estimation error, and antenna beam width.
- We design a comprehensive workflow between the cloud, the mobile ES, and the ER, enabling the self-management of energy for indoor IoT devices.
- We propose a two-stage positioning system that meets the accuracy requirements for RF-based localization, while also ensuring ease of deployment.

The remainder of this chapter is organized as follows. Section 8.2 introduces the system and outlines the functional architecture of the mobile ES and ER. Section 8.4 provides a detailed explanation of our proposed indoor positioning system and workflow of the energy management system. Sections 8.5 and 8.6 present a comprehensive hardware implementation of the mobile ES and the ER, respectively. Section 8.7 evaluates the performance of our prototype and demonstrates its application in a smart door lock scenario.

8.2 RF-Based Energy Management System

As shown in Fig. 8.1, in an indoor environment, an energy management system that includes a cloud server, a mobile ES, and some ERs is used to manage the energy of the IoT devices. These devices periodically send their remaining energy level to a server through the roaming mobile ES. When the server detects that a device's energy level falls below a predefined threshold, it informs the mobile ES of the device's ID. The mobile ES then moves to the proximity of the device, guided by location coordinates obtained from an indoor positioning system. By following a designed workflow, the ES determines the azimuth and elevation angles, allowing it to continuously charge the device. Once all the devices' charging demands are met, the mobile ES returns to its initial position.

The core components of this system include the mobile ES, the ER, and the two-stage positioning system. The following subsection will introduce the ES and the ER, while the positioning system will be described in detail in Sect. 8.4.

8.3 Functional Architecture

Mobile Energy Station The mobile ES is responsible for mobility, energy transmitting, and communication relay. The detailed functional architecture is elucidated in Fig. 8.2.

First and foremost is the mobility function. The mobile ES is equipped with a dynamic real-time 2D mapping capability, utilizing Simultaneous Localization and Mapping (SLAM) technology [22]. It can generate maps in real time and calculate obstacle-based paths. When encountering new obstacles, the robot updates the map and recalculates the path accordingly. To achieve motion control, the system utilizes state feedback. To collect state data, we have incorporated several modules, including a laser radar (LiDAR), a photoelectric encoder, an Inertial Measurement Unit (IMU), and a motor drive unit.

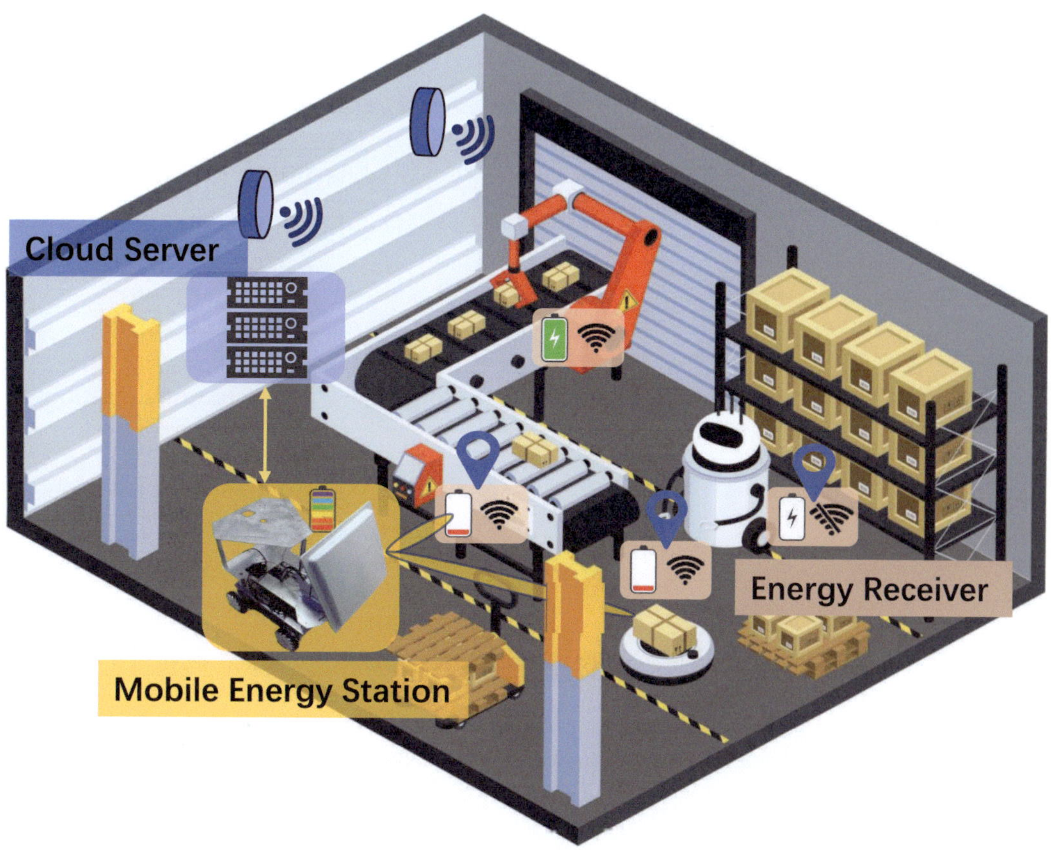

Fig. 8.1 RF-based energy management system

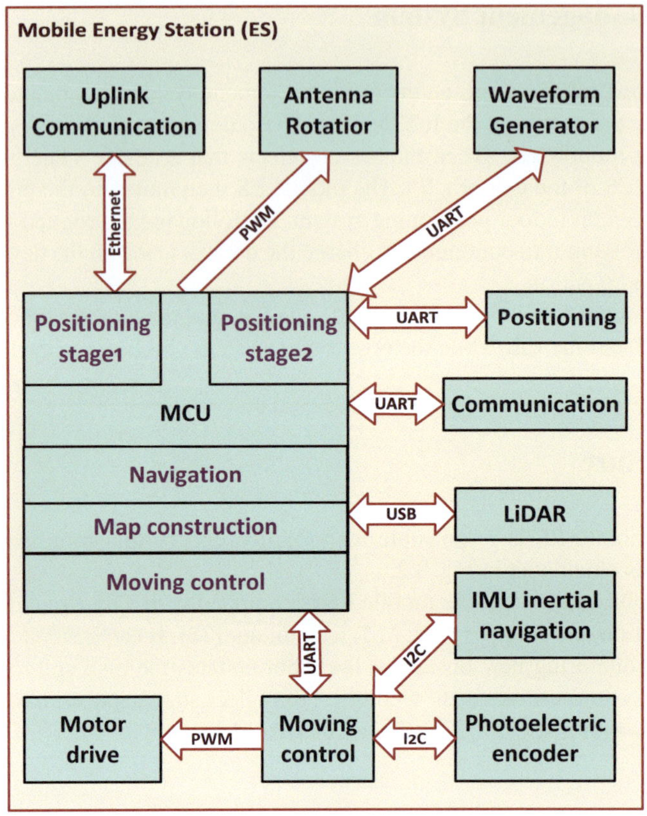

Fig. 8.2 Functional architecture of the mobile energy station

Fig. 8.3 Functional architecture of the energy receiver

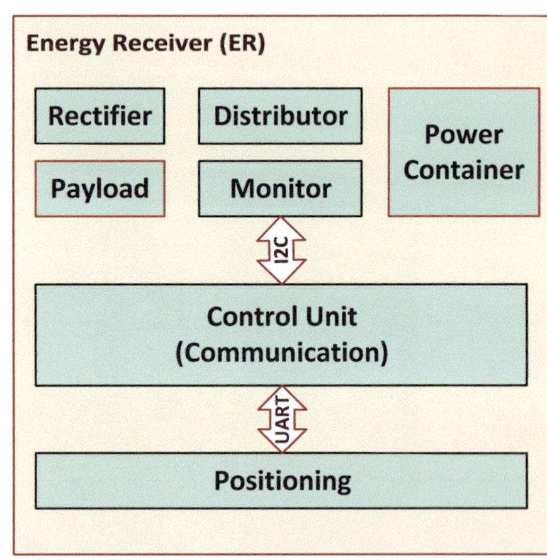

The other functions of the mobile ES are to provide continuous energy transmission and serve as a communication relay, converting the Wi-Fi protocol into a more energy-efficient ZigBee protocol. Therefore, the mobile ES consists of a Main Control Unit (MCU), a waveform generator, an antenna rotator, an uplink communication module, and a communication module. Additionally, we have included a positioning module to implement the two-stage positioning approach proposed in our system.

Energy Receiver As depicted in Fig. 8.3, as a module embedded between the payload circuit and the energy container, the ER is primarily responsible for handling RF energy, monitoring remaining energy, and broadcasting frames for positioning.

Firstly, the ER processes RF energy. Since RF energy cannot be directly utilized, a rectifier is employed to convert it into storable DC energy. Additionally, to ensure that the load circuit can be sufficiently supported, energy is likely to be extracted from the energy container. A distributor is used to regulate the energy distribution among the charging circuit, energy container, and payload.

Secondly, the ER provides power-level feedback. A charging monitor is employed to measure the charging power and battery level. These measurements are then transmitted through the communication module.

Lastly, in our proposed two-stage positioning system, the ER module needs to transmit data frames with constant tone extension (CTE) for positioning. Therefore, a dedicated positioning module is necessary.

Cloud Server The cloud plays a critical role in the system by providing control and management functions for IoT devices. It allows for effective monitoring of the device's energy status by setting different low-battery alert thresholds. This enables the mobile ES to be notified to charge the battery promptly, ensuring normal operation. Additionally, the cloud can clean and analyze the data reported by IoT devices, offering more precise data analysis and decision support. For instance, real-time status monitoring, detection of abnormal situations, and timely action can be achieved.

8.4 Two-Stage Positioning System and Workflow

8.4.1 Two-Stage Positioning System

We designed a two-stage positioning approach to help ES determine ER's position. In the first stage, the absolute position of the target is determined. However, there may be errors in this position estimation. To address this, the second stage is designed to determine the relative direction of the receiver with respect to the mobile ES. The two-stage positioning is depicted in Fig. 8.4.

1. *Positioning Stage 1*: Positioning Stage 1 aims to provide an initial estimation of the ER's location by using a signal propagation model and the principles of trilateration. The ES's movement is guided by this estimation, reducing the distance between itself and the target.

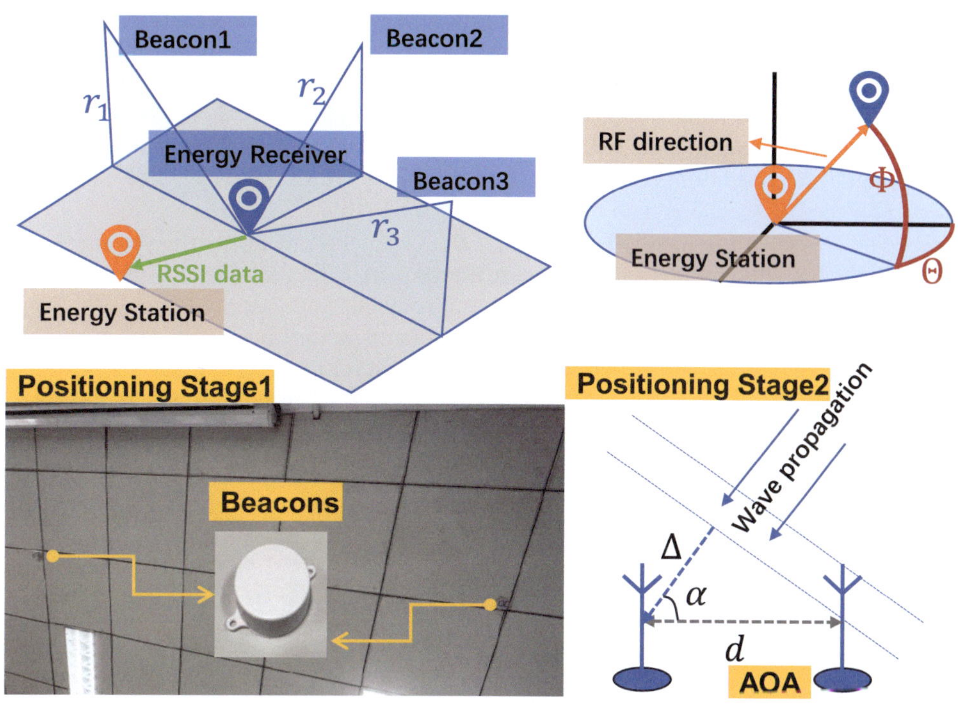

Fig. 8.4 Two-stage positioning system

In our approach, three fixed beacons are strategically positioned on the room's ceiling, as depicted in the top left of Fig. 8.4. These beacons emit signals with equal power gain into the environment. The ER captures these signals, yielding different Received Signal Strength Indication (RSSI) values for each beacon. The RSSI data is then transmitted to the mobile ES. By utilizing the propagation model, the distance between the ER and each fixed beacon, denoted as r_i, $i = 1, 2, 3$, can be calculated. Once three distances have been obtained, they can be used to determine the receiver's coordinates in 3D space through geometry analysis.

One notable advantage of positioning stage 1 is its energy consumption. Since the receiver is only required to receive broadcast information from the fixed beacons and not transmit any data, energy consumption is minimized. However, it is worth noting that due to various factors, such as refraction and reflection caused by indoor obstacles, the accuracy of positioning Stage 1 typically ranges around 1 meter [23].

2. *Positioning Stage 2* : Since the coordinates obtained from positioning stage 1 are not accurate, it is necessary to introduce the second stage of positioning.

   As shown in the bottom right of Fig. 8.4, assuming we utilize two receiver antennas when the incoming signal impinges upon these antennas at an oblique incident angle, it engenders a disparity in phase ψ. This discrepancy arises due to the propagation of the wavefront at the velocity of light c, thus requiring additional time to reach the antenna positioned farther away. Considering the wavelength of signal λ is a known constant, the phase shift Δ can be interpreted as $\psi \frac{\lambda}{2\pi}$. And the angle of arrival α can be readily ascertained from this spatial disparity by arccos $\left(\frac{\Delta}{d}\right)$.

   In practice, the constant wavelength λ is achieved by employing frames with constant tone extension (CTE) broadcasted by the ER. The mobile ES, equipped with an antenna array, captures the signals and obtains inphase quadrature (IQ) results by controlling the sampling frequency and order of the output signals from each antenna. These IQ results are then sent to the MCU for further processing, where data in the form of azimuth and elevation can be obtained.

8.4.2 *Workflow*

The working procedure of an RF-based energy management system is depicted in Fig. 8.5. A single workflow consists of the following four steps:

1. *Deployment Stage*: During the deployment stage, all the devices are registered in the database. Low battery alarm threshold and max battery energy level are configured. Beacons are deployed for the purpose of positioning. ERs are connected in a ZigBee network under the management of the coordinator [24], and the access of the uplink module to the Internet signifies the completion of the deployment phase.
2. *First Positioning Stage* : When the remaining energy drops to the configured threshold, the mobile ES will broadcast a wake-up signal. Then the positioning module will be activated in the target ER, evaluating the RSSI of the fixed beacons deployed on the ceiling. The RSSI information is sent to the mobile ES for processing. The mobile ES will then plan a trajectory and depart for the ER's location with the assistance of LiDAR.
3. *Second Positioning Stage* : When the mobile ES arrives at the destination, it enters the second positioning stage. The positioning module in the ER will continuously broadcast frames with CTE for AOA estimation. Broadcasting is ended with a shutdown command from ES after outputting the azimuth and elevation results.
4. *Wireless Energy Transmitting Stage* : The antenna is then rotated to the estimated direction. And the robot will continuously transmit RF energy to the ER. Meanwhile, the ER being serviced continuously sends the remaining energy level to the mobile ES every 30 seconds, based on which ES decides when to stop transmitting RF energy. Besides, feedback of charging power is utilized to indicate abnormal charging situations. Finally, devices will send an updated message to the cloud server.

When there are multiple devices that need service, the mobile ES will plan a trajectory for all devices by repeatedly conducting stages 2, 3, and 4.

8.5 Implementation of the Mobile Power Beacon

The hardware implementation of the mobile ES is depicted in Fig. 8.6, which includes a waveform generator, an antenna rotator, a main control unit, an uplink module, a positioning module, and some sensors for state feedback.

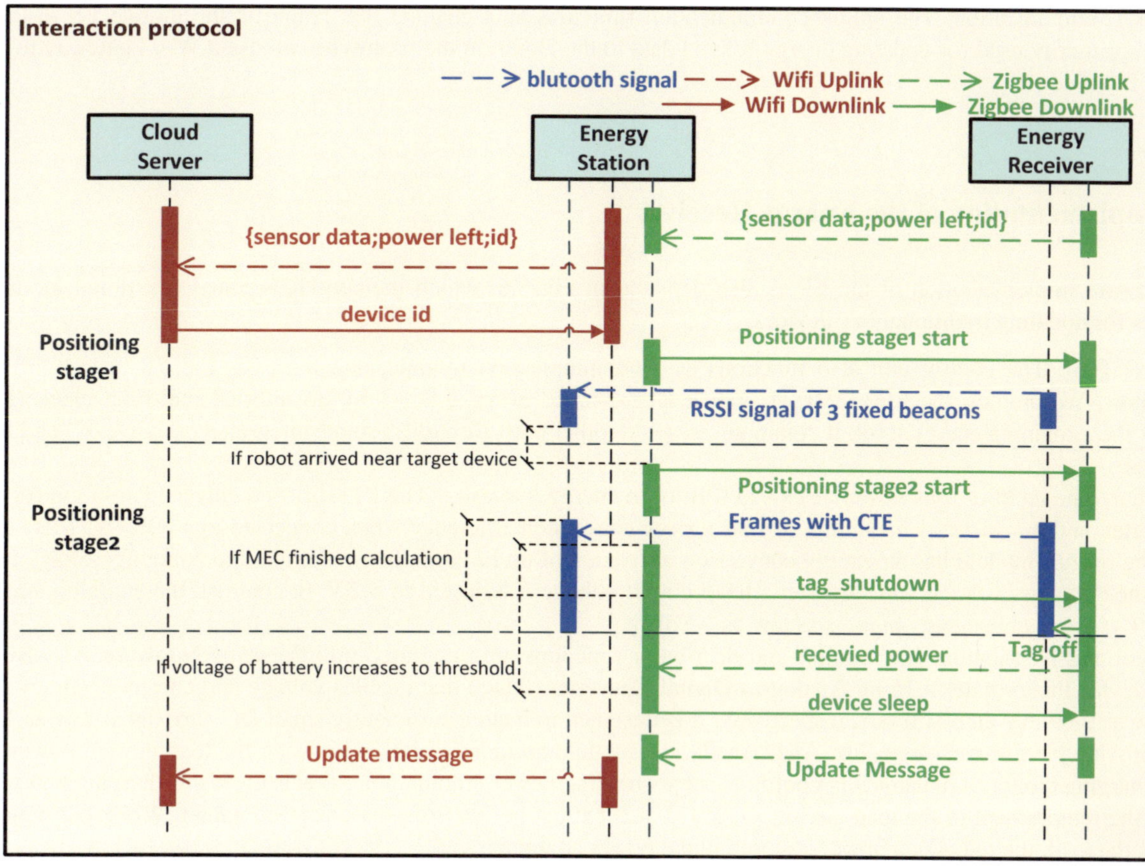

Fig. 8.5 Workflow of RF-based energy management system

- Main Control Unit (MCU): The MCU utilizes Invidia's Jetson Nano TX1 with chip architecture ARM Cortex-A57 which maximizes performance and energy consumption. It integrates advanced 3D graphic processing capabilities and has powerful parallel computing capabilities. TX1 is employed to work simultaneously for calculating the position coordinates and AOA information of the ER. Besides, it builds maps and plans paths based on processing real-time data from LiDAR. Meanwhile, MCU collects status feedback from sensors, such as speed and attitude.
- Waveform Generator: Waveform generator uses Semtech's LoRa modulation module LLCC68 and a 20 dB power amplifier with an adjustable amplification factor alone with a 915 M directional antenna. After receiving bitstream data through UART, LLCC68 modulates the bitstream to a frequency shift key (FSK) waveform with a spreading factor of 5 at 915 MHz [25], which is then amplified by the power amplifier and transmitted to the antenna. This module mainly operates in stage 4.
- Communication Module: We adopt a low power cost, short distance communication protocol ZigBee. The EFR32MG21 of Silicon Labs is adopted and typically operates in EM2 mode with an energy consumption of 4.5 uA. This module is awakened by an external interrupt controlled by the MCU and broadcasts the specific ID to wake up the target ER. It receives the beacon signal strength packets and real-time power feedback from the target terminal and transmits the control signal to synchronize the workflow. Besides, an onboard 2.4 GHz omnidirectional antenna is implemented for the communication.
- Antenna Rotator: The antenna rotator utilizes a servo motor and a simple Pulse Width Modulation (PWM) input signal to rotate to a given angle precisely. The control pulse period for this system is 20 ms, with pulse widths ranging from 0.5 to 2.5 ms, corresponding to positions from -90° to 90°. By adjusting the PWM pulse width, the motor's rotation angle can be accurately controlled.
- Positioning Module: The positioning module utilizes an EFR32BG22 of Silicon Labs and a 4×4 antenna array board, SLWRB4185A. This module is used to receive frames with CTE from the ER, allowing for the IQ values received by each element of the SLWRB4185A to differ. The IQ values are then transmitted to the MCU via UART for further calculation.

- Uplink Communication: The uplink communication functions as a channel that connects the mobile ES to the cloud server, primarily used for updating energy-related data to the cloud. In the prototype, we use a Wi-Fi gateway to connect to the Internet.

8.6 Implementation of the Energy Receiver

The hardware implementation of the ER is also depicted in Fig. 8.6, which includes RF-related functional modules and interfaces for updating traditional IoT devices.

- Control Unit: The control unit also functions as a communication module, and it assists the ES to synchronize the workflow. We employ the EFR32MG21, which is awakened every 2 hours by an internal real-time clock (RTC) to report the remaining energy level. It communicates with other parts through an inter-integrated circuit (I2C) and a UART interface.
- Rectifier: The rectifier uses Powercast's P1110B as an energy harvester. The P1110B is a universal solution for RF-DC converter, and it achieves a maximum of 75% energy conversion efficiency when connected to a 50Ω 915 MHz antenna, and the overall module has an energy conversion efficiency of up to 70%. To ensure that the charging voltage is higher than the container voltage, the harvester's fixed output voltage is adjusted to 4.2 V. In contrast, the charging current can vary. The acceptable energy input is as low as −5 dBm.
- Monitor and Distributor: The monitor and distributor functions are integrated into the same hardware, TI's BQ25155. This device incorporates a 16-bit Analog-to-Digital Converter (ADC) that enables voltage and current accuracy to reach as low as 0.09 mV and 0.06 mA, respectively. It is designed to handle sufficiently small DC currents as low as 1.25 mA and provide precise measurements. Additionally, the module provides isolation between the load circuit, power source, and charging circuit. It dynamically coordinates the charging power among them. Precisely, if the charging power cannot meet the requirement of the load circuit, BQ25155 will extract power from the power container. While the input power meets the load, the redundant power is stored in the power container.
- Positioning Module: Similar to the positioning module in the mobile ES, we employ EFR32BG22. To improve the harvesting efficiency, the positioning module is supposed to stick as close as the receiving antenna.

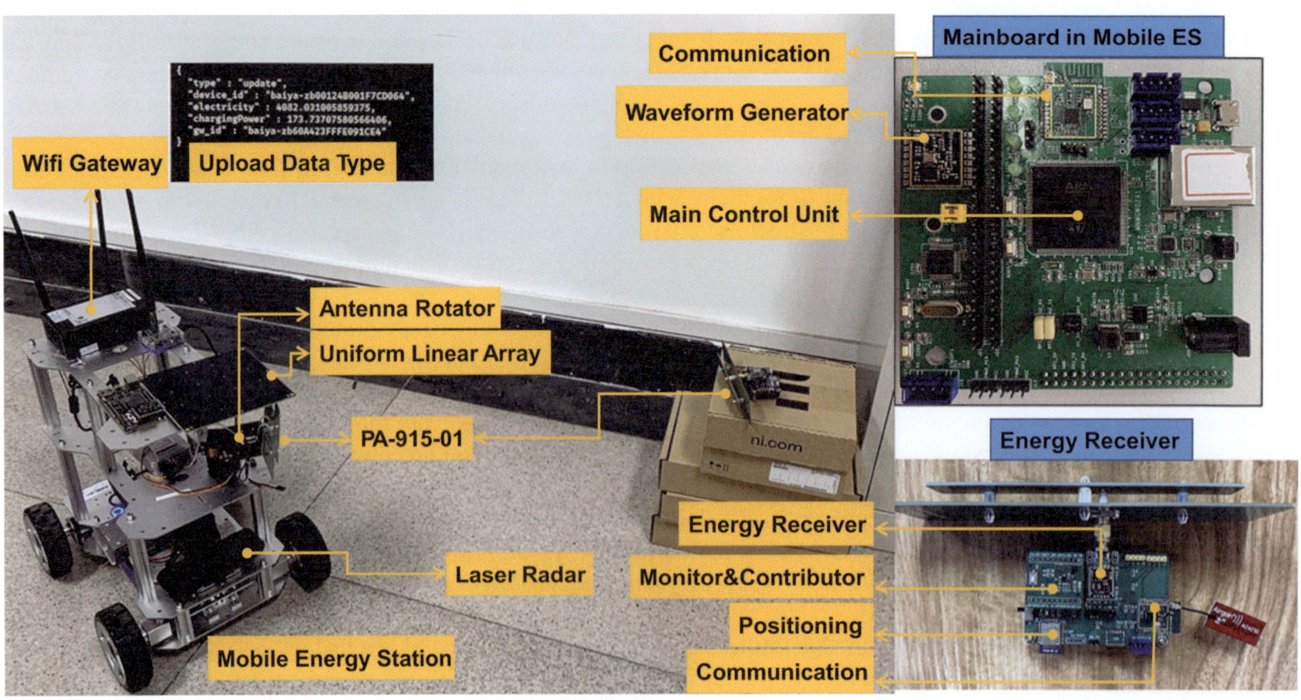

Fig. 8.6 Hardware implementation of energy station and energy receiver

- Power Container and Payload: These two modules are the interfaces for updating traditional IoT devices. It is required to connect the traditional device's power source and payload to the energy container and payload interfaces in the ER, respectively. A new load circuit is generated in this way, and the new circuit is under the management of the contributor. It is important to note that the default configuration of the ER is designed for universally used 4.2 V lithium batteries, and an additional boost circuit is required for higher voltages.

8.7 Performance Evaluation

8.7.1 Performance Analysis of First Positioning Stage

Due to various factors, such as environmental conditions and noise, indoor positioning accuracy in the first stage is challenging to quantify.

We denote the real location vector of the target as R and the estimation location vector as $\hat{R}$ which complies with normal distribution, and the positioning distance deviation $e = ||\hat{R} - R||$ also conforms to a normal distribution, according to the central limit theorem [26]. The histogram in Fig. 8.7 represents the probability of estimation errors, with σ_e and μ_e denoting the standard deviation and mean of the errors, respectively.

We evaluated the indoor system's error using 1000 sampling results, with a mean value $u_e = 1.0545$ m. The extremely small standard deviation signifies the stability of the first positioning stage. It is worth noting that a more complex environment may lead to an increase in average error.

8.7.2 Performance Analysis of Second Positioning Stage

We conducted a performance evaluation of the second positioning stage, with the ER's height set at 1.3 m and the antenna array height set at 1 m. The antenna array was placed parallel to the ground. We change the horizontal distance d between the ES and the ER, and the azimuth angle deviation from the actual value is denoted as e_d. We performed 50 estimations at

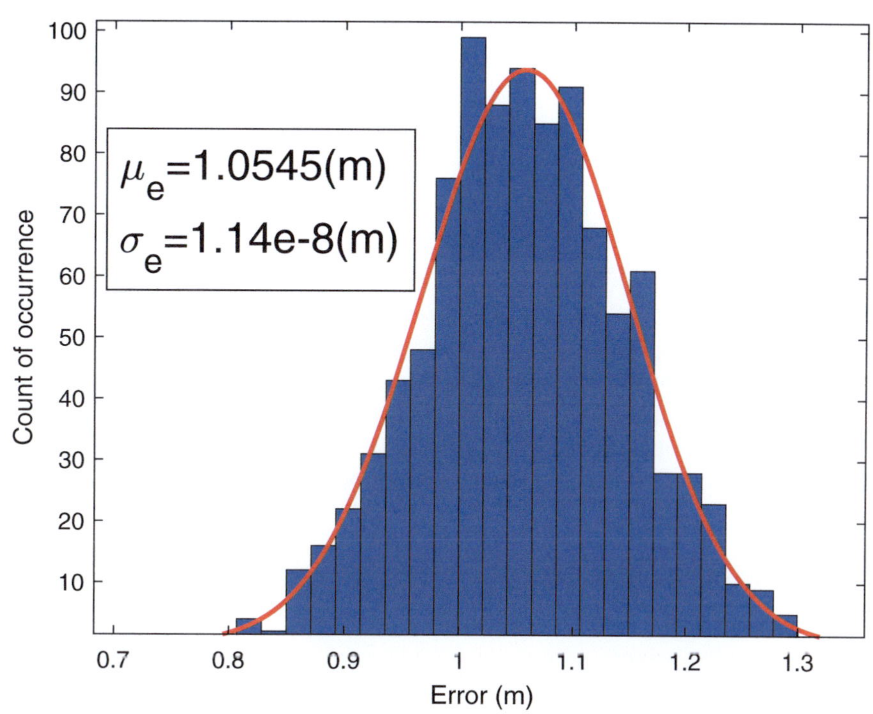

Fig. 8.7 Probability histogram of the measurements error

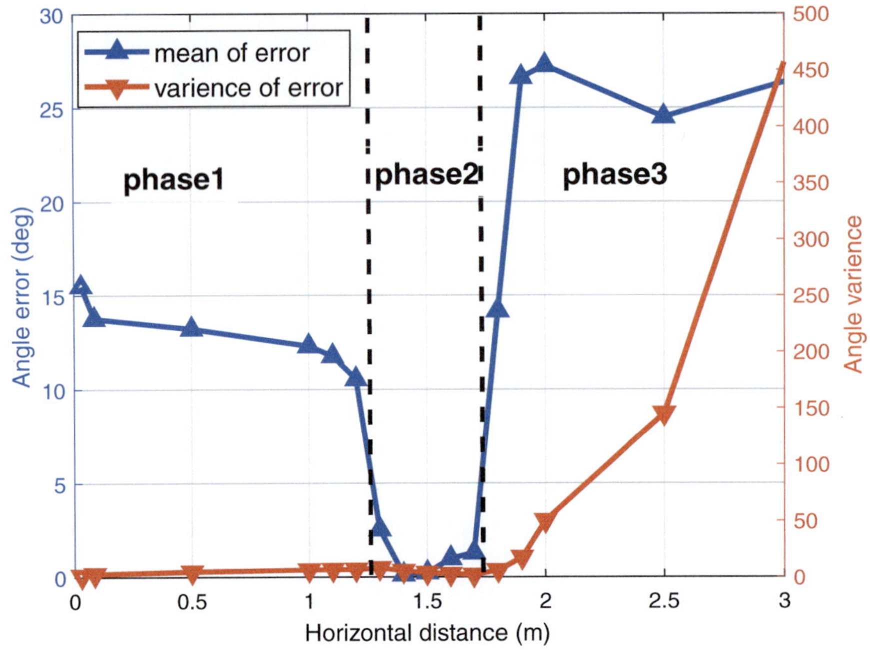

Fig. 8.8 Variance and mean of $\mathbf{E_d}$ with d in range 0–3 m

each e_d; then $\mathbf{E_d} = e_d[1], \ldots, e_d[50]$ is obtained. We further calculate the mean and variance of $\mathbf{E_d}$, and they are plotted in Fig. 8.8. The results revealed that both too close and too far horizontal distances reduce the performance of the second stage positioning.

From Fig. 8.8, the variance curve shows that when the mobile ES is located within a range of 0–1.7 m from the ER, the variance of the azimuth angle is relatively small. However, the variance increases sharply when the distance exceeds 1.7 m, indicating unreliable azimuth angle estimation. This phenomenon is probably attributed to the elevation angle.

Considering signal processing, there is a performance metric called the Cramér–Rao bound (CRB), which indicates the lower bound of the variance of the azimuth angle estimation [27]. And CRB $\propto 1/\tan\phi$, where ϕ denotes the elevation angle between the ES and the ER. CRB $\rightarrow \infty$ as $\phi \rightarrow 0°$, resulting in a large variance and unreliable estimation results. In our system, when ϕ is less than

$$\phi_l = \arctan\frac{1.3-1}{1.7} = 10.008°, \tag{8.1}$$

the variance tends to increase sharply.

Next, we analyze the mean angle error of the estimations at each distance. From Fig. 8.8, the curve for angle error can be divided into three phases with respect to d:

- Phase 1 (0–1.2 m): The mean angular error falls within the range of 5–10°. As mentioned earlier, the slight variance indicates that the large error is not due to variance but other factors. One probable reason is the preconditions of the AOA algorithm, which assumes far-field signals. When the distance is within 0–1.2 m, the CTE signals do not satisfy the far-field assumption of the parallel beam, rendering the AOA algorithm ineffective.
- Phase 2 (1.2–1.7 m): In this distance range, the system operates outside the near-field region, and the AOA algorithm becomes effective with a relatively small variance. From the perspective of error consideration, this range represents the optimal distance for WET.
- Phase 3 (1.7–3 m): Within this range, there is significant variance. And the AOA estimations are not reliable due to significant fluctuations.

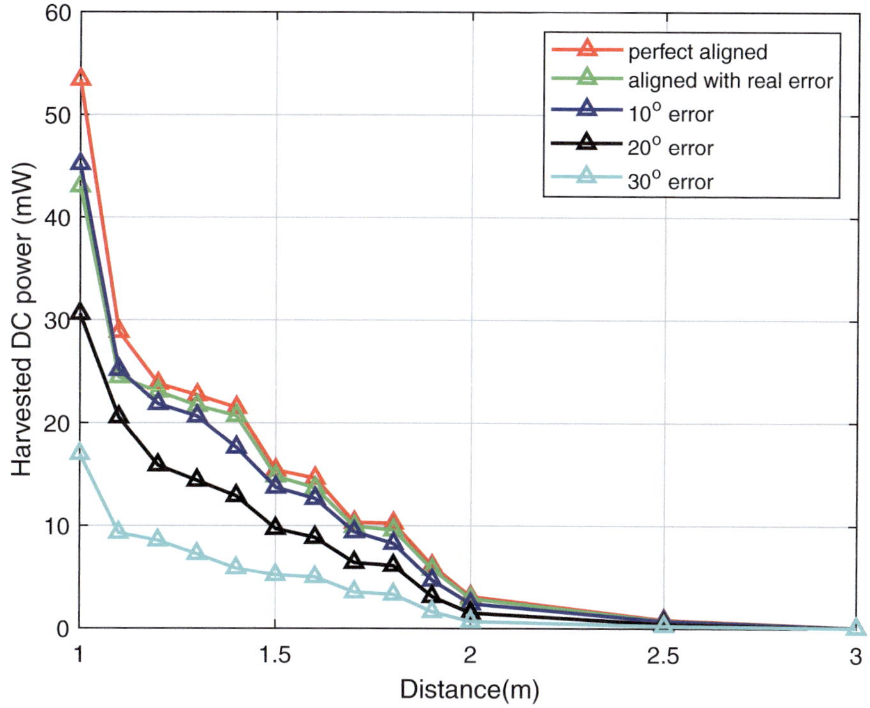

Fig. 8.9 Received power of perfect alignment and self-locating alignment (PA-915-01)

8.7.3 *Harvesting Performance Evaluation*

We employ a linearly polarized antenna, PA-915-01, which exhibits a half-power point at a vertical angle of 120° and a horizontal angle of 62°. Additionally, it maintains a maximum gain of 4.5 dBi when horizontally and vertically deviating by 10° and a residual gain of 4.6 dBi.

In the evaluation, we set transmitting power to 10 W, and the resulting curve is represented in Fig. 8.9. It can be observed that with the existence of errors, the receiving power remains within a deviation of no more than 10%. Significant losses have not been encountered. This is due to the fact that the deviation angle of the beam remains within the high-gain range of the PA-915-01 antenna.

Assuming that the two localization stages are independent random processes, the average error in indoor localization during the first stage is approximately 1.0545 m, consequently causing the error in the second positioning stage to reach approximately 12.5°. As a result, the ER is capable of stably receiving at least 43 mW of energy with autonomous operation.

8.8 A Realistic Application

As depicted in Fig. 8.10, we update the traditional door lock with our designed ER. The ER uses omnidirectional flexible patch antennas with a maximum gain of 4 dBi.

In the mobile ES, we employ a different RFID circularly polarized directional flat panel antenna, which boasts narrow horizontal and vertical half-power beamwidth of 40° (allowing for a unilateral deviation of up to 20°) and a maximum gain of 12 dBi, controlling for the same transmission power at 10 W.

We charted the received power for perfect alignment, a 10° deviation, a 20° deviation, and under actual error, as shown in Fig. 8.11.

As shown in Fig. 8.11, on the one hand, we can see that using a narrower beamwidth antenna yields a 30–50% increase in received power. At a distance of 1.9 m, when the angle error is above 24°, the ER has a sharp decrease, and this is because both distance and estimation error are in terrible conditions.

Besides, evidence suggests that a shorter distance does not necessarily guarantee a higher energy harvesting efficiency. We emphasize the received power at 1.2 and 1.3 m in Fig. 8.11. Although the distance at 1.2 m is closer, the angle estimation

Fig. 8.10 An application for the smart door lock

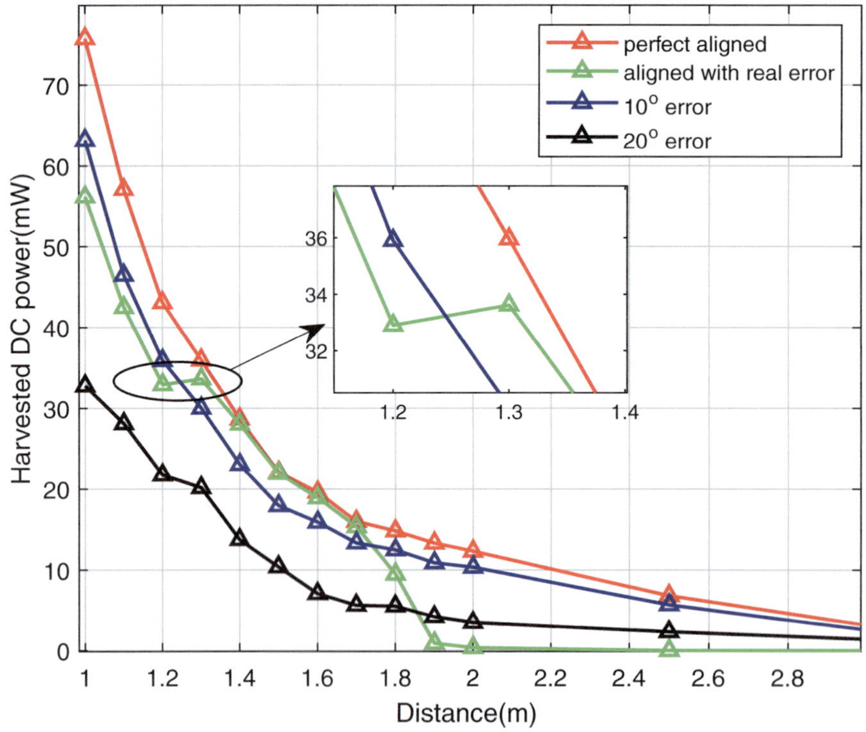

Fig. 8.11 Power harvested in smart door application (RFID circularly polarized directional flat panel antenna)

error is greater, leading to the fact that the received energy in 1.2 m is less than that of 1.3 m. This phenomenon may be more apparent with narrow beamwidth antennas.

Based on the performance of the proposed two-stage positioning system evaluated in Sects. 8.7.1 and 8.7.2, the receiver is capable of stably receiving at least 58 mW of energy with autonomous operation.

8.9 Exercises

1. Using the free-space propagation model, the RF power received by the receiver can be simply predicted. The frequency of the RF signal is 915 MHz, the transmitter's transmitting power is 10 W, and the gain of the transceiver antenna is 6 dBi. How much RF power can the receiving antenna receive?
2. What types of antennas are used in DEIN systems and what are their characteristics?
3. Which are the main components of a typical RF harvesting enhanced device? Describe the role of each of these components.
4. Discuss the reasons for using AOA technology in DEIN systems assisted by mobile base stations, and consider which other technologies can be used in engineering applications to achieve the same goal.
5. Discuss several typical structures of rectifiers, and explain their characteristics, advantages, and disadvantages separately.
6. In a practical engineering scenario, the receiving side expects to receive 100 mW DC power at a distance of 3 m. How to design this wireless RF energy transmitting system?

References

1. Hu, J., Kun, Y., Guangjun, W., & Lajos, H. (2018). Integrated data and energy communication network: A comprehensive survey. *IEEE Communications Surveys Tutorials, 20*(4), 3169–3219.
2. López, O. L. A., Alves, H., Souza, R. D., Montejo-Sánchez, S., Fernández, E. M. G., & Latva-Aho, M. (2021). Massive wireless energy transfer: Enabling sustainable IoT toward 6G era. *IEEE Internet of Things Journal, 8*(11), 8816–8835.
3. Clerckx, B., Huang, K., Varshney, L. R., Ulukus, S., & Alouini, M.-S. (2021). Wireless power transfer for future networks: Signal processing, machine learning, computing, and sensing. *IEEE Journal of Selected Topics in Signal Processing, 15*(5), 1060–1094.
4. Huang, K., & Zhou, X. (2015). Cutting the last wires for mobile communications by microwave power transfer. *IEEE Communications Magazine, 53*(6), 86–93.
5. Zhan, B., Xia, W., Xiong, C., & Liu, S. (2021). Simulation analysis of coupling coil of 13.56MHz magnetic coupling resonant wireless energy transmission system. In *2021 22nd International Conference on Electronic Packaging Technology (ICEPT)* (pp. 1–6).
6. Dean, J., Coultis, M., & Van Neste, C. (2021). Wireless sensor node powered by unipolar resonant capacitive power transfer. In *2021 IEEE PELS Workshop on Emerging Technologies: Wireless Power Transfer (WoW)* (pp. 1–5).
7. Ma, D., Lan, G., Hassan, M., Hu, W., & Das, S. K. (2020). Sensing, computing, and communications for energy harvesting IoTs: A survey. *IEEE Communications Surveys & Tutorials, 22*(2), 1222–1250.
8. López, O. L. A., Kumar, D., Souza, R. D., Popovski, P., Tölli, A., & Latva-Aho, M. (2022). Massive MIMO with radio stripes for indoor wireless energy transfer. *IEEE Transactions on Wireless Communications, 21*(9), 7088–7104.
9. Tjukovs, S., Eidaks, J., & Pikulins, D. (2018). Experimental verification of wireless power transfer ability to sustain the operation of LoRaWAN based wireless sensor node. In *2018 Advances in Wireless and Optical Communications (RTUWO)* (pp. 83–88).
10. Liu, S.-B., Zhang, F.-S., Boyuan, M., Gao, S.-P., & Guo, Y.-X. (2022). Multiband dual-polarized hybrid antenna with complementary beam for simultaneous RF energy harvesting and WPT. *IEEE Transactions on Antennas and Propagation, 70*(9), 8485–8495.
11. Shen, S., Kim, J., Song, C., & Clerckx, B. (2021). Wireless power transfer with distributed antennas: System design, prototype, and experiments. *IEEE Transactions on Industrial Electronics, 68*(11), 10868–10878.
12. Tang, H., Wu, Q., Chen, W., Wang, J., & Li, B. (2021). Mitigating the doubly near-far effect in UAV-enabled WPCN. *IEEE Transactions on Vehicular Technology, 70*(8), 8349–8354.
13. Xie, L., Cao, X., Xu, J., & Zhang, R. (2021). UAV-enabled wireless power transfer: A tutorial overview. *IEEE Transactions on Green Communications and Networking, 5*(4), 2042–2064.
14. Fan, X., Hu, J., & Yang, K. (2023). UAV-aided data and energy integrated network: System design and prototype development. *China Communications, 20*, 1–13.
15. Wei, H., & Huang, M. (2020). Intelligent vehicle positioning method based on GPS/Lidar/Derivative data fusion. In *2020 IEEE 3rd International Conference on Information Systems and Computer Aided Education (ICISCAE)* (pp. 153–157).
16. Tiemann, M., Schmuelling, B., & Clemens, M. (2019). Control signaling in wireless power transfer for electric vehicles through ultra-wideband. In *2019 Fourteenth International Conference on Ecological Vehicles and Renewable Energies (EVER)* (pp. 1–5).
17. Guo, X., Ansari, N., Hu, F., Shao, Y., Elikplim, N. R., & Li, L. (2020). A survey on fusion-based indoor positioning. *IEEE Communications Surveys & Tutorials, 22*(1), 566–594.
18. Mohanti, S., Bozkaya, E., Naderi, M. Y., Canberk, B., Secinti, G., & Chowdhury, K. R. (2021). WiFED mobile: WiFi friendly energy delivery with mobile distributed beamforming. *IEEE/ACM Transactions on Networking, 29*(3), 1362–1375.
19. Buyle, C., Cox, B., der Perre, L. V., & De Strycker, L. (2021). A multi-band solution for interacting with energy-neutral devices. In *2021 55th Asilomar Conference on Signals, Systems, and Computers* (pp. 329–333).
20. Koonen, T., Mekonnen, K. A., Huijskens, F., Pham, N.-Q., Cao, Z., & Tangdiongga, E. (2020). Fully passive user localization for beam-steered high-capacity optical wireless communication system. *Journal of Lightwave Technology, 38*(10), 2842–2848.
21. Joshitha, K. L., Jayalakshmi, T., Bhavani, R., & Preethibala, S. (2018). Automatic charging of multiple smartphones using light beam. In *2018 IEEE International Conference on System, Computation, Automation and Networking (ICSCA)* (pp. 1–4).
22. Liao, Z., Hu, Y., Zhang, J., Qi, X., Zhang, X., & Wang, W. (2022). SO-SLAM: Semantic object SLAM with scale proportional and symmetrical texture constraints. *IEEE Robotics and Automation Letters, 7*(2), 4008–4015.

23. Hassanien, A., & Vorobyov, S. A. (2011). Transmit energy focusing for doa estimation in MIMO radar with colocated antennas. *IEEE Transactions on Signal Processing, 59*(6), 2669–2682.
24. Pan, G., He, J., Wu, Q., Fang, R., Cao, J., & Liao, D. (2018). Automatic stabilization of Zigbee network. In *2018 International Conference on Artificial Intelligence and Big Data (ICAIBD)* (pp. 224–227).
25. Lavric, A., & Popa, V. (2017). Internet of things and LoRa™ low-power wide-area networks: A survey. In *20117 International Symposium on Signals, Circuits and Systems (ISSCS)* (pp. 1–5).
26. DasGupta, A., & DasGupta, A. (2010). Normal approximations and the central limit theorem. In *Fundamentals of probability: A first course* (pp. 213–242).
27. La Manna, M., & Fuhrmann, D. R. (2017). Cramér-Rao lower bounds comparison for 2D hybrid-MIMO and MIMO radar. *IEEE Journal of Selected Topics in Signal Processing, 11*(2), 404–413.

Index

J. Hu, K. Yang, *Integrated Data and Energy Transfer in Wireless Networks*, Textbooks in Telecommunication Engineering,
https://doi.org/10.1007/978-3-031-60514-7